D0072816

Controversy in Victorian Geology

Controversy in Victorian Geology

The Cambrian-Silurian Dispute

James A. Secord

PRINCETON UNIVERSITY PRESS
PRINCETON, NEW JERSEY

Published by Princeton University Press, 41 William Street,
Princeton, New Jersey 08540
In the United Kingdom: Princeton University Press,
Guildford, Surrey

Library of Congress Cataloging in Publication Data will be
found on the last printed page of this book

ISBN 0-691-08417-3

This book has been composed in Linotron Palatino

Clothbound editions of Princeton University Press books
are printed on acid-free paper, and binding materials are
chosen for strength and durability

Printed in the United States of America
by Princeton University Press
Princeton, New Jersey

For Anne

Contents

Illustrations

N.B. Many of the explanatory diagrams included in the illustrations listed above represent my own interpretations of verbal descriptions in the primary sources. The maps and sections that are taken directly from nineteenth-century originals have usually been redrawn and simplified so as to bring out features of special importance to the controversy. The originals should be consulted in all cases of doubt.

Preface

IN RECENT YEARS the early Victorian period has become a focus for some of the most innovative and exciting work in the entire history of science. As yet, however, technical developments in scientific knowledge have frequently been studied in isolation, only tenuously related to the wider culture in which they were originally produced. By examining in depth a single major controversy, this book attempts to bring a fully contextual perspective to bear on the analysis of a typical Victorian science. It is intended both as a good story and as a contribution to the cultural history of actual scientific practice.

It is a pleasure to thank the many individuals and institutions whose generous help has made this book possible. I wish first of all to thank my parents, Jane and John Secord, who have avidly encouraged my academic work from the very beginning. My interest in the Cambrian-Silurian controversy was initially sparked as an undergraduate at Pomona College, where Donald McIntyre and his colleagues teach the earth sciences with a lively appreciation of their wider significance. In the years that followed, my approach to history was shaped in the stimulating environment of the Program in History of Science at Princeton University. I owe a particular debt to two outstanding scholars. Charles Gillispie skillfully directed the early stages of my writing and research, while Gerald L. Geison introduced me to many of the techniques employed in this study. Both have continued to serve as valued sources of advice, encouragement, and friendship ever since. On this side of the Atlantic, Roy Porter has provided unflagging enthusiasm and inspiration for over a decade now. Long before I took up this project, I began to benefit from John Thackray's familiarity with the source materials of early British geology. More recently, Jack Morrell has lavished time, wit, and critical expertise towards improving the book. I am particularly glad to record my intellectual debt to the writings of Martin Rudwick, and to acknowledge his kindness in allowing me to read several draft chapters of his narrative of the Devonian controversy shortly before my own manuscript went to press. Thanks also go to

Hugh Torrens for an invaluable line-by-line scrutiny of the entire text, and to Patrick Boylan, Janet Browne, Adrian Desmond, Colin Forbes, Jeanne Pingree, Phil Powell, David Price, Adrian Rushton, Simon Schaffer, Cecil Schneer, Crosbie Smith, and Robert Stafford for help of various kinds. I can only hope that the present volume goes some way towards meeting the very high standards that have been set in this area of research; any mistakes that remain are of course my own.

The writing of this book, together with a number of related projects, would not have been possible without financial support from the Fulbright-Hayes Commission (1978-1979); the Social Science Research Council (U.S.A.), (1978-1979); the National Endowment for the Humanities (1979-1980); and the North Atlantic Treaty Organization (1981-1982). My earliest research in London was made even more pleasant by the welcome at Imperial College given by Marie and Rupert Hall, while at a later date, William F. Bynum arranged for a fruitful affiliation with University College. I am especially grateful to the Master and Fellows of Churchill College, Cambridge, for electing me to a Junior Research Fellowship in 1982, the early part of which I devoted to completing a revised draft of this study. The College has made every effort to further my work; I would like particularly to thank Michael Hoskin for his encouragement. Andrew Tristram spent several late nights at the word processor, and Nicky Milsom and other members of the staff helped in a variety of ways. Fiona Cooper and Linda Mullen assisted with various stages of the typing, and the cost of this was partially offset by a small grant from the College's research fund.

For access to materials in their collections, I wish to acknowledge the assistance of numerous archivists and librarians. Permission to quote from unpublished manuscripts has been given by the American Philosophical Society; the Library, St. Andrews University; the Director, the British Geological Survey; the British Library; the British Museum (Natural History); the Syndics of the Cambridge University Library; Edinburgh University Library; the Geological Society of London; the Archives of Imperial College, London; the Mitchell Library, Sydney; the National Museum of Wales; the Curator of the Sedgwick Museum, University of Cambridge; the Master and Fellows of Trinity College, Cambridge; the University Museum, Oxford; and Yale University Library.

One contribution of my wife, Anne, to the present work is as large as it is easy to specify: while occupied with a full-time job, she has given up a great many of her free hours to draw all of the maps and line diagrams. Her constant support, sympathetic criticisms, and common sense have had equally important, if less tangible, effects on the character of the argument and the quality of the prose. This book, finished at last, is for her.

Manuscript Sources

APS American Philosophical Society, Philadelphia. Darwin/Lyell (D/L), S. P. Woodward (W).

BGS British Geological Survey, London. Official Archives, Murchison.

BL British Library, London. Murchison.

BMNH British Museum (Natural History), London. Owen (O), Sowerby (S).

CUL Cambridge University Library. Sedgwick, Featherstonhaugh, Fisher, Greenough, Harkness, Hughes.

DRO Devon County Record Office, Exeter. Buckland.

EUL Edinburgh University Library. A. Geikie, Lyell, Murchison.

GSL Geological Society of London. Official archives, Murchison, Sedgwick.

ICL Imperial College, London. Huxley (H), Ramsay (R).

ML Mitchell Library, Sydney. McCoy (all references are to frame numbers on microfilm CY499).

NMW National Museum of Wales, Geology Department, Cardiff. De la Beche.

SM Sedgwick Museum, Department of Earth Sciences, University of Cambridge. Sedgwick.

TC Trinity College Library, Cambridge. Whewell, Sedgwick.

UMO University Museum, Oxford, Department of Geology. J. Phillips.

YUL Yale University Library, New Haven. Dana, Silliman.

Controversy in Victorian Geology

Introduction

> Two famous geologists, earnest and true,
> Through Wales ran a race to find something new.
> The one came in first and a world did discover,
> The other came last, for he got a roll over.
> Silurian beds we in myriads number,
> Cambrian strata 'stat nominis umbra,'
> S says M knows not his beds when he's got 'em,
> That his system is base and his base has no bottom;
> Whilst M makes appeal to the sense of mankind
> Whether he should be stifled, 'cause S lagg'd behind.
>
> —Edward Forbes[1]

In 1833 the daughter of Lord Northampton became tired of an interminably dull discussion between William Buckland, Roderick Murchison, and Gideon Mantell about a motto for the Geological Society of London. "Pa—I've got it," the young girl told her father, and as Murchison later remembered, "she had drawn as our Crest a '*mole*' & under it the motto 'I bore.' "[2] Historians of science, understandably anxious to avoid a similar censure, face a difficult problem in dealing with the practice of geology in the nineteenth century. Much effort is expended on a few figures who contributed to theoretical controversies: Charles Lyell, Charles Darwin, and the diluvial geologists are the subjects of a considerable and growing literature. Some authors are now endeavoring to show how atypical, and in some respects uninfluential, these men may have been in the mainstream development of geology.[3] But as in other areas of the history of science, the mainstream itself has been little studied. The great mass of books, maps, and memoirs, including those produced by figures of the first rank, remain unused, unread, and little understood. Even a rapid glance at this forgotten material suggests that

[1] [Forbes] 1852a; italics in original removed.

[2] Murchison, Journal "Scientific Life, Alps, London, Silurian, 1826-38," GSL: M/J7, p. 207.

[3] Bartholomew 1976, 1979; M. Greene 1982; Page 1969; Porter 1976. Rupke 1983 carries the devaluing of Lyell to an extreme. For criticisms of some of these views, see L. G. Wilson 1980.

emphasizing the philosophical and theoretical aspects of the subject gives a radically misleading view. The principal preoccupations of the scientific community, as evidenced by the sheer quantity of available sources, await examination; the practical activity of geology in the Victorian era remains unexplored.

And here the mole is an appropriate mascot. Embracing a self-conscious empiricism, the gentlemanly Fellows of the Geological Society of London claimed that their science should be centered, not on a cosmological theory, but rather on a method and an activity—geological fieldwork, the tracing of rock strata over particular areas. Geology was virtually identified as the elucidation of the stratigraphical succession, a taxonomic enterprise involving three dimensions and focused on maps and sections. Time, process, and causation usually entered the picture only at the very end of an investigation: the reconstruction of the history of the earth and the succession of life upon it were long-term motivating goals of only secondary importance for most everyday practice. The greatest accomplishment of English geology was not Lyell's uniformitarian theory or any of the alternative schemes interpretive of terrestrial history. Rather, Victorian geologists took the leading role in establishing the fundamental order of the world's rocks.

Despite the central importance of this accomplishment, the establishment of a taxonomy for the strata has not attracted extensive historical attention. Like most elements of the classificatory sciences, the creation of the geological column has traditionally been dismissed as a noncontroversial accumulation of factual information. In part this picture contains an element of truth. From the special viewpoint of stratigraphy, the British rocks were unquestionably better known in 1850, say, than in 1830. As I hope to show, empirical findings acceptable to all investigators working within the principal tradition of research played a pivotal role in shifting the consensus on classification. The cumulative and progressive aspects of science cannot be ignored. They require close historical study.

Nevertheless, to see the establishment of the geological column (or any other grouping of objects in nature) as the gradual unveiling of a transcendent reality would be far too simple a view. For the rocks themselves are silent; the classification that orders them is something *made*. Geologists of the early nineteenth century recognized this problem themselves when they proclaimed as their task the ordering of a "chaotic" sequence of strata. Although geology had emerged as a coherent scientific discipline in the previous century, many interpretations of particular phenomena remained possible. The difficulties are reflected in the myriad controversies that crowd the pages of Victorian scientific periodicals. Almost all of the major divisions of the geological column were settled only after long and acrimonious debate. Lyell argued with Edward Charlesworth about the boundaries of the Tertiary subdivisions; William Fitton and Gideon Mantell fought over the position of certain beds within the Upper Cretaceous; a host of leading researchers battled over the Devonian on two separate occasions during the century. As one modern undergraduate textbook puts it, the received stratigraphical column "grew like Topsy."[4]

This book is about the most famous of these controversies, the dispute over the proper boundary between the Cambrian and the Silurian divisions in the oldest fossil-bearing rocks. The controversy grew out of the work of two leading geologists of the Victorian era: Roderick Impey Murchison, an independently wealthy gentleman living in London, and Adam Sedgwick, professor of geology at the University of Cambridge. In a preliminary classification of the older rocks in Britain during the 1830s, Murchison founded a "Silurian system" among the strata of the Welsh Borders, and Sedgwick described a "Cambrian system" in the ancient strata of North Wales. The debate emerged during the following decades as it became clear that these two groupings were in large part synonymous. Across an extensive overlap, Cambrian and Silurian offered alternative classifications for rocks of identical age. The resulting controversy eventually drew in a large part of the contemporary geological community, with John Phillips, Charles Lyell, Daniel Sharpe, Henry De la Beche, Andrew Ramsay, Edward Forbes, and numerous other leading naturalists playing important roles. It ended in Britain at the close of the century with the adoption of an intervening "Ordovician system" to include part of the disputed strata.

The Cambrian-Silurian controversy has always occupied a central place in the history of geology as one of the great set-pieces of the science's "golden age." The dispute has been used to point cautionary tales about scientific method, condemned as an instance of arbitrary power politics in science, pictured as a chapter in the extension of the geological time scale or the search for the origin of life, and interpreted as the outcome of technical mistakes by one or another of the principals.[5] But almost without exception, even the

[4] Dott and Batten 1976: 46. Many of these debates are mentioned in Zittel 1901; also helpful (if not always dependable) is Berry 1968.

[5] Earlier historical discussions of the present controversy include Geikie 1875; Clark and Hughes 1890, esp. 2: 508-563; Hunt 1878; Dana 1890; Berry 1968: 80-99;

most insightful of these accounts ultimately depend upon retrospectives written by Sedgwick, Murchison, and their late Victorian biographers. Upon closer examination these have proved to be utterly untrustworthy, fully reflecting the charged circumstances in which they were written. By making use of an untapped wealth of manuscript and printed sources relating to the debate, I hope to have lifted the veil of myth from some of the most important scientific research in the nineteenth century.

At a deeper level, a study of the controversy provides an ideal opportunity for a thorough reassessment of the day-to-day workings of a Victorian science. Such an investigation achieves its significance through—rather than in spite of—its involvement with the mundane materials of "normal science."[6] For if Victorian geologists saw strata classifications and maps as their greatest achievements, then it only makes sense that we should first turn to these in placing the subject in its wider historical setting. A close examination of such materials opens up a host of crucial issues. Most significantly, recent work in cultural anthropology and the sociology of knowledge has shown that the conceptual framework that brings the natural world into a comprehensible form becomes especially evident when a scientist constructs a classification. Previous experience, early training, institutional loyalties, personal temperament, and theoretical outlook are all brought to bear in defining particular boundaries as "natural." Emphasizing this point, Mary Douglas and others have suggested that scientific classifications hold a wealth of cultural significance.[7] Throughout this book I argue that geological nomenclatures defined separate intellectual properties, recognized through the coloring of maps and signalled by the names "Cambria" and "Siluria." These territories acquired a host of meanings which had to be accounted for in defining classifications. Particularly in the later and more vehement stages of the controversy, these underlying meanings come forward with particular force.

Holland 1974; Rudwick 1976c; Thackray 1976; Balan 1979: 384-392; Speakman 1982: 74-85; together with many popular and semipopular works in geology and history of geology. Challinor 1969 is unusually accurate and grounded in a close reading of the original published papers; unfortunately it is brief and centered on a discussion of a single Welsh county.

[6] "Normal science" was characterized by Kuhn 1970; its significance is developed and extended in Barnes 1982. It will be obvious that I am speaking here of normal science as an activity pursued under a broad consensus, not one that has achieved total agreement.

[7] See Douglas 1973, Bloor 1982, Collins 1981b, and J. Dean 1979.

The dispute over Cambria and Siluria, although typical of the "unrestricted sciences" in the complexity of its details, can be briefly summarized in broad outline.[8] The events can be shaped into four distinct phases, each of which involves changing technical problems and theoretical emphases and thereby brings different analytical issues to the fore. More generally, the dynamic character of the debate underlines the fact that scientific controversies pass through distinctive stages, with many different patterns evident throughout the history of science. While possessing its own characteristic features, the Cambrian-Silurian dispute also illustrates a number of general issues relating to the study of scientific controversy.

The first phase involves a study of the problems and opportunities presented by collaboration in science. The early decades of the nineteenth century had seen the subdivision of the younger rocks of Great Britain, generally known as the "Tertiary" and "Secondary." But underlying these were the so-called Grauwacke, or Transition strata, of Wales and western England, an ancient series only poorly understood. Early in the 1830s Sedgwick and Murchison—at this time close friends—entered Wales from opposite ends to unravel the sequence of these rocks. In 1835 they gave their respective groups of strata names: Sedgwick's "Cambrian" designation for the slates of North Wales recalled the Latin name for the Principality, while Murchison's "Silurian" memorialized a warlike tribe that had inhabited the Welsh Borders in Roman times. In examining these preliminary researches, I devote special attention to the elements that eventually led to an open break. On a methodological level, the later dispute grew out of differences in the circumstances in which the principals practiced geology. Murchison, trained in a tradition based in London and Oxford, emphasized the fossil remains found in his Silurian system, whereas Sedgwick always maintained the structural outlook characteristic of the science as it developed in the institutional environment of Cambridge. During these early years, such contrasts remained implicit, and difficulties were resolved through careful observance of the rules of geological etiquette. For example, a highly developed sense of scientific property maintained a strong separation between their respective research areas, producing a boundary that was simultaneously geological and territorial. Through the mores developed by the scientific community for fostering cooperation, an uneasy peace was maintained throughout the 1830s.

[8] Pantin 1968 describes the "unrestricted sciences."

The second phase of the debate brought this convenient arrangement to an end. During the early 1840s, other investigators quickly showed that Cambria and Siluria included many strata and fossil species in common. This episode not only illustrates the problems involved in agreeing on certain kinds of apparently straightforward facts, but also raises the issue of dissent in science.[9] Throughout this period the dispute was in many ways a one-man affair, the story of Sedgwick's attempts to replace the widely accepted Murchisonian classification with alternatives that he believed to be more methodologically sound. Other geologists found his constantly shifting alternatives difficult to follow, however, and almost to a man they seconded Murchison in sweeping the Silurian colors down to the base of the fossiliferous rocks. Previously, the establishment of this consensus has been pictured as the result of an arbitrary exercise of power by Murchison. I argue that it actually reflects the commitment of most British geologists to the wide-ranging program of international correlation opened up by the use of fossils. For Murchison, as for many others, this program was closely associated with active support for an emerging ideology of imperial expansion. Increasingly frustrated by the lack of interest in his alternatives, Sedgwick reasserted the existence of a separate Cambrian grouping in 1846, first in a limited version (based partly on a hitherto-unnoticed use of the theory of organic progression as a tool for classification) and soon afterwards by simply extending his Cambrian upwards to encompass the entire so-called Lower Silurian fossil fauna. These revivals brought the controversy into print for the first time. Sedgwick opposed almost all his contemporaries on the single most important issue Victorian geologists faced, the method for constructing a "natural classification" of the strata.

The growth of professional careers in science has recently been the subject of much historical debate. The third phase of the Cambrian-Silurian controversy hinges on this issue and suggests that the importance of professionalization for scientific practice has been considerably underestimated. As a result of the advent of a large-scale government Geological Survey, the outline of the geological succession in Wales was established with a comprehensiveness difficult for individuals to challenge except in details. Opportunities for the older generation of gentlemanly specialists[10] underwent radical changes, and with many of the factual questions in the debate apparently settled, attention centered on history as well as geology.

[9] Dolby 1976.

[10] Rudwick 1985 and 1982b: 189, adapted from Morrell and Thackray 1981.

Both Sedgwick and Murchison wrote elaborate retrospectives invoking the events of the preceding twenty years in arguments over priority in discovery and responsibility for error. Even the poem by Edward Forbes quoted above was written to win support for Murchison; Sedgwick was not amused, and condemned it as "offensive doggerel."[11]

As the controversy during the 1850s turned explicitly towards priority, it continued to foster important new research. One of the most significant findings—a gap in the original Silurian succession at May Hill—added renewed urgency to Sedgwick's case for a separate Cambrian. Another series of discoveries also provided an opening for positions intermediate between the extremes of the increasingly estranged principals. In Wales, America, Scandinavia, and Bohemia, large groups of fossil species were sought and found at the base of the geological column. In the view of Charles Lyell, John Phillips, John Salter, and Charles Darwin, these provided entirely new grounds for the Cambrian. The sparsely fossiliferous strata studied by Sedgwick so long before finally had a fauna of their own. Both Sedgwick and Murchison utterly rejected this compromise, but there can be no doubt that their dispute had generated new scientific knowledge.

The fourth phase of the controversy illustrates the ways in which a priority dispute can continue long after the deaths of the original participants and will be dealt with more briefly here. During this period the growing acceptance of various mediating positions led to the adoption of a threefold division for the Lower Palaeozoic. Most of the lower half of Murchison's Silurian was given neither to the Cambrian nor to the Silurian, but rather to a newly coined system, the Ordovician. Based so clearly upon developments arising out of the controversy itself, the Ordovician (named by Charles Lapworth in 1879 after another Welsh tribe) was finally accepted by the Geological Survey in 1901. Seventy years after the entry of Murchison and Sedgwick into Wales, the debate between Cambria and Siluria was essentially over in Great Britain, and the order was settled in its current form: Cambrian, Ordovician, and Silurian.[12]

It is indicative of the state of research into the history of science that the Cambrian-Silurian dispute has never been the subject of a

[11] Sedgwick 1855: xci.

[12] It was finally settled on an international level only after the Second World War. Consequently, a full inquiry into the closure of the controversy is beyond the scope of this book, which aims to shed light on the practice of a Victorian science in its context.

full historical study. Traditionally, historians have focused almost exclusively on developments in high-level theory. Viewed from such a perspective the history of science becomes a sweeping succession of conceptual breakthroughs made by a handful of heroic individuals—rather than a practical intellectual activity, what Jerome Ravetz has called a "craft."[13] With a few outstanding exceptions, most notably Adrian Desmond's work on vertebrate palaeontology, Mary P. Winsor's investigation of invertebrate taxonomy, and Martin Rudwick's recent study of the Devonian controversy, there are no explicit models for dealing historically with debates in the classificatory sciences.[14] What follows are the guiding principles that direct the present account.

To begin with, I believe that past science can best be approached from the standpoint of general social history, as the product of particular people working in particular places in particular circumstances. In carrying this precept into practice, the sociology of knowledge has proved immensely helpful, because it makes no *a priori* judgments about the boundaries of science or canons of rationality in the past: these are to be established by the historian through empirical enquiry.[15] It can also be applied to questions discussed both inside and outside the boundaries of a scientific community (despite common assumptions to the contrary, sociological explanations in history need not always invoke issues of politics or class). And most important, it provides a rationale for explaining all beliefs sympathetically, without regard to whether they have later proved true or false. My account differs in two ways from those produced by the most thoroughgoing adherents of a sociological approach. First, some authors in this tradition have virtually denied all possibility of imputing intentions to historical figures. Despite

[13] Ravetz 1971. Within the history of geology, I have in mind works like Bowler 1976 and Rudwick 1976a. Along with Leonard Wilson's 1972 biography of Lyell, these contrast with earlier historical literature in their attention to technical arguments; however, they typically do so with an eye on philosophical questions concerning the history of life.

[14] Desmond 1979, 1982; Winsor 1969, 1976; Rudwick 1979a, 1985. Also important in this context are Rudwick 1974 and J. Dean 1979. For surveys of some of the relevant literature, see Secord 1985a and Allen 1983.

[15] Barnes 1974, 1982; Bloor 1976, 1982; Fleck 1979; for important commentaries, see Hesse 1980 and Whitley 1983. A useful survey of sociologically informed writings in the history of science is given by Shapin 1982, an essential starting place for work in this field. Historical case studies include a valuable collection of essays edited by Barnes and Shapin 1979; Neve 1980 and Rudwick 1980 provide illuminating essay reviews.

potential pitfalls, however, an inquiry into intentions seems absolutely essential if individual human personality is not to disappear as an element in science.[16] Second, rather than speaking of the "construction" of knowledge, I have frequently used the participants' own language and spoken of "discovery." Of course, perceptions of the natural world are always mediated by interpretive conventions, and in this sense all knowledge has a social dimension. But in a historical account of a developed science, the existence of these conventions must after a certain point be taken for granted. To do otherwise would severely hinder historical empathy and understanding. In the present dispute, for example, the criteria for making a competent observation in the field were never in doubt. Instead, controversy focused on the relative importance of observations and how they should be placed in broader arrays.

At the same time, I have made every effort to take the technical details of the dispute just as seriously as they were by the participants. Victorian geology depended very much on particulars, and the type specimens so important for a classificatory science were found as outcrops and exposures in the field. The geology of a small region like Wales was potentially of international importance. As a result, much of what follows is devoted to individual beds of limestone and sandstone, and to the fragmentary fossil specimens found in them. Although these cannot explain the existence of the controversy, they constituted its basic materials; they were what the dispute was about. Such details cannot be glossed over or relegated to appendices; one of the principal jobs of the historian of science is precisely to explain how knowledge about nature is created. Just as historians of the physical sciences are turning their attention to experimentation and measurement, those interested in the natural sciences are beginning to investigate the special problems of description and classification.[17]

Reliance on narrative is an equally essential feature of close attention to context. Throughout the controversy, one is faced with rap-

[16] Shapin and Barnes 1979 explicitly denies the validity of imputing intentions; criticisms of this view are in Rudwick 1980. For an account giving the recovery of intentions a central place in intellectual history, see Skinner 1969.

[17] Medawar 1969 pointed to the need for detailed studies of scientific practice, a call since taken up both by historians (e.g. Holmes 1974, 1981, and Gooding 1982, among many other authors) and by sociologists, anthropologists, and ethnographers of science. A pioneering study is Latour and Woolgar 1979; see also the sampling of essays available in Knorr et al. 1980, and current research published in the journal *Social Studies of Science*.

idly shifting classifications developing in response to alternative positions and new discoveries. Understanding the changing fortunes of these classificatory systems requires that the constraints of circumstance be taken into account. Thus when the debate is viewed as an issue developing in time, the early acceptance of Murchison's views and Sedgwick's reasons for opposing them can be seen in clear perspective. There is no need to take sides. Retrospectives written during the debate can be interpreted in their own right rather than taken as authoritative accounts of earlier work. One can also see the dispute gradually widen to draw in an ever-growing circle of participants whose changing views can be charted and understood. Most important of all, a contextual narrative shows that the controversy was a productive affair that helped to forge the keys to its own solution.

While maintaining a broadly narrative approach, I have not adopted the strictly "nonretrospective" technique being pioneered by Martin Rudwick in his study of the Devonian controversy.[18] The kind of exacting attention to temporal order that he recommends is essential in discussing certain issues, such as the emergence of a consensus about technical findings and the making of a specific discovery. But for other issues addressed here it is only of secondary importance, and can even be misleading. Concerns for property and territory, underlying motives for studying science, changes in scientific language, the emergence of new kinds of professional careers—all are best analyzed outside a strict narrative framework. Rudwick has applied a "historical microscope" to an entire scientific debate; more conventionally, I have sometimes used a microscope, sometimes a magnifying glass, and at other times a wide-angle lens, depending on the object in view.

In one respect I have purposely limited my scope. This book is an analysis of a Victorian scientific controversy, not a survey of Lower Palaeozoic geology in the nineteenth century. The geological community in Britain was a tightly knit group of men concentrated around a cluster of institutions, approaches, and concerns. Its leaders viewed the names of the stratigraphical column as a peculiar province of the English nation, and although foreign geologists suggested alternatives to Cambrian and Silurian, only rarely were these taken into consideration. Attitudes towards a parallel controversy in the United States illustrate this insularity. During this debate, which involved rocks of the same age, Ebenezer Emmons la-

[18] Rudwick 1985.

bored to establish a "Taconic system" as the oldest group of fossiliferous rocks. Although the two controversies eventually intertwined in the second half of the century, no British geologist during the period covered here ever contemplated adopting any classification offered by an American. Similarly, I have found only one contemporary British reference—and that a mocking one—to the French geologist Alcide d'Orbigny's suggestion that some of these strata be termed the "Murchisonien."[19]

By uncovering the structure and evolution of a famous scientific debate, this book explores the ways in which established technical procedures, previous training, social pressure, and personal pride joined together to shape a classification. My method is intensive rather than extensive; concrete and particular rather than abstract and general. My strategy is to provide an interpretive picture of a single episode from as many perspectives as possible—to describe, in the anthropologist's sense of a "thick description," the full range of meanings revealed by a scientific debate.[20] Viewed in this way, the closely textured materials of a technical dispute open a broad window on science and its place in culture.

[19] On the Taconic, see Schneer 1969b and 1978. Emmons even offered to withdraw his term in favor of the Cambrian in 1845 (at a time when Sedgwick was advocating "Protozoic" instead); see Emmons to Sedgwick, 28 Nov. 1845, CUL: Add. ms 7652IF103. For Murchisonien, D'Orbigny 1849-1852, *1*: 27 and a mention in E. Forbes 1854b.

[20] Geertz 1973. For historical applications, see Stone 1979 (and the reply in Hobsbawm 1980).

CHAPTER ONE

Controversy and Classification

GEOLOGY enjoyed a remarkable popular success in Victorian England. Crowds thronged to the geological section at the annual meetings of the British Association for the Advancement of Science; Hugh Miller's works sold like fashionable novels; geological imagery graced poems, plays, and common speech. Without the slightest touch of intended irony Tennyson placed geology next to astronomy as a "terrible muse."[1] But as a subject of serious research the science was pursued even in its Victorian heyday by only a small group of men. This coterie of active researchers centered its activities in the Geological Society of London, founded in 1807 as the first specialist society devoted to the exploration of the mineral structure of the globe. From the end of the second decade of the century onwards the Society's leaders exercised almost complete control over creative innovation in the earth sciences in Britain, directing an enterprise focused on the determination of stratigraphical order. The sixty-year battle over the boundary between the Cambrian and Silurian systems of strata was fought out almost entirely within an elite of the Geological Society, whose social organization and scientific practices conditioned the course and indeed the very possibility of such a debate.

The world of the metropolitan geologists revolved on a seasonal calendar: winter and spring were highlighted by spirited discussions at meetings of the Geological Society, and summer and autumn were spent in mapping and classifying strata. Accordingly, we begin in the meeting rooms at Somerset House, and then turn to the "grand and sublime scenes of nature" to be found in the field.

A FORUM FOR DEBATE

By the time Sedgwick and Murchison began the work that was to lead to the dispute, the Geological Society had established highly

[1] "Parnassus," in Tennyson 1899, *8*: 203. Geology's popularity is discussed in many works, esp. D. Dean 1981 and G. Davies 1969: 200-205. Allen 1976 gives a su-

effective procedures for validating new knowledge about the strata. In particular, the organization's meetings and publications provided geologists with carefully circumscribed forums for debate, a process of mutual criticism that culminated in the famous discussions following the reading of papers before the assembled Fellows.[2] These discussions, which occurred at fortnightly intervals, were widely hailed as the most exciting in scientific London. Many of the Fellows were eloquent speakers, possessing the enviable gift of infusing interest into the driest topics. On occasion the arguments echoed through the halls of Somerset House until the small hours of the morning. The quality of the meetings led even the misanthropic Charles Babbage to exclude the organization from his strictures of 1830 on the supposed decline of science in England. "It possesses all the freshness, the vigour, and the ardour of youth in the pursuit of a youthful science," he wrote, "and has succeeded in a most difficult experiment, that of having an oral discussion on the subject of each paper read at its meetings."[3] As Babbage's remarks suggest, such debates (introduced in 1822-1823) represented a considerable innovation for a scientific society in the early nineteenth century. In the wake of contemporary social and political unrest, many considered the geologists rather bold for allowing disputation so central a place in scientific discourse. The sprightly disagreements characteristic of the early Royal Society had faded during its eighteenth-century somnolence, and by the time the Geological Society was founded, many Fellows of the Royal actually maintained that discussions of any sort were precluded by their charter. As a result, throughout most of the nineteenth century the Royal Society meetings suffered a deserved reputation for dullness.[4] Members of the Astronomical Society (founded in 1820), confident of the certainty of their subject, believed it unsuited to discussion. The Linnean Society, dating from 1788, avoided controversy even more assiduously. In order to *prevent* their meetings from becoming a forum for specialized debate, the subjects of forthcoming papers were never announced in advance. A policy more diametrically opposed to that of the Geological Society is difficult to imagine. Thomas Bell, who finally introduced active discussions at the Linnean during his

---

perbly readable introduction to the world of natural history in the nineteenth century.

[2] For more on the Geological Society, see Morrell 1976, Woodward 1908, Porter 1977, Weindling 1979, and Laudan 1977.

[3] Babbage 1830: 45.

[4] Lyons 1944: 254-255; Weld 1848, 2: 472.

long presidency in the 1850s, recalled predictions of "the ruin of the Society" and fears that the "meeting room would become the arena of almost gladiatorial combats of rival intellects."[5]

Many men of science discountenanced debates because their immediacy precluded the patient reflection believed requisite for careful scientific work. In the individualistic and competitive world of Victorian science, tempers might flare in open discussion. Sceptics pointed across the Channel for a lesson in the dangers of unrestricted debate, for the conduct of scientific controversy in Europe appeared bitter and vindictive to English eyes. Disputes like those between Georges Cuvier and Étienne Geoffroy Saint-Hilaire and between Léonce Élie de Beaumont and Louis-Constant Prévost bore the taint of jobbery, with intimate ties to favoritism and the unfair use of patronage. Unrestricted controversy, it was thought, could easily lead English naturalists into "the dreary wild of politics" that continually distracted their Continental counterparts.[6] Given this wariness of controversial excess, it is hardly surprising that even the geologists occasionally betrayed nervousness about the inflammatory character of verbal combat. William Fitton, a fiery-tempered Irishman and no stranger to controversy himself, stressed soon after the innovation had been introduced "the self-command that renders both agreeable and instructive the conversations (I will not call them discussions—much less debates) with which it is now our practice to follow up the reading of memoirs at our table." Fitton's euphemistic dodging of even the hint of controversy is evident, and other presidents were similarly at pains to point to the dispassionate atmosphere of the meetings.[7] While the geologists studiously avoided political controversy, the argumentative style of their debates possessed something of a parliamentary air. A Fellow usually made only one or two contributions to a discussion, often in prepared speeches of some length. Speaking styles and acceptable methods of delivery were generally more formal than those characteristic of most modern scientific meetings, although jokes, innuendoes, and repartee often added spice to the proceedings. Like their parliamentary counterparts, the geologists were conscious of performing for an audience accustomed to brilliant oratory.[8] Several

[5] Quoted in Gage 1938: 30, 52. For the Astronomical Society, see Babbage 1830: 46.

[6] The Cuvier-Geoffroy debate is described in Appel 1976; also of interest is A. Boué to Murchison, 7 Mar. 1868, GSL: M/B20/9. For "the dreary wild," see Sedgwick 1834: xxix.

[7] Fitton 1828: 61; also similar comments in Whewell 1838: 648.

[8] Herbert 1977: 159-178; "performing" in eighteenth-century science has been discussed more extensively: see Shapin 1974 and Schaffer 1983.

leading Fellows—including George Greenough, George Poulett Scrope, and Henry Warburton—were members of both Parliament and the Geological Society. Even the parallel rows of facing benches at the Geological Society mimicked the arrangements at Westminster, although the soporific Society of Antiquaries, which had a similar seating plan, demonstrated that this was no guarantee of controversial vigor.[9]

The English geologists prided themselves above all on the sportsmanlike quality of their scientific sparring (Fig. 1.2). In a presidential address to the Geological Society of Dublin, J. Beete Jukes complimented the sister society in London on the temperance of its discussions:

> Geologists have ever been remarkable, perhaps above every class of scientific men, for the cordial union, the hearty good fellowship, which has knit them together into a band of brothers. Their contentions and dissensions have almost ever been kept down to mere means of eliciting the spark of truth by the collision of various intellects, or at most have been displays of personal strength and skill, knightly combats in all honour and love, preceded and ended by the cordial shake of the hand, which is the manly habit even of our common pugilists.[10]

Or as Babbage wrote in a similar vein, "the continuance of these discussions evidently depends on the taste, the temper, and the good sense of the speakers." At the end of his life Sedgwick recalled the early members as "robust, joyous, and independent spirits, who toiled well in the field, and who did battle and cuffed opinions with much spirit and great good will."[11] On occasion, of course, questions of personality and politics arose in English geological controversy: Jukes's own comment, for example, came as a direct response to the disagreement between Sedgwick and Murchison, eventually "the most noted instance of the *odium geologicum* which the history of British science has yet offered."[12] But for the most part, geologists lauded themselves for patching up their disagreements as gentlemen and remaining friends.

The Geological Society could encourage debate within its walls in part because it was so thoroughly dominated by a small group of leading researchers. Although the membership of the Society to-

[9] Allen 1976: 60 mentions the parliamentary connection. For an illustration of the Antiquaries' rooms, see Needham and Webster 1905, facing p. 239.

[10] Jukes 1854: 108. Similar comments in Trimmer 1844: 434.

[11] Babbage 1830: 45-46; Sedgwick 1855: xc.

[12] Geikie 1875: *1*: 387.

Fig. 1.1. Meeting of the Geological Society. The figure on the left (in this instance the treasurer giving a gloomy financial report) addresses the president in the chair; what is presumably a geological map of England and Wales hangs on the wall behind the seated Fellows.

talled well into the hundreds throughout the century, at any one time the central elite numbered nearer a dozen. Like any elect, these leading Fellows were acutely aware of the precise membership of the circle of "believers." The same names crop up as officers, as participants at meetings, as contributors to Society publications, and as members of the governing Council. When one considers the friendships, joint papers, inside jokes, constant interchange of letters, and exclusive dining club, it comes as no surprise to learn that some

Fig. 1.2. *Phillipsia* attacking *Griffithides*: John Phillips and Richard Griffith cross swords in a dispute over the affinities of Irish Palaeozoic fossils, specimens of which also serve as their armor in controversy. Murchison, the leading expert on the strata involved, acts as umpire and wears the tabard of his Silurian system.

outsiders accused a clique of controlling the organization's affairs. "It is a great difficulty," wrote Charles Moxon in 1842 of the meetings, "to find any assemblage beyond the president, curator, secretary, Messrs. Fitton, De la Beche, and one or two others, and their friends." In a similar but more positive vein, the youthful Andrew Ramsay once compared the Geological Society to a family.[13] During the 1830s and 1840s, the period of the Society's greatest success, the leading members constituted an urban social and scientific elite. Like Lyell, Darwin, Fitton, Murchison, and Greenough, most possessed independent incomes of varying size and security. A few, notably Buckland from Oxford and Sedgwick from Cambridge,

[13] *Geologist* 1842, *1*: 162; Ramsay, Diary, entry for 10 Mar. 1847, ICL(R): KGA Ramsay 1/8 f. 28r. The dominance of a small group of individuals in determining the consensus on any particular issue in the Geological Society is illuminated by Harry Collins's concept of a "core-set": see Rudwick 1982a, 1985, and Collins 1981b.

were beneficed academics, holders of the few university chairs in England devoted to the natural sciences. Almost none were professional men of science in the economic sense except for the few involved in the official Geological Survey, which was founded in 1832 but did not achieve substantial size until in the 1840s.[14] Whatever their source of income, all had pledged themselves to a geological vocation. For aspirants to this status in the higher and lower reaches of the social scale, this could be decidedly difficult (or even impossible) to do. "It is a *tyrannical* study," the aristocratic Charles Bunbury complained of his troubles in becoming an adept in the science, "which requires the devotion of a man's whole time and thoughts, to do anything great in it." On the opposite end of the spectrum the Coventry ribbon weaver Joseph Gutteridge lamented that lack of means and leisure precluded serious geological research by the working class. The relatively homogenous social background of those at the center of British geology ensured a very broad consensus on religious and political issues. One scarcely expected to find freethinking secularists and Scriptural literalists frequenting the rooms of a scientific society in Somerset House.[15] Given basic agreement on these issues among the leading Fellows, the give-and-take of constant debate could form the principal pattern of discourse in English geology without splitting the already small group of active geologists into warring factions of party or class.

Most of those at any particular meeting of the Society were not, however, members of this elite; they came instead simply to listen and learn. A combination of various partial records suggests that attendance ranged anywhere from twenty-five to fifty Fellows and their guests, with especially exciting papers (like one by Louis Agassiz on the ice age) drawing up to a hundred or more. Such a crowd probably exceeded the seating capacity of the meeting room. Most of the men attending—women were not admitted until the present century—had only a general interest in geology and were there to hear the speeches, comments, and rebuttals from the half-dozen or so on the front benches who were really knowledgeable about the topic under discussion. Sir Robert Peel, John Ruskin,

[14] For these points, see Morrell 1976; Porter 1973, 1978; Secord 1986.

[15] Morrell and Thackray 1981, esp. pp. 21-29. One radical who did attend Geological Society meetings was Robert Edmond Grant, professor of zoology at University College London; see Desmond 1984a. The quotations are taken from Bunbury to Horner, 14 Apr. 1848, in Bunbury 1890-1893: *Middle Life*, *1*: 357; and (for Gutteridge) Chancellor 1969: 207-210.

John Lockhart, and other Victorian notables occasionally attended the debates, and professors like Sedgwick or Robert Grant of University College often brought their most promising pupils.[16]

The importance of debate in judging and establishing new geological knowledge in Victorian England can scarcely be overstressed. The discussions provided entertainment for the nongeological visitor, instruction for the novice, and peer evaluation for the expert few. As Murchison told the assembled Fellows in 1833, nearly a decade after debate was introduced, "The ordeal . . . our writings have to pass through in the animating discussions . . . within these walls, may be considered as the true safeguard of our scientific reputation."[17] The debates set methodological standards and offered opportunities for announcing new results, establishing priority claims, and achieving consensus on nomenclature and classification. Rather than representing a breakdown in a process of rational agreement, debate within the Society was the anticipated norm. Thus delivery before the small community of geologists at Somerset House, not publication, was usually the most important moment in the history of a paper. Darwin summed up the situation to a friend while finishing his book on the geology of South America:

> As for your pretending that you will read anything so dull as my pure geological descriptions, lay not such a flattering unction on my soul for it is incredible. I have long discovered that geologists never read each other's works, and that the only object in writing a book is a proof of earnestness, and that you do not form your opinions without undergoing labour of some kind. Geology is at present very oral, and what I here say is to a great extent quite true.[18]

Notably, the Society strictly prohibited any published mention of the discussions, presumably to avoid the indignity of having their disagreements aired in print. The secretary placed abstracts of the formal communications to the Society in official *Proceedings*, in the weekly *Athenaeum* and *Literary Gazette*, and in the monthly *Philo-*

[16] Information derived from the minute books of the Geological Society (which, with the exception of the annual meetings each February, list only the Fellows bringing visitors), together with a variety of anecdotal evidence. The Agassiz meetings are described in Woodward 1908: 136-144.

[17] Murchison 1833a: 464.

[18] Darwin to J. M. Herbert, [1846], in F. Darwin 1887, *1*: 334-335.

sophical Magazine. Full versions were usually published (albeit with painful slowness) in the quarto *Transactions of the Geological Society* or, after 1845, in the more expeditiously printed *Quarterly Journal*. But any editor who included the smallest scrap of a discussion received an angry letter from the officers.[19] Thus controversy in its main forum was ephemeral, and the importance of debate for early Victorian geology is scarcely evident in the materials that reached the hands of a publisher. As a result of the rule of silence, the oral history of scientific disputes must be reconstructed from manuscript letters and diaries.

The limits on the reporting arose for several reasons. Many felt that the freedom of the discussions depended on their being kept in relative confidence. Outsiders could attend, but only at the invitation of a Fellow, and consequently the meetings possessed something of the character of an exclusive debating society. In addition, geologists wished to maintain a public image of their science as a stable, ordered enterprise, an "inductive science" akin to astronomy or physics. An undue reputation for controversy would tarnish this image and make geological knowledge seem uncertain and provisional, an easy target for Scriptural geologists and others outside the pale. The language of geology also became increasingly technical and esoteric in the course of the century, and opportunities for errors in reporting by nonspecialists multiplied accordingly. This concern for accuracy led to an understandable desire for strict control over new geological information reaching the public. The annual meeting every February, at which the president read a lengthy address on the achievements of the preceding year, provided the principal occasion for an official evaluation of the state of knowledge. Another mark of the same tendency is found in general periodicals, such as the *Edinburgh*, the *Quarterly*, and the *Athenaeum*; almost all their reviews of geological books were written by members of the inner circle of specialized experts.[20]

Similar concerns motivated the Geological Society's attitude towards the refereeing and publication of scientific papers after they had been read at a meeting. Any paper delivered before the Society became its property, and only rarely was an author allowed to make

[19] E.g. W. J. Hamilton to C. Moxon, [1842], in Woodward 1908: 146. Moxon was editor of the *Geologist*.

[20] Fitton, Scrope, Lyell, Sedgwick, Edward Forbes, Richard Owen, and many other Fellows contributed reviews; for details, see Houghton 1966–. Most of the authors of the anonymous reviews used in the present study have been identified through this source.

alterations on his own accord before publication. (This policy was less strict before the late 1830s, perhaps because of inordinately long delays in publishing the *Transactions*.) For an individual with second thoughts about a paper, the only recourse was a request to the Council for changes or possible withdrawal. The Council, on the other hand, was perfectly free to require substantial alterations or shortening, and it frequently did so after taking advice from referees. Anything on the far reaches of theory was usually suspect, especially if unaccompanied by extensive factual documentation. Personal animus was strictly taboo, and reference to previous studies being corrected or rejected was usually kept to a minimum. In spite of the restrictions, however, the Geological Society undoubtedly offered a far more liberal publication policy than did the Linnean or Zoological societies; here a parallel can be drawn with their differing attitudes towards verbal disputation.[21] Although geologists sometimes chafed under their Society's strictures on the introduction of overtly controversial matter into published papers, most restrictions operated as unwritten laws maintained by mutual consent. The same self-restraint that characterized the discussion of papers also regulated their appearance in print. As a result, often only extensive familiarity with contemporary geological literature makes it possible to place particular contributions within the process of scientific debate. What appear today to be dry stratigraphical treatises were understood at the time as provocations or rebuttals in controversy.

Occasionally the etiquette governing publication and discussion was thrown aside in the heat of controversy, and debate passed from the circle of specialists into a more public arena. In such cases geologists appealed to wider audiences at meetings of the British Association, in general scientific periodicals like the *Philosophical Magazine*, in books and other independent publications, or even in popular illustrated weeklies and newspapers. Some of these alternatives, particularly the British Association, could also be used for staking claims to priority in advance of the winter meetings of the Geological Society. As arenas for disputation, they had their own rules and limitations while permitting wider publicity and a much greater latitute of controversial discourse.[22] They were generally employed for serious debate only after the usual means of resolving disputes within the geological community had failed, when indi-

[21] Herbert 1977: 172-176.

[22] For the British Association (and much else besides) see Morrell and Thackray 1981. MacLeod and Collins 1981 offer a useful collection of essays.

viduals required controversial resources unavailable in a discussion among specialized experts.

The Stratigraphical Enterprise

Transported back in time to one of the Geological Society debates, a modern student of Victorian intellectual history would certainly be surprised, perhaps disappointed, or even (as were Prince Albert's equerries) lulled to sleep.[23] There can be no doubt that the study of the earth broached questions of profound importance for the intellectual life of the nineteenth century: the reality of a recent flood, the uniformity of natural processes, the age of the world, the progressive history of life, and the origin of man.[24] But these issues, although understandably emphasized in the secondary literature on the history of geology, occupied popular reviews, prefaces, and presidential addresses and were seldom discussed during the average Geological Society meeting. Most British geologists were stratigraphers, and the debates at Somerset House usually dealt with questions of classification—with the order and succession of the materials of the earth's crust. Thus, more than theorists like Lyell or Darwin, the protagonists in the Cambrian-Silurian controversy represent the principal tradition in British geology in the early nineteenth century. The word "stratigraphy" itself arose during the second half of the century when the subject became a specialty *within* geology.[25] Before that time all "geologists" were assumed to be students of the strata.

Roy Porter has shown how geology emerged as a coherent discipline at the end of the eighteenth century with rock strata as its central focus. In stratigraphy, as in Abraham Werner's "geognosy" or Alexander von Humboldt's "pasigraphie," the fundamental relation to be determined was superposition: the placement of strata in sequential order, one above the other. Such an ordering of strata, securely fixed in a limited geographical area, could then serve as a "type" for similar sequences elsewhere, and the network of scientific order could gradually be extended across the globe on the basis of palaeontological and lithological similarities. Bands of limestone,

[23] Woodward 1908: 167-168.

[24] Important studies of the great theoretical controversies include Gillispie 1959, Rudwick 1976a, Bowler 1976 and 1984, W. F. Cannon 1960a, Hallam 1983, Rupke 1983, Ruse 1979, and L. G. Wilson 1972.

[25] See "stratigraphy" in the *Oxford English Dictionary*. Challinor 1978 gives historical information on many geological terms.

sandstone, and conglomerate gained meaning from their place in this worldwide sequence of strata, from their positioning in a classification. The elite of the Geological Society spent most of their time tracing and mapping strata so as to put them in order; the immediate goal was not a history of the earth but rather a classification of the stratigraphical column.[26]

After the last Geological Society meeting of the spring season, the leading researchers gathered up their hammers and their wives and set off on extensive stratigraphical tours. For an understanding of the practical activity of geological classification, we must follow them from the rooms of the Society into the field. With the development of the Romantic taste for wild nature and the creation of a network of inns, itineraries, roads, and guides, travel had become a fashionable pursuit for the leisured classes. Geological fieldwork arose in part as a scientific counterpart of these developments. Specialized handbooks and maps guided the carriages of the leading Fellows much as Gilpin and Pennant had directed their nongeological predecessors. Stratigraphy, like sketching scenery, keeping a journal, or collecting beetles, gave serious purpose to tours that might otherwise have seemed frivolous or unimproving. In recommending "travel—travel—travel" as his three pieces of advice to the beginning geologist, Lyell gave counsel his contemporaries were only too willing to follow.[27]

The geological traveller hoping to establish the sequence and precise distribution of an unstudied group of strata usually began with a broad overview of the area to be investigated. An observer skilled in inferring underlying geological structure from surface topography could often accomplish much of his work from a single well-chosen mountaintop. With a general conception of his field area in mind, the geologist next planned a series of parallel traverses more or less at right angles to the outcrops of the strata. Walking across the upturned edges of the geological succession, he marked in a notebook the directions of dip and strike of each stratum encountered. The angle of dip, usually estimated by sight, measured the maximum inclination of tilted strata from the horizontal; the point of the compass towards which the rock beds inclined was known as the direction of dip. The strike of the strata, always at right angles to the direction of dip, represented the line of intersection of the tilted plane of the strata with the horizontal. Along the line of his

[26] Porter 1977: 202-215; Rudwick 1982a.

[27] C. Lyell 1830-1833, *1*: 56-57; for geology and Romanticism, see D. Dean 1968; Nicolson 1959; Porter 1978: 820-822.

Fig. 1.3. *Punch*'s geologists on a British Association excursion.

traverse the geologist also noted the lithological characters, thicknesses, and fossil contents of each outcrop. The resulting traverse (also known as horizontal) sections united isolated exposures and observations into a connected slice of the succession, cutting across the countryside like a trench. Such sections were best exposed along stream beds and coastal cliffs, two of the favorite haunts of the Victorian geologist.[28]

In addition to these traverse sections, geologists very occasionally traced out an individual stratum along the entire length of its outcrop. Mapped in this way, a bed with distinctive lithology could serve as a marker horizon, with its appearance in the separate traverse sections allowing them to be coordinated into a general picture of regional geological structure. In more detailed mapping, virtually all the individual rock formations were followed out at full length, and the details of every outcrop in the region recorded. But such a procedure was relatively rare in the exploratory investigations characteristically pursued by the gentlemen geologists.

These stratigraphical relations of the rock beds with one another, often called their "physics," provided the foundation for establishing the geological succession in a particular area. This area could then become a type, to serve as a standard for spreading stratigraphical order to other regions. Of course, in this process of exten-

[28] Details of field methods can be found in most contemporary manuals, e.g. De la Beche 1831, 1833 (esp. pp. 598-606), although obviously these cannot communicate the important tacit elements in fieldwork training. See also C. Darwin 1977b, and the helpful discussions in Rudwick 1976b and Donovan 1966.

sion the evidence of superposition provided unequivocal testimony, but direct links of this kind were frequently unavailable. In most cases, isolated sequences of strata had to be matched with the standard succession using other evidence. During the 1830s, most geologists relied on a combination of simple lithological and palaeontological characters for such correlations. Although an individual bed of sandstone or limestone could thin out unexpectedly, lithology alone remained constant enough over very short distances to serve as a useful first approximation, especially within the confines of a single country. Fossils, however, had become established both as crucial elements in any finished local description and (more contentiously) as the sole criterion for correlations between strata of different countries. The development of this emphasis was associated with the names of William Smith in England and Georges Cuvier and Alexandre Brongniart in France, although none of these advocates of the palaeontological method (nor their immediate followers) demanded more than a few specimens for establishing a correlation. Sometimes even these few were difficult to find and hard for the nonspecialist to recognize; in other cases, fossils were so abundant as to constitute in themselves a prominent part of the lithology. In actual practice geologists resorted to an eclectic combination of criteria to correlate strata, using fossils, lithology, and the available evidence of physics to form a unified picture of the succession. The balance struck among these diverse elements varied according to individual training and inclination, as a host of contemporary controversies evidence only too well.

As geology came of age in the early decades of the nineteenth century, a specialist corps of paid palaeontologists emerged, typified by James Sowerby and his family of fossilists in London. Later, such men tended to become unrivalled experts on a single group (corals or molluscs, for example) and identified specimens from the field for an appropriate fee. In general, they came from humbler social backgrounds than the major metropolitan stratigraphers. An equally important part in the stratigraphical enterprise was played by the many amateur cultivators of science found throughout the provinces. They constituted the audience for geological publications and the British Association, and many occupied highly respectable positions in their local communities. In an elaborate system of paternalistic exchange, the provincial cultivators furnished the "great guns" of the Geological Society with much of their observational ammunition. As we shall repeatedly see, all the participants in the present controversy relied heavily upon the empirical findings of local men.[29]

The wealth of detail that went into any geological work is one of the most difficult aspects of the science to convey in a brief compass. William Whewell, Master of Trinity College in Cambridge, wrote that geology was in itself almost a liberal education, "one of the best schools of philosophical and general culture of mind." It required what the Prussian geologist Leopold von Buch called "the necessary talent of combination," the ability to generalize from a multitude of facts seemingly unimportant in themselves.[30] Unlike physics, geology could claim no great conceptual difficulty or need for higher mathematics; rather, its technical complexity lay in the burdensome quantity of particulars that had to be brought to bear on even the most straightforward problems. In the course of a single summer's fieldwork a geologist might fill his notebook with thousands of details, each tied to the circumstances of an individual locality. Guided by expectations of an underlying consistency and simplicity, the geologist then interpreted particular exposures of the strata to find their place in a general classificatory scheme.

A careful observer could determine the succession with relative ease in areas like the Isle of Wight, where the rocks were well exposed and relatively undisturbed. These localities served as training grounds for nineteenth-century geologists. However, anyone who has trudged across a Welsh bog or a valley deep with alluvium can testify that this task was rarely so simple. Strata could be obscured by vegetation, buried under layers of soil or gravel, altered by cleavage or jointing, tilted upside down, or pulverized by faulting. Even the best maps and sections represented extrapolations from a few data points: an abandoned quarry, a canal or road cutting, the stones brought to the surface by a farmer's plough. As a result, the existence of any order whatsoever in the materials of the earth's crust was far from obvious or commonsensical, and the "language" of the strata could only be read by the trained initiate who coordinated an assemblage of scattered outcrops into a coherent classification. The ambiguity and complexity of the evidence made disputes almost inevitable even when geologists agreed on the criteria to be used in ordering the strata.[31] The most straightfor-

[29] See Allen 1985, Morrell and Thackray 1981, Reingold 1976, Rudwick 1982b and Sheets-Pyenson 1982 for discussion of the role of experts and collectors; and Secord 1985c for a brief characterization of intellectual paternalism in Victorian science.

[30] Buch to Murchison, 20 Apr. 1846, GSL: M/B33/6; Whewell 1837, 3: 524.

[31] Rudwick 1976b; Harrison 1963. For an extreme example of the opposite view,

ward maps and classifications could in this way become polemical weapons in the Geological Society debates.

Geology, as a science that established the geographical extent and spacial orientation of rock strata, depended on an extensive repertoire of illustrative materials—colored maps, strata charts, sectional diagrams, and pictorial views—all intended as shorthand expressions of the order perceived in nature. Of these visual techniques, the geological map best captures the essence of the science in the Victorian era. As Sedgwick once wrote, the geologist was like a mechanical engineer, dissecting the earth to reveal the interconnections of its parts.[32] Maps, to carry the analogy further, served as the blueprints of the stratigraphical enterprise. Guided by the neatly bounded taxonomic units that the stratigraphical sequence provided, geologists could spread scientific order across the earth's surface. This integration of locality and taxonomy, expressed on a map, represents the fundamental character of nineteenth-century geology. As Greenough, the first president of the Geological Society, wrote in 1840:

> Words following words in long succession, however ably selected those words may be, can never convey so distinct an idea of the visible forms of the earth as the first glance of a good Map. . . . In the extent and variety of its resources, in rapidity of utterance, in the copiousness and completeness of the information it communicates, in precision, conciseness, perspicuity, in the hold it has upon the memory, in vividness of imagery and power of expression, in convenience of reference, in portability, in the combination of so many and such useful qualities, a Map has no rival.

Greenough, who had been trained in Germany at Göttingen, was a thoroughgoing Humboldtian who included geology under an all-embracing geographical vision. Under his leadership the fledgling organization had sensibly focused its earliest collaborative efforts on a geological map of England and Wales. Considerably more accurate in most areas than Smith's remarkable solo effort of 1815, this map continued to serve as a basis for further studies long after its appearance in 1820.[33]

---

which nonetheless contains much useful information on mapping technique, see Greenly and Williams 1930.

[32] Sedgwick 1843a: 238. Rudwick 1976b provides a general discussion of geological illustration; [Laudan] 1974 investigates the early history of mapping in the science.

[33] Rudwick 1963; for a different view, Laudan 1977. The quoted passage is from Greenough 1841: lxxvi.

The emphasis on mapping gave geological knowledge a uniquely territorial dimension. The competitive world of early Victorian science viewed trespass on another man's subject matter as a serious offense, and to impinge unduly upon a topic already taken up in earnest by another could readily be considered bad manners or even theft.[34] In geology, the importance accorded to maps associated such considerations directly with landed property and, as in the present case, with territorial imperialism. Similarly, classifications also received an unusually high value as scientific property because they could achieve concrete visual representation on a map; practitioners typically thought of rock formations as geographical areas rather than as periods of time, and accordingly the territory occupied by particular strata became the personal preserve of individual geologists. From beginning to end, the controversy over Cambria and Siluria revolved around questions of acreage, territory, property, and boundaries.

A reduction of Greenough's 1820 map appropriately formed the frontispiece of William Daniel Conybeare and William Phillips's *Outlines of the Geology of England and Wales*, a work that epitomizes the monographs and memoirs produced by the inner circle at the Geological Society. This handbook, published in 1822, served as a concrete exemplar of stratigraphical practice and defined the course of much subsequent research. As we shall see, its influence over the research of Murchison and Sedgwick was profound and lasting. Conybeare and Phillips devoted most of their book to a description of the British sequence, beginning with the recent unconsolidated deposits and ending at the base of the Old Red Sandstone. Notably, they named the principal groups of strata "supermedial," "medial," and "submedial"—in other words, according to their order of superposition (Fig. 1.4, column c). In the text, strata are usually described as being above, below, or parallel to one another, rather than as younger, older, or contemporaneous; they are objects in themselves rather than signs of a succession of changing environments or the passage of time. Only in a brief introductory section is the *history* of the earth discussed at all.[35]

Conybeare and Phillips's handbook indicates that geologists in the early nineteenth century usually thought of sand*stone* and lime*stone*. Only rarely, and typically in popular works and public lec-

[34] For the competitive and individualistic character of Victorian science, see Morrell 1971.

[35] Conybeare and Phillips 1822. Kuhn 1970: 187-191 develops the notion of exemplars in science.

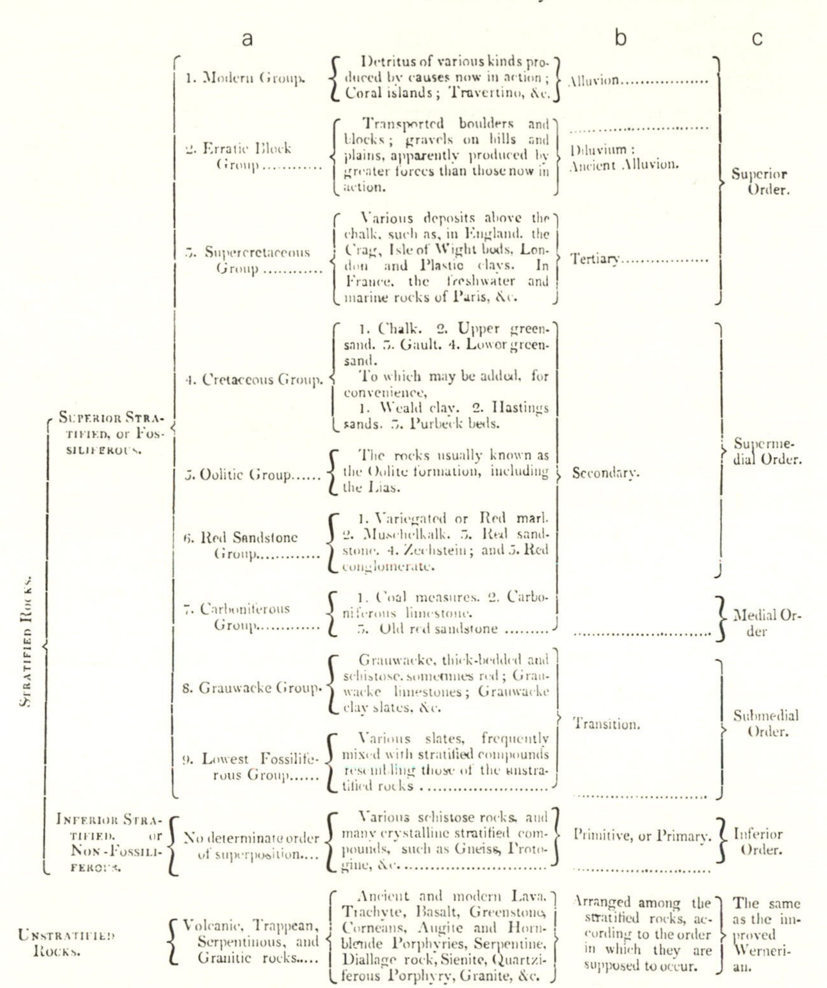

	a		b	c
	1. Modern Group.	Detritus of various kinds produced by causes now in action; Coral islands; Travertino, &c.	Alluvion..................	
	2. Erratic Block Group............	Transported boulders and blocks; gravels on hills and plains, apparently produced by greater forces than those now in action.	Diluvium: Ancient Alluvion.	Superior Order.
	3. Supercretaceous Group............	Various deposits above the chalk, such as, in England, the Crag, Isle of Wight beds, London and Plastic clays. In France, the freshwater and marine rocks of Paris, &c.	Tertiary..................	
	4. Cretaceous Group.	1. Chalk. 2. Upper greensand. 3. Gault. 4. Lowor greensand. To which may be added, for convenience, 1. Weald clay. 2. Hastings sands. 3. Purbeck beds.		
Stratified Rocks. Superior Stratified, or Fossiliferous.	5. Oolitic Group......	The rocks usually known as the Oolite formation, including the Lias.	Secondary.	Supermedial Order.
	6. Red Sandstone Group............	1. Variegated or Red marl. 2. Muschelkalk. 3. Red sandstone. 4. Zechstein; and 5. Red conglomerate.		
	7. Carboniferous Group............	1. Coal measures. 2. Carboniferous limestone. 3. Old red sandstone	....	Medial Order
	8. Grauwacke Group.	Grauwacke, thick-bedded and schistose, sometimes red; Grauwacke limestones; Grauwacke clay slates, &c.	Transition.	Submedial Order.
	9. Lowest Fossiliferous Group......	Various slates, frequently mixed with stratified compounds resembling those of the unstratified rocks	....	
Inferior Stratified, or Non-Fossiliferous.	No determinate order of superposition....	Various schistose rocks, and many crystalline stratified compounds, such as Gneiss, Protogine, &c.	Primitive, or Primary.	Inferior Order.
Unstratified Rocks.	Volcanic, Trappean, Serpentinous, and Granitic rocks.....	Ancient and modern Lava, Trachyte, Basalt, Greenstone, Corneans, Augite and Hornblende Porphyries, Serpentine, Diallage rock, Sienite, Quartziferous Porphyry, Granite, &c.	Arranged among the stratified rocks, according to the order in which they are supposed to occur.	The same as the improved Wernerian.

Fig. 1.4. Strata classifications commonly used in England in 1831: (*a*) De la Beche; (*b*) "Improved Wernerian" as accepted by many geologists, although Transition could also include the Carboniferous and the Primary, or both; (*c*) Conybeare and Phillips's *Outlines*.

tures, did they picture ancient beaches surrounding a primordial sea. Maps and sections, not "local histories," were the primary goal.[36] As Whewell noted in his *History of the Inductive Sciences* of 1837, almost all geological work was descriptive and phenomeno-

[36] For "local histories" (Jonathan Hodge's term), see W. F. Cannon 1976: 112.

logical. The discussion of mechanisms and causes, those features that made geology the type of his special class of "palaeotiological sciences," had only just commenced. In Whewell's opinion both the so-called uniformitarian and catastrophist views of geological causation were premature and there existed as yet no truly "theoretical or philosophical history" in the science at all. Thus geology, like its sister palaeotiologies, remained incomplete. For a discussion of the chief principles used in most contemporary geological studies, Whewell referred his readers to an earlier section of the *History* dealing with botanical classification. "We may observe also," he wrote in another passage, "that descriptive geology, such as we now speak of, is one of the classificatory sciences, like mineralogy or botany; and will be found to exhibit some of the features of that class of sciences." The essentially static character of Whewellian "descriptive geology" is often missed. Philip Lawrence, for example, conflates it with what he calls "historical geology," a category actually approximating to the elusive goal of palaeotiology.[37] A view similar to Whewell's was taken by Sedgwick, who opened his lectures for over half a century by locating geology among the sciences. Although he hoped that the study of the earth would draw from physics and mathematics, he linked the discipline closely with the rest of natural history. His published syllabuses show that this was no mere prefatory flourish, for more than two-thirds of the remaining lectures are devoted to stratigraphical classification.[38] Even Lyell's *Elements of Geology*, which might be expected to paint a vivid picture of changing environments and ancient forms of life, consists largely of a dry listing of strata and the fossils used to identify them. As it gradually expanded into a comprehensive textbook, the *Elements* in later editions became even less dynamically oriented. Of course, in debating the uniformity of natural processes in the wake of Lyell's earlier and considerably more famous *Principles of Geology* (1830-1833), geologists had begun to explore a philosophical dimension unique to the palaeotiological sciences.[39] We shall see that John Phillips and De la Beche were particularly impressed by the need to apply the new approach to the interpretation of the older strata, and the Cambrian-Silurian controversy shows that theoretical constructs like progressionism and uniformitarianism had a role in the framing of classifications.

[37] Lawrence 1978: 102; Whewell 1837, *3*: 479-622, at p. 491.

[38] Sedgwick 1821: 4.

[39] C. Lyell 1838, 1830-1833. My view differs here from that of L. G. Wilson 1972 (see esp. 506), which nonetheless provides the only extensive study of the *Elements* and its place in Lyell's work.

When seen exclusively in light of these developments in theory, geology has often been hailed as the first truly historical science. But the preceding outline of contemporary field methods and aims indicates that it better belongs under the broad rubric of classificatory natural history. David Oldroyd and other authors, in developing the perspective of Michel Foucault into a claim that the basic orientation of the earth sciences in this period was fundamentally "historicist," have thus ignored the bulk of geological practice. As Paul Farber and Mary P. Winsor have suggested, the continuities with the taxonomic vision usually associated with the Enlightenment are striking throughout the natural sciences in the nineteenth century. While naturalists found order in the living world in terms of a divine plan or ideal archetype, geologists ordered the earth's crust spacially and structurally, through the essentially static medium of a geological map. Most geologists thus classified strata in much the same way as botanists classified plants or palaeontologists fossils: the hammer joined the butterfly net, the collecting jar, and the herbarium as an emblem of natural history. Even stellar astronomy could be part of the classificatory enterprise. The period that witnessed the rise of Kew Gardens and the creation of the London Zoo also produced the Geological Society and the Geological Survey.[40]

In geology, the emphasis on classification resulted in part from a narrowing of horizons, the rejection of the grandiose "theories of the earth" so common during the previous century. Porter has shown that sensitive questions involving the origins of man, of new species, and of the planet itself had little or no part to play in the newly defined science. In the turbulent decades following the French Revolution, the focus on strata offered a relatively safe alternative to earlier and more speculative ways of perceiving the planet.[41] But the tacit agreement to thus limit geology entailed more than an escape into an ideologically neutral empiricism. Geology, like the other elements of natural history, had powerful supports from the social order of nineteenth-century England: for all its internal quarrels, classificatory stratigraphy spoke to the public as a science that established position and place. The fundamental appeal of such an activity for a society increasingly conscious of class

[40] Compare Farber 1982a, Farber 1982b, and Winsor 1976 with Oldroyd 1979 and Foucault 1970. Janet Browne's (1983) discussion of geology as "a science of processes" would seem basically allied to the latter perspective, as would Rupke 1983 and a number of older works, e.g. Toulmin and Goodfield 1965: 141-170; Gillispie 1959. J. Greene 1959 discusses the replacement of a static by an evolutionary view of nature; Schaffer 1980 discusses stellar astronomy and natural history.

[41] Porter 1977, 1979.

and an elite troubled by dissent from the "lower" orders is unmistakable. In 1841 *Punch* made the point ridiculously concrete by compiling a stratigraphical column of social classes, from "people wearing coronets" at the summit to "tag-rag and bob-tail in varieties" at the very base (Fig. 1.5).[42]

To see the rocks as a layered sequence of strata was at first far from obvious. Only at the end of the eighteenth century had naturalists agreed that a coherent succession might be discovered amidst the shattered ruins of the earth's crust; perceiving order of any kind in the face of such apparent confusion required considerable training and frequently a wholesale rejection of common sense. In complicated districts like Wales the Victorian geologist could readily experience the exhilaration of grappling with "primordial chaos," of forcing disruption and disorder into the confines of a classification. More fundamentally, in placing the component parts of the earth's crust into relation with one another, geologists hoped to reveal the outlines of the divine plan for the world by reading the "pages" of the "book" of nature. While taking the extreme antiquity of the earth for granted, almost all the major figures of the period believed that the strata exhibited traces of divine planning and foresight. Buckland and Murchison, for example, showed that the very structure of the English coal basins proclaimed the benevolent wisdom of the Creator.[43] In their view, geological classifications were emphatically not artificial groupings for convenience, but attempts to uncover divisions preexistent in nature. In writing of the natural sciences more generally, Martin Rudwick has emphasized that "classification was not a means to an end, a clue to evolutionary relationships, for example: it was itself an end, the end of knowing the true order of Nature."[44] As a part of natural history, geology took a prominent role in this search for order. It has long been recognized that in their English versions the theories of catastrophism, progressionism, neptunism, and diluvialism all postulated divine intervention in the history of nature to lesser or greater degrees, in part as antidotes to religious disbelief and civil disorder. But the influence of such concerns penetrated to a more basic level as well, to the day-to-day activity of the geologist in the field. The Creator's benevolent maintenance of the natural and social order was celebrated

[42] [Anon.] 1841; Perkin 1969; J.F.C. Harrison 1979. For theoretical perspectives on the relationship between classifications of nature and society, see Bloor 1982, which offers an important updated reading of Durkheim and Mauss 1963.

[43] W. Buckland 1837, *1*: 524-547; Murchison 1834c.

[44] Rudwick 1976a: 208.

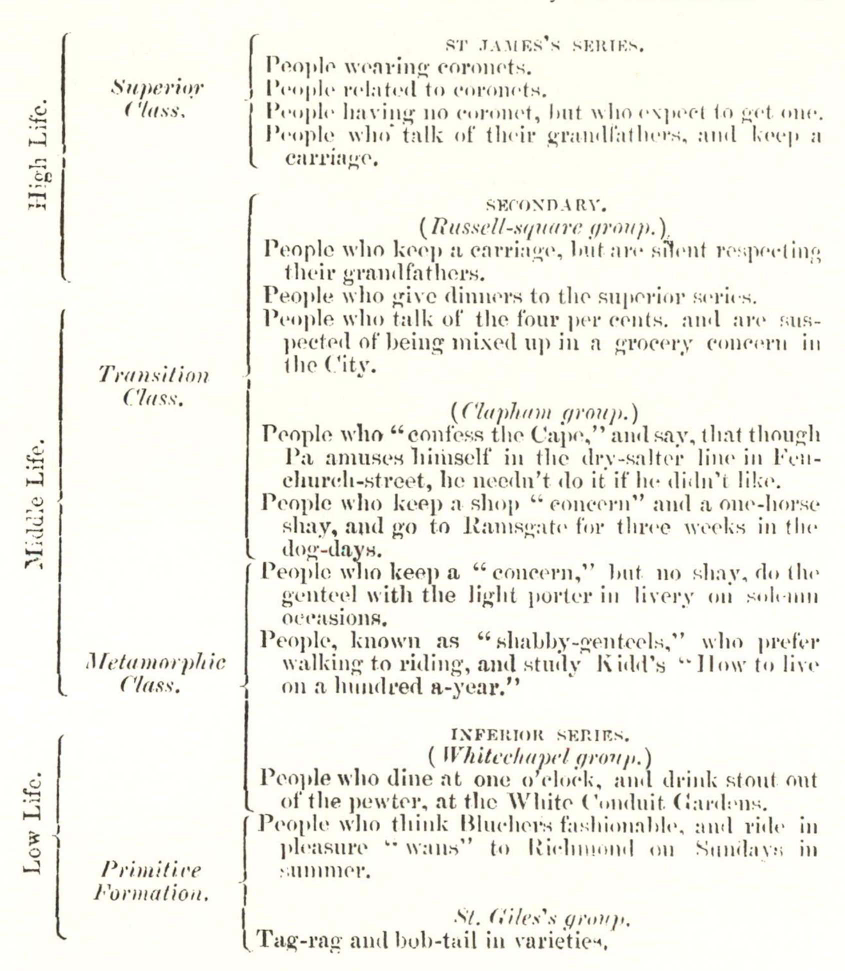

High Life.

Superior Class.

ST JAMES'S SERIES.

People wearing coronets.
People related to coronets.
People having no coronet, but who expect to get one.
People who talk of their grandfathers, and keep a carriage.

SECONDARY.
(*Russell-square group.*)

People who keep a carriage, but are silent respecting their grandfathers.
People who give dinners to the superior series.
People who talk of the four per cents. and are suspected of being mixed up in a grocery concern in the City.

Middle Life.

Transition Class.

(*Clapham group.*)

People who "confess the Cape," and say, that though Pa amuses himself in the dry-salter line in Fenchurch-street, he needn't do it if he didn't like.
People who keep a shop "concern" and a one-horse shay, and go to Ramsgate for three weeks in the dog-days.

Metamorphic Class.

People who keep a "concern," but no shay, do the genteel with the light porter in livery on solemn occasions.
People, known as "shabby-genteels," who prefer walking to riding, and study Kidd's "How to live on a hundred a-year."

Low Life.

INFERIOR SERIES.
(*Whitechapel group.*)

People who dine at one o'clock, and drink stout out of the pewter, at the White Conduit Gardens.

Primitive Formation.

People who think Bluchers fashionable, and ride in pleasure "wans" to Richmond on Sundays in summer.

St. Giles's group.

Tag-rag and bob-tail in varieties.

Fig. 1.5. *Punch*'s "Geology of Society."

not only in geological theories, but also, if less directly, in ordinary strata charts, museum catalogues, and stratigraphical maps. In many ways, the modern stratigraphical column bears traces of the man-centered vision of the Victorians: the divisions become finer and more nuanced as one approaches the present, and the advent of the human species originally defined the commencement of the last major geological era.[45]

[45] The connections at the level of theory are brought out by the stimulating analysis in Gillispie 1959, esp. pp. 217-228, and excellent articles by Brooke (1979) and W. F. Cannon (1960b). The ideological underpinnings of the standard Victorian classifica-

The search for order in the strata was well advanced by the time Murchison and Sedgwick entered Wales in 1831. Before turning to a discussion of their early work, I shall briefly outline the accomplishments of the stratigraphical enterprise up to that date. Three principal classifications used by the Fellows of the Geological Society are shown in Figure 1.4, derived from a chart in a textbook by De la Beche. Conybeare and Phillips's *Outlines* in 1822 had summarized knowledge of the geological record from the most recent unconsolidated deposits of eastern England to the base of the Old Red Sandstone in the west. These rocks, the "Tertiary" and "Secondary" of most classifications, were relatively flat and undisturbed and possessed a clear sequence of lithological types and many fossils. The Old Red Sandstone at their base, which occupied much of Herefordshire and southern Scotland, was a gritty reddish rock often classified as a lower extension of the Carboniferous Limestone. Far below the Old Red, at the base of the entire stratigraphical column, geologists had found "Primary" rocks: granites and gneisses thought to result either from metamorphism of preexisting strata or from the cooling of the originally incandescent globe.

Between these Primary rocks and the base of the Secondary lay a vast thickness of strata occupying much of the mountainous hill country of Britain, particularly Wales, the Welsh Borderland, large parts of southern Scotland, Cornwall and Devon, and the Lake District (Fig. 1.6). British geologists customarily referred to these older strata either as "Transition" rocks or as the "Grauwacke." At the end of the eighteenth century Abraham Werner had called them the Transition series because of their transitional character between two states of deposition in his universal ocean. Although by 1831 almost all British geologists had rejected the neptunistic theory implied by "Transition," they continued to use the term as a practical way of referring to rocks with a particular position in the sequence. The alternative usage "Grauwacke" (or "Greywacke" in its anglicized version) was a German mining term for a coarse dark sandstone characteristic of the older strata. In many of the mountainous districts where these rocks were found, their physics could be interpreted only with great difficulty, for they were folded, faulted, and altered by cleavage and low-grade metamorphism. With the exception of a few fossiliferous bands known from the top of the sequence, the Transition strata possessed relatively few organic re-

---

tions of the geological column are especially apparent when these are compared with a radically different alternative; see Desmond 1984b: 402.

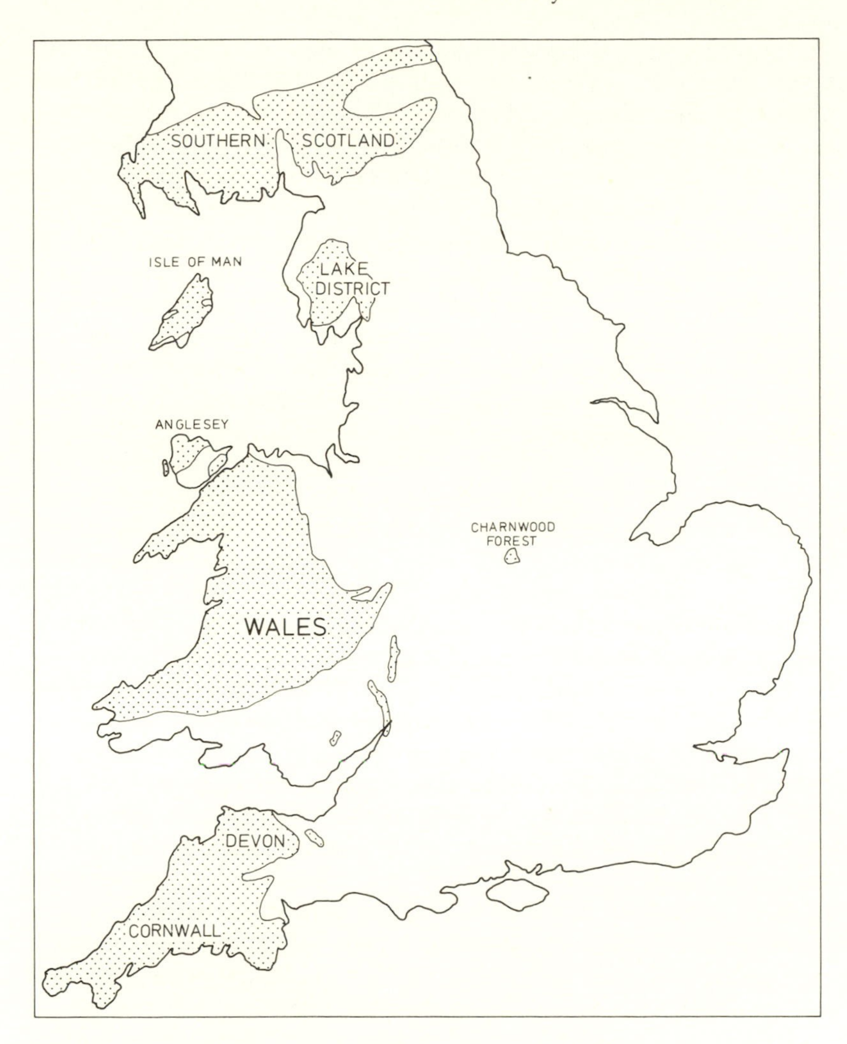

Fig. 1.6. Areas of Primary and Transition rocks in 1822.

mains and could be correlated with similar rocks in other countries only in a very general way. However, numerous fossil species were regularly used as indicators of the presence of rocks of this age as a whole: trilobites, brachiopods, corals, and other invertebrates had all been found in them and were listed in the usual works. Compared with the Secondary and Tertiary strata, though, the Transition rocks had received little attention since the general shift from mineralogical to palaeontological criteria for correlation. In 1822

Conybeare and Phillips could apologize for their omission of the strata below the Old Red on the grounds that they were actually *better* known than the overlying beds. But even as the *Outlines* was being written, the work of John MacCulloch, Arthur Aikin, Leonard Horner, Thomas Weaver, and the older Continental geologists was going out of date owing to its relative lack of attention to fossil remains. As Hugh Torrens rightly points out, the scope and quality of the previous generation's work has often been underestimated, but the fact remains that most of its accomplishments were either unpublished or unintegrated into any general stratigraphical scheme. By the standards of mapping and classification maintained by the leading Fellows of the Geological Society in 1831 the rocks below the Old Red Sandstone desperately required reinvestigation.[46]

In many ways the problem of unravelling the ancient strata was characteristic of those being set by the geological community in the early nineteenth century. It involved classification, it required the extension and development of field techniques already applied in other contexts, and perhaps most significantly it was accepted as a worthy subject of inquiry by all concerned. As a result the older rocks offered substantial opportunities for scientific achievement and reward. Moreover, broad agreement about the important tasks facing geology made controversy on a subject like the older rocks potentially fruitful, for rather than arguing past one another, as their eighteenth-century predecessors had so often done, the geologists of the nineteenth debated from the common perspective of the stratigraphical enterprise. They rallied around a set of common goals—the classification and mapping of strata—rather than a single theory, and within the confines of a specialized coterie of gentlemen debating at Somerset House, scientific knowledge accumulated through a process of continuous criticism.

[46] Conybeare and Phillips 1822: preface; Torrens 1983. The opinion that the older strata deserved a new look is evidenced by contemporary addresses to the Geological Society, e.g. Fitton 1828: 58-59. The development of the stratigraphical column has yet to be studied in detail. For some discussion, see Zittel 1901.

CHAPTER TWO

Collaboration and Contrasts

THE SEEDS of the Cambrian-Silurian controversy lie in collaboration rather than in conflict, for Sedgwick and Murchison began their study of the older rocks as close friends. The classificatory geology pursued by the leading Fellows of the Geological Society relied heavily upon such collaborative efforts. In the eighteenth century isolated individuals had undertaken field studies, and some of the earliest maps of the nineteenth—those of William Smith and John MacCulloch in particular—were largely individual projects as well.[1] But even these men collaborated more than is generally recognized, and a network of exchange for observations and specimens was relatively well established. Specialization, increasing standards, and the large number of details to be mastered, soon demanded collaborative ventures of ever greater scope. The foundation of the Geological Society in 1807 brought some of the most significant of these projects into being. Its first public manifestoes spoke self-consciously of the need to unite the scattered labors of those engaged in the study of the earth; both Greenough's map and Conybeare and Phillips's *Outlines* were compilations representing the accumulated efforts of the metropolitan elite.[2] In such an atmosphere, joint touring in the field came as naturally as the good-humored camaraderie of the meetings. The frequency of collaborative research underlines the strength of the social and intellectual ties that bound the leading members of the Society, even in the course of their continual debates. Many of these scientific partnerships warmed into personal friendships: fellow geologists attended on family crises, consoled in times of loss, and acted as godfathers to each other's children.[3]

[1] Discussions of early fieldwork are available in Porter 1978 and Neve and Porter 1977; for Smith, see J. M. Eyles 1969; for MacCulloch, see V. A. Eyles 1937 and Cumming 1985. L. G. Wilson (1972: 126) shows that Charles Lyell helped MacCulloch with "a blank on his grand map."

[2] The manifesto at the beginning of the *Transactions* is [anon.] 1811; see also (for the map) Rudwick 1963, and the unpaginated introductory notice in Conybeare and Phillips 1822.

[3] Buckland's son Adam was also called "Conybeare Sedgwick" after his two godfathers; see Gordon 1894: 105, and Clark and Hughes 1890, *1*: 511.

The names Adam Sedgwick and Roderick Murchison are linked not only in a celebrated quarrel, but also in the most famous partnership in the history of geology. Before dividing on the issue of the Cambrian-Silurian boundary, the two men collaborated on important papers over a twenty-year period. They joined forces against De la Beche on the age of the Devon strata and against Ami Boué and his Continental compatriots on matters relating to the geology of the Alps. Moreover, they were close personal friends from their first field trip together in 1827 to the final break in 1856. As with any joint venture in science, the partnership resulted from an intersection of needs and interests at a particular period of their careers. The researches in Wales that led to the founding of the Cambrian and Silurian systems, while begun separately, became in many respects an extension of these earlier cooperative efforts.

The Origins of a Friendship

To the end of his life, men spoke of Sedgwick as if he had been quarried in his native Yorkshire dale of Dent rather than merely born there. He always retained a special fondness for the place of his birth in 1785, and the quick-tempered passion of the North Country dalesman remained a prominent element in his personality. "The undisguised nature of the man," wrote Cornelia Crosse in her recollections, "was vehement, genuine, very kindly, and highly explosive."[4] Sedgwick's father was parish curate and headmaster of the local grammar school that young Adam, the third of seven children, attended in his earliest years. Further university training was something of a family tradition, and at the age of twenty Sedgwick entered Trinity College at Cambridge, which remained his home uninterruptedly for the next sixty-nine years. Older than the other students, a sizar, and a provincial who had scarcely ventured twenty miles from his parish, Sedgwick felt ill at ease at first, but soon began to make his way and graduated B.A. in 1807 as Fifth Wrangler. He inclined towards a legal career after graduation, but his father's financial position demanded the more secure and immediate income that the clerical profession would provide. Towards this end he commenced further study of classics and mathematics (finding neither to his special taste) and won a college fellowship in 1810. Like most Cambridge Fellows of the early nine-

[4] Crosse 1892, *2*: 36; for Sedgwick's early life, Clark and Hughes 1890, *1*: 1-165, Speakman 1982.

Fig. 2.1. The young Sedgwick at Cambridge; later characterized as "a fine, commanding-looking man with a wide brow & raven hair. . . . He has wonderful fluency & abounds in imagination and wit."

teenth century, Sedgwick merely marked time in the following years: his election to the Fellowship, appointment as a mathematical lecturer in 1815, and ordination in 1817 were all steps on the road to a beneficed living.

The 1818 election for the Woodwardian chair of geology formed the turning point in Sedgwick's life. The duties of the professorship, established by the irascible naturalist John Woodward almost a century before, included lectures and the care of a collection. Like most eighteenth-century academics, previous holders of the chair had largely neglected these provisions.[5] There seemed at first little prospect that the new appointee would greatly improve matters, and although enthusiastic about his new position, Sedgwick only gradually abandoned his plans to use the post as a stepping stone to a comfortable country parish. On taking up the chair—one of the few devoted to the natural sciences at Cambridge—the new professor knew only slightly more of geology than did the average educated Englishman. He had read Georges Cuvier's preliminary *Essay* and several articles in the popular reviews, had attended Edward Clarke's lectures at Cambridge on mineralogy, and on at least one occasion (in 1816) had been introduced as a visitor to the Geological Society. But all this made him no more than a beginner. As his principal opponent for the chair complained, "Sedgwick is put up by a large College, merely as a *man of talent*, who *can* soon fit himself for his office."[6] Yet the election was by no means corrupt. During the early part of the century, specialized qualifications carried less weight at Cambridge than general culture and qualities of character, which Sedgwick evidently possessed in abundance. In any event the electors' choice proved a wise one. He took to the field almost immediately and spent the summer of 1818 learning the methods of geology firsthand and collecting specimens for the museum at Cambridge. Students heard the first of what soon became a celebrated course of lectures in the Easter term of 1819. He soon joined the Geological Society and after two or three field seasons began presenting papers of considerable interest.

Thus by the time that Murchison entered the Society in 1824, Sedgwick was a tried veteran like Buckland, Greenough, or Conybeare.[7] Murchison commenced his scientific studies at the relatively advanced age of thirty-three, just as Sedgwick had, and came to the subject via an equally circuitous route. Born in 1792, he was the eldest son in a wealthy Scottish family; his father was a surgeon who had purchased property in Rosshire after having amassed a fortune

[5] Rothblatt 1968 analyzes Cambridge academic life and Sedgwick's part in changing it. Clark and Hughes 1890, *1*: 166-198 give a history of the chair.

[6] G. C. Gorham to his father, 17 May 1818, in Clark and Hughes 1890, *1*: 157; also *1*: 152-165 and *2*: 349 for the election. Sedgwick's attendance at a Geological Society meeting is noted in Woodward 1908: 39 and Speakman 1982: 56.

[7] Murchison's early life is recounted in Geikie 1875, *1*: 1-95; for the difference in generations, pp. 114-115.

Fig. 2.2. Murchison in 1836. Five years earlier the self-confident young physicist James D. Forbes had described him as follows: "A Rossshire proprietor of gentlemanly manners but too much Sang froid. 12 years in the Army. Afterwards a foxhunter. Then became geologist to which he has devoted himself with laudable perseverance. . . . He is a diligent observer & I believe a good naturalist but does not appear possessed of much originality."

(under somewhat dubious circumstances) in India. In later life Murchison loved to dwell upon his Highland ancestry, but with the exception of a few months' babyhood on the family estate in Rosshire and two years in Edinburgh, he lived in England and spoke with an English accent. Intending to become a soldier, a dream from his earliest youth, Murchison enrolled in the military college at Great Marlow in 1805 and within a few years fought in some of the most famous campaigns of the Peninsular Wars. In 1809 his future looked even brighter, as his maternal uncle, General Mackenzie of Fairburn, named the young lieutenant as an aide-de-camp. But rather than a royal road to rapid promotion, the appointment led to idle years of garrison duty in northern Ireland, far from the excitement of the wars raging on the Continent. Murchison's remaining hopes for further advancement were finally dashed in 1815, "the battle of Waterloo having submerged all my ambition, as well as that of the great Napoleon." In that year he also met and married the heiress Charlotte Hugonin, an attractive and well-educated woman about three years his senior. Sedgwick, in contrast, remained "mired in celibacy" throughout his life, prohibited from taking a wife by the terms of his professorship.[8]

The failure of his military career bitterly disappointed Murchison, and he even contemplated exchanging his martial uniform for clerical robes. Eventually, though, the retired soldier adjusted to the carefree life of a North Country gentleman, as long antiquarian rambles on the Continent alternated with foxhunting, partridge shooting, and an endless round of entertaining. But these pursuits, however agreeable, were not to prove Murchison's métier. The couple were living far beyond their means (apparently on the proceeds of speculative investments), and a severe financial squeeze in 1823 forced them to give up house, horses, and subscription to a pack of hounds. Murchison later remembered this as a well-planned prelude to his taking up science, but practicalities of pence and pounds almost certainly dictated the change of scene. After a year spent with Charlotte's parents, the Murchisons—still possessed of a substantial independent income—set up house on a reduced scale in London. Only then did Murchison settle on geology as an appropriate outlet for his immense energy and enthusiasm, doubtless hoping to follow his friend the chemist Humphry Davy in proving that a sporting man could also be a scientific one. Not all his old ac-

[8] For Napoleon: Geikie 1875, *2*: 332; for Charlotte Murchison: [F. Buckland] 1869; for Sedgwick: Clark and Hughes 1890, *1*: 181-185, 386.

quaintances were so sanguine about the shift of interests: one army friend lamented that Murchison, once such "an excellent fellow and a most agreeable comrade," eventually became capable of attending only to an absurd pomposity labelled "the Silurian System."[9]

Murchison knew even less of geology in 1824 than Sedgwick had upon his appointment to the Woodwardian chair. Nevertheless, with characteristic determination he rapidly made amends by attending lectures at the Royal Institution, by joining the Geological Society, and by meeting the leading researchers. He soon went into the field, and in his early papers described strata in regions already familiar to him, including the Tertiary geology of Petersfield in Hampshire, where his wife's parents lived, and the sedimentary rocks of the Scottish Highlands. Two years after the move to London, Murchison was widely recognized as a rising star of British geology, "an independent gentleman having a taste for science, with plenty of time and enough of money to gratify it."[10]

For both Murchison and Sedgwick, geology promised the rewards of a lifework. Though the financial rewards were minimal, geology was at least a vocation worthy of a gentleman, offering intellectual satisfaction and social sanction commensurate with the established professions.[11] In the absence of direct economic benefits, both men looked to geology as a source of accomplishment, a means of social advancement, and a measure of self-worth. These aspirations greatly heightened the importance of property and priority in their scientific careers and provided their later dispute with much of its force.

Sedgwick and Murchison must have met soon after the latter's "conversion" to science, presumably through mutual involvement in the Geological Society.[12] But it was not until the summer of 1827 that friendship really blossomed. During the previous field season Murchison had been unable to interpret the geological succession on the island of Arran, and he asked Sedgwick (already England's principal authority on the older rocks) to accompany him on a tour of the red sandstones of Scotland. "Dinna forget the land of cakes, whiskey, & hospitality, where I hope to act as your aide-de-camp," Murchison wrote early in 1827, "which having been my old trade

[9] [Campbell] 1906, *1*: 349; more generally, Geikie 1875, *1*: 73-95.

[10] Geikie 1875, *1*: 129.

[11] As Porter (1978: 817-825) points out, most considered science *superior* to a paid profession.

[12] See Sedgwick to Murchison, [28 Mar. 1827], GSL: M/S11/1. For their first meeting, Clark and Hughes 1890, *1*: 299-300; Geikie 1875, *1*: 124.

before I took to the road, enables me to say without presumption that I am not a bad caterer."[13] The Scottish tour of that summer was the first of many in succeeding years. It was only the pressure of his own engagements that led Sedgwick to refuse an invitation to join Murchison and Lyell on their trip to the French Auvergne in 1828,[14] but during the next summer he accompanied Murchison on an extensive expedition to the eastern Alps. These travels resulted in bulky memoirs in the Geological Society *Transactions*, including reinterpretations of Scottish geology opposed to the pioneering work of John MacCulloch and what Whewell termed a "Mount Blanc" of a paper on the Alps.[15]

The hundreds of letters that passed between London and Cambridge during the long partnership produced one of the major Victorian scientific correspondences, paralleling that between Darwin and Joseph Hooker in botanical geography or George Stokes and William Thomson in physics.[16] The "Dear Sir" of the earliest exchanges soon gave way to the familiar "Dear Sedgwick" and "Dear Murchison," while at the end of their letters they pledged themselves "yours to the earth's centre" and "yours to the top end of his hammer." By the end of the Scottish tour the two geologists had already become fast friends, and although the bulk of their correspondence focused on technical issues in stratigraphy, a prominent place was also reserved for personal, political, and social news.[17] From the beginning, however, Murchison felt more strongly than Sedgwick the need for a collaborator and a friend. The retired soldier, particularly during the early years in London, had relatively few acquaintances of long standing. Sedgwick's sense of humor and oratorical brilliance attracted his attention immediately. "Perhaps you know," Lyell wrote, "that he idolizes even more than the Cantabs 'the first of men,' as Adam is usually styled there."[18] On scientific matters Murchison valued the views of Sedgwick above all others and went to great lengths to gain his approval on doubtful

[13] Murchison to Sedgwick, [pmk. 12 May 1827], CUL: Add. ms 7652IIID88 (copy).

[14] Sedgwick to Murchison, 7 Apr. 1828, in Clark and Hughes 1890, *1*: 320.

[15] Sedgwick and Murchison 1835a, 1835b, 1835c, 1839. "Mount Blanc" is taken from Whewell to Murchison, 21 Oct. 1831, GSL: M/W4/5.

[16] For the Darwin-Hooker and Stokes-Kelvin correspondences respectively, see J. Browne 1978 and D. B. Wilson 1976. Two very different studies of collaborative work are available in Guerlac 1976 and Pinch 1980.

[17] Sedgwick to Murchison, 25 June 1828, 5 June 1832, printed in Clark and Hughes 1890, *1*: 323, 391. About one-quarter of the extant Sedgwick-Murchison correspondence is printed in their biographies.

[18] Lyell to Scrope, 9 Nov. 1830, in K. M. Lyell 1881, *1*: 309-311, at p. 310.

issues. Sedgwick's own reasons for entering into such a close union are somewhat less evident. In part it was a matter of literary inertia. Almost from the beginning of his career Sedgwick believed himself "constitutionally incapable of much sedentary exertion" and valued the younger man's stamina in the study as much as his indomitable energy in the field.[19] Murchison's knowledge of traveller's German and skill at bargaining at inns were welcome assets in geologizing outside the British Isles. While abroad, Sedgwick could collect specimens for the Cambridge collections and compare the successions on the Continent with those found at home. He found Murchison an excellent companion, as did Lyell and others, although for really close friendships he usually turned to his circle at Cambridge. For Murchison, Sedgwick was "the man of my heart," but Sedgwick confided his inmost thoughts to Charles Ingle and William Ainger, two friends he had made while still a student.[20] Even with all the joint papers and tours, Sedgwick and Murchison remained highly distinctive, sharply drawn personalities. Friendship grew out of their collaboration in science rather than existing independently.

These caveats aside, the association was remarkably close, persistent, and scientifically productive—in every respect exemplifying the strong bonds of male friendship characteristic of the nineteenth century. As correspondents, travelling companions, and mutual critics, the two men possessed contrasting virtues that complemented one another ideally. "I am indeed confident," Murchison wrote in 1852 at the bitter height of their controversy, "we shall slide down the hill of life with the same mutual regard which animated us formerly when climbing together many a mountain both at home and abroad."[21] This reconciliation never took place, however, and their dispute ultimately destroyed not only a famous collaboration but also a great Victorian friendship.

Programs of Research

The main story of the development of the Cambrian and Silurian systems begins in 1831, when Sedgwick and Murchison commenced classifying the older rocks of Wales. Given their earlier col-

[19] Sedgwick to C. N. Wodehouse, 12 Oct. 1837, in Clark and Hughes 1890, *1*: 499.

[20] See the numerous letters to these two men in Clark and Hughes 1890; for "the man of my heart," Murchison to Sedgwick, 19 Jan. 1838 [not sent], in Craig 1971: 498-499. Similar qualifications on their friendship, although too strongly stated, are given in Clark and Hughes 1890, *1*: 300-301.

laboration, it has always seemed something of a mystery that they began from opposite ends of the country in researches that were at first entirely independent. In the end the consequences of this lack of coordination became all too obvious: as Sedgwick's biographer John Willis Clark put it, separate entries meant separate field areas and separate field areas meant separate geological "systems."[22] Although Clark's lament involves an element of wishful thinking, it does suggest the importance of events in 1831 for the later controversy. Just as the early collaboration represents a conjunction of interests, so does the separation in 1831 indicate a divergence. The coauthored publications and famous friendship, while illustrating the compatability of their approaches, have obscured important differences in their reseach aims. These separate concerns make it clear why the two geologists did not plan a joint attack on Wales.

By the summer of 1831 Sedgwick's examination of the region was long overdue. Conybeare and Phillips's *Outlines* of 1822 had "Part I" boldly printed on the title page, and the preface optimistically promised a second volume covering the Primary and Transition rocks. After the death of William Phillips in 1828 Sedgwick had assumed most of the responsibility for this long-awaited work, which would necessarily include chapters on Welsh geology. From that point onwards he undertook almost all of his own fieldwork as preparation for the *magnum opus*. The book's format was to be patterned upon its predecessor, with each formation introduced in descending order and each occurrence described from north to south. Conybeare outlined the state of research in a long letter to Sedgwick in 1828. Certain areas, he noted, had already been "done" according to the standards of geological fieldwork described in the previous chapter. John S. Henslow had written a superb paper on Anglesey during his tenure of the mineralogy chair at Cambridge, Arthur Aikin possessed manuscript information on Shropshire, and the literature on Cornwall and Devon could be coordinated after a brief tour. Sedgwick had already finished the Lake District and he planned to visit southern Scotland in the near future.[23]

This listing included almost all the Primary and Transition districts in the British Isles (Fig. 1.6). The largest blank was Wales. Conybeare suggested a unified approach; he would do South Wales

[21] Murchison 1852c: 184.

[22] Clark and Hughes 1890, *1*: 377; also Geikie 1875, *1*: 191.

[23] Conybeare to Sedgwick, 24 Apr. 1828, in Clark and Hughes 1890, *1*: 324-325. Aikin's work is fully described in Torrens 1983; the study of Anglesey is Henslow 1822.

in concert with De la Beche and Sedgwick, while Henslow might be persuaded to extend his investigation across the Menai straits to the mountainous regions of Snowdonia. The actual writing could then be shared by Sedgwick and Conybeare. As it happened, most of this anticipated assistance failed to materialize. The first volume of the *Outlines* had been a compilation of the labors of the entire Geological Society, but the second would have to be almost entirely the work of one man. A considerable effort in the field was required, including investigations of complex regions that had hitherto attracted little attention. Most important, the succession in each isolated district had to be unified into a classification applicable throughout Great Britain. Even with prevailing standards of accuracy, allowing large districts to be mapped in weeks, Sedgwick faced a formidable task.

Sedgwick had hoped to fill in some patches of North Wales in 1830, but torrential downpours drove him out of the field.[24] When he finally began during the following summer, it was without great enthusiasm. "What a horrible fraction of a geological life sacrificed to the most toilsome & irksome investigations belonging to our science!" he complained to Murchison. "When I finished Cumberland I hoped some one else would have done N. Wales—but I have been disappointed."[25] The book on the older rocks had already consumed far more time than Sedgwick had anticipated, and he was anxious to turn to less arduous tasks. Many contemporaries wondered if he had the stamina to complete an entire volume. Lyell doubted that it would ever be finished, and remarked in his journal that Sedgwick "has not the application necessary to make his splendid abilities tell in a work. Besides every one leads him astray."[26] This telling passage referred not only to the press of university administration and teaching on the Woodwardian professor, or to his active involvement in Whig politics and scientific administration. The "every one" leading Sedgwick astray undoubtedly included Murchison, for almost all of their collaborative research actually diverted Sedgwick from completing the *Outlines*. Murchison usually initiated the joint excursions, and the trips (actual or proposed) to Scotland, Germany, the Alps, and the Auvergne grew almost entirely from his own research problems. Frequently he had already visited an area by himself, had studied the stratigraphy,

[24] Clark and Hughes 1890, *1*: 365; he also contemplated a trip to North Wales in 1828 (*1*: 322).

[25] Sedgwick to Murchison, 20 Oct. 1831, in Clark and Hughes 1890, *1*: 382.

[26] K. M. Lyell 1881, *1*: 375.

and then wished to return with Sedgwick's critical eye. But as the 1820s drew to a close, these enticing invitations to international travel were clearly keeping Sedgwick from his own more pressing project. In 1828 Murchison pleaded with his friend to join him on the Continent; "pray even do it," he stressed, "before you bring forth that long-expected second volume on the Geology of England and Wales."[27] Murchison scarcely intended to keep him from completing the *Outlines*. But there can be no doubt that the joint work was diverting Sedgwick from his own research, and his refusal of yet another tour signalled a determination to complete the synoptic view of the older rocks. In 1831 Sedgwick entered North Wales with this task firmly in mind.

For Murchison the situation was completely different. Contrary to his own later testimony and the statements in dozens of histories and textbooks of geology, he had no particular plans for a major study either of the Transition rocks or of the geology of Wales when he first crossed the Welsh border in 1831. He had no idea that this would be the first of seven visits, no idea that he would transform the study of the older rocks, and no idea that he would eventually overtake Sedgwick as their leading specialist. All of these consequences were at first unseen, for the decision to spend several years classifying the Transition rocks of Wales and the Welsh Borders grew unexpectedly from a tour undertaken with completely different aims from those eventually accomplished. Given their divergent aims, the two men could not conceivably have joined forces for an attack on the older rocks of Wales. Far from planning an "invasion of grauwacke" parallel to Sedgwick's, Murchison was initially engaged in another enterprise altogether, and only in the light of subsequent events did their goals seem so similar.[28]

Left to his own ends Murchison would almost certainly have preferred yet another season of geologizing on the Continent. But it appears probable that his mother-in-law had finally enforced an old demand that Roderick and Charlotte refrain from such extensive travels abroad lest her death should intervene before their return. Murchison thus faced the presumably unwelcome prospect of an entire summer within Britain. But there were compensations. Charlotte, an amateur naturalist in her own right and a trained artist of some skill, could be by his side tagging specimens and sketching landscapes. Murchison's plans included a general survey of British

[27] Murchison to Sedgwick, 18 Aug. 1828, in Clark and Hughes 1890, *1*: 340.

[28] For a full study of this tour, see Secord (forthcoming).

geology; with Greenough's map in hand he could clear up unsolved stratigraphical difficulties across the length and breadth of the island. By 1831 the boundaries of this map needed many important corrections, one of the most notoriously inaccurate lines being that between the Old Red Sandstone and the Grauwacke in Wales and the Welsh Borders (see Fig. 2.3). In the north of England, visits to old hunting companions could be combined with mapping of the Secondary strata on their estates. Other aspects of the tour demonstrate Murchison's special interest in the theoretical program of Lyell, a close personal and scientific associate at this time. For example, in many areas (again especially in Wales and the Borders) Murchison could investigate the relation between the uplift of strata and the underlying processes of igneous action. Also like Lyell, he hoped to find seashells on the summits of mountains and other indications of extensive geological changes in recent times. As the summer drew to a close Murchison even proposed spending the next season, not among the Welsh strata later so closely linked with his name, but rather on the opposite coast in an elaborate Lyellian study of the confused Tertiary sequence known as the "Crag."[29]

These concerns, as evidenced in field notebooks and contemporary correspondence, show that Murchison's return to Wales in 1832 and the succeeding summers necessitated a major shift in his research plans. He had ended the summer intending to revisit East Anglia and instead went west again to Wales to study rocks at the bottom of the geological succession. In 1831 he had been particularly concerned with the igneous rocks of the Principality, but by 1832 his interest had shifted to those of sedimentary origin. It should already be evident that Murchison and Sedgwick commenced the study of Welsh geology with completely different aims; behind the collaborative tours of previous summers were two individuals pursuing separate programs of research.

Perfecting the Geological Record

In continuing the coverage of the *Outlines* below the Secondary formations, Sedgwick looked to Wales for a solution of a difficulty he had faced repeatedly in his earlier work. The record of the rocks in all the areas he had previously investigated seemed incomplete: between the base of the Old Red Sandstone and the commencement of the underlying Grauwacke, a body of strata of unknown

[29] The Crag debates are summarized in L. G. Wilson 1972: 461-495.

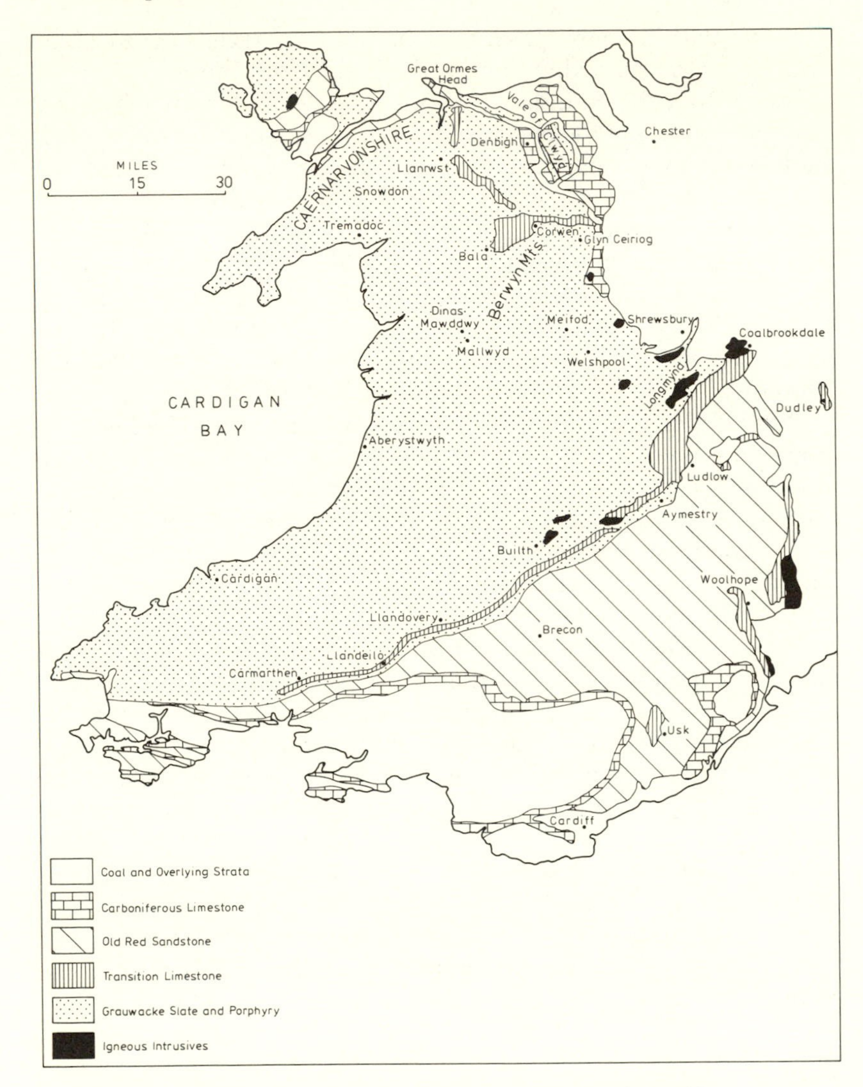

Fig. 2.3. Geological map of Wales, based on Greenough's 1820 compilation produced for the Geological Society. Wales, which had been investigated by Aikin, Greenough, and Buckland, was one region where this map had a demonstrable superiority over that of William Smith.

thickness was missing. As a result of this unconformity, the relations of Sedgwick's older rocks to the overlying strata remained unclear. His slaty grauwackes in Scotland and the north of England were strata set adrift, as it were, without a secure mooring to the known succession of overlying Secondary rocks. However, their approximate position was inferable from previous studies of the older strata of Wales, particularly by Greenough, Buckland, and Conybeare. These authors had described smoothly conformable sequences between the Grauwacke and the Old Red Sandstone in several localities. Such sequences were generally called "passage beds," for they bridged a gap commonly found in the record of the rocks.[30] On a theoretical level, the problem of continuity in the stratigraphical record was one of the most important issues confronting geologists in the early nineteenth century, just as it continues to be today. But given the established picture of the structure of Wales and the Welsh Borders, Sedgwick had every expectation of finding passage beds linking the uppermost Grauwacke and the bottom of the Old Red.[31] While not absolutely essential for his field research, such a continuous sequence would give his sections for the *Outlines* a more secure basis than they had hitherto possessed.

Sedgwick thus started in 1831 with little doubt that he would bridge the troublesome gap. Preparing himself for his task, he began his field studies in August at a well-known district of Transition Limestone near the rising industrial town of Dudley (for localities, see Fig. 2.3). These highly fossiliferous rocks were known to occupy a position somewhere just below the Old Red Sandstone, although the presence of a slight unconformity prohibited the assignment of a precise horizon. Sedgwick hoped to use the Dudley beds as a type for similar calcareous strata he expected to find in North Wales.[32] Greenough's map indicated a clear succession near the northern coast, with patches of Old Red and Transition strata by the Vale of Clwyd and Great Ormes Head passing occasionally into underlying calcareous slates. Accompanied for a few days by the young Cambridge graduate Charles Darwin, Sedgwick started by tracing the

[30] Buckland and Conybeare 1824: 221, 280-284; Greenough 1820.

[31] Fitton 1829: 121-123. Discussions with Hugh Torrens and Martin Rudwick have convinced me that Sedgwick and Murchison would not have been at all surprised to find conformable passages in Wales. The problem of continuity in the geological record is treated more generally in Rudwick 1976a: 189-191.

[32] The tour is described in Sedgwick to Murchison, 13 Sept. 1831, printed in Clark and Hughes 1890, *1*: 377-379; details of the Dudley visit are in Sedgwick to Murchison, [?25 Jan. 1836], GSL: M/S11/96a & b.

boundaries of the Carboniferous Limestone scarps in these areas. The two men soon found that the Old Red and Transition strata underneath the Carboniferous were disturbed or nonexistent. Greenough's map was highly misleading. "The *Old Red* all round Orm Head &c. &c. is a *pure fiction*," Sedgwick wrote Murchison. "At least I can't see a trace of it. There is not a particle of it between Denbigh and the Isle of Anglesea."[33] As a result of their mapping, the distribution of formations along the coast was thoroughly altered, and the Vale of Clywd now appeared in the trough-like form of a syncline. This tour provided Darwin with his only training in the geological investigation of an unknown area before the departure of the *Beagle*; and although he never mentioned these remarkable findings in his autobiography, they surely must have shown him just how open for discovery the new science could be. Even within the British Isles, the structure of large regions could still be radically revised.[34]

For his part, however, Sedgwick faced the same annoying break that had troubled his work in Scotland and the Lake District. Since the Old Red and Transition limestones were missing in North Wales, he was unable to work from the known succession down into more ancient strata. As the only alternative he plunged to the bottom of the sequence and began his sections in the Primary rocks to the west of Snowdon. After an abortive attempt to prepare himself by examining Anglesey with Henslow's paper in hand, he returned to the mainland and commenced mapping in earnest. He established his base line among the ancient slates and porphyries on the western slopes of Snowdonia and made several long traverses into the overlying strata to the east.[35] By October (when the weather finally drove him out of Wales) Sedgwick could claim a certain measure of success. Although his work had proceeded more slowly than he would have liked, he had nearly finished a geological map of the rugged Caernarvonshire mountains in two months. As yet, his failure to find a conformable succession remained only a local difficulty. Completion of a structural map of Wales occupied most

[33] Sedgwick to Murchison, 13 Sept. [1831]; italics are as in original at GSL: M/S11/51.

[34] Darwin's notebooks have been published in Barrett 1974; Sedgwick's are as follows: No. XXI August 1831 and No. XXII August 30 to September 15, 1831, SM. Transcripts of parts of these by Owen T. Jones are also available. See also Clark and Hughes 1890, *1*: 379-381; De Beer 1974: 38-40.

[35] Sedgwick to Murchison, 13 Sept. [1831], GSL: M/S11/51; Henslow 1822.

of his attention, and a base line of some sort was all that this relatively straightforward task required. After all, Sedgwick had studied Lakeland geology despite a virtually identical unconformity in that region.[36]

But the failure to find a conformable succession did have major consequences for the future. In subsequent years Sedgwick emphasized his initial inability to find passage beds below the Old Red, largely because this link (in the form of the Silurian system) had by then assumed such massive significance. As might be expected, a far more radical process of autobiographical myth-making informed Murchison's retrospective accounts of his concurrent success in obtaining this objective. According to his version of events, repeated by all subsequent commentators, Murchison intended from the start of his 1831 tour to look for passage beds below the Old Red. South of Builth, along the banks of the River Wye, he found in a series of gently dipping mudstones what retrospectively became the "first true Silurian section," the site of a discovery that filled a gaping hole in the succession. As so often in the history of science, this romantic account telescopes a complex series of researches into a single moment of blinding illumination. But the reconstruction of his actions in light of later events went further still, affecting even the very essence of the "discovery" itself. Because of his wish to investigate igneous uplift and make corrections to Greenough's map, he had happened on one of the few areas in Britain where such a conformable series could readily be seen. The passage beds along the Wye, far from being unexpected as he later claimed, were precisely what any experienced geologist would have anticipated. What actually drew Murchison's attention was the undisturbed and highly fossiliferous character of these Transition strata. As became clear within a few days, an even more impressive sequence could be found to the north near Aymestry and Ludlow, where the Rev. Thomas T. Lewis, the local curate, had worked out the succession in convincing detail. Once back in London Murchison made a striking change in plans and decided to return to Wales during the following summer to work out the geology of the Transition strata more completely. In other words he had "discovered" not a series of passage beds, but rather a potential new direction for his research, one safely distinct from the dynamical geology and studies of Tertiary strata that Lyell was making so completely his own. By

[36] See the comments in Sedgwick 1873: xv.

comparison the older rocks were almost unexplored. Only Sedgwick was seriously investigating them, and he wanted all the help he could get.[37]

After his decision to follow up his work on the Grauwacke, Murchison returned to hammer the sections with his accustomed enthusiasm. Lewis and his friends gladly aided in this prestigious enterprise by collecting specimens, tracing strata, and watching for new outcrops. At the close of the 1832 field season a distinct series of formations extended from the base of the Old Red down to the slaty rocks to the north that Sedgwick had begun to describe. Most important, Murchison's strata were laden with fossils, which suggested that they could serve as a basis for long-distance correlations. The gap that Sedgwick had so often confronted had been bridged far more thoroughly than ever before.

Given the importance of the 1831 tour both for his career and for the history of science, it is not surprising that Murchison should view its issue as the outcome of a deliberate search. Although glad of his help, Sedgwick after a few years understandably saw this question from a different perspective. He complained of his friend's revisionism after reading a manuscript account of the discovery penned in 1836. "Starting then without any anticipation of what has turned up," he wrote, "you stumbled on a rich field and have since gathered an ample harvest after enormous labours—This kind of statement takes not away one jot from your discoveries—As the passage now stands, you assume a prescience I don't believe real—and you set down as little better than a pack of asses every one who had preceded you—."[38] Of course, these remarks are touched by more than a shade of jealousy, as by this date the definitive quality of Murchison's work was already evident. Sedgwick recognized that the discovery account had been aimed not only at predecessors like the provincial clergyman Lewis and the metropolitan specialist in Shropshire geology Arthur Aikin,[39] but also against himself. This made it all the more necessary that Murchison put his accomplishment in proper perspective. Otherwise Sedgwick feared appearing in the roll call of scientific worthies as little more than the leading "ass" in the pack.

Initially Sedgwick had welcomed Murchison's change of plans as

[37] Secord (forthcoming). The rivalry between Murchison and Lyell developing at about this time is described in Page 1976. Lewis's work is described in J. C. Thackray 1977 and [Fitton] 1841.

[38] Sedgwick to Murchison, [?25 Jan. 1836], GSL: M/S11/96a & b.

[39] For Lewis: J. C. Thackray 1977; for Aikin: Torrens 1983.

affording unexpected assistance in a heavy task. But coupled with his inability to find a complete succession in North Wales, the events of 1831 also set the stage for confrontation. From the outset, the position of Murchison's beds was precisely known. Sedgwick, on the other hand, was forced to employ a base line deep in the Primary rocks, and his groups of strata remained unconnected to the standard sequence by continuous sections. However accurate his structural interpretations of various districts were, their relations to one another and to the overlying beds remained relatively ambiguous, although a general placement deep in the Grauwacke seemed beyond question. The exact relationship between the respective groups of Sedgwick and Murchison awaited determination. As yet only a minor problem, it would have to be faced by them together.

Styles and Contexts

Differences in Sedgwick's and Murchison's methods and techniques parallel the contrast in their research aims. Within the larger tradition of field studies explored in the first chapter, two separate geological styles can be characterized, just as an art historian might distinguish the work of two masters painting in different workshops within the confines of a developed artistic tradition. Although the methods of Murchison and Sedgwick were similar enough to permit joint projects like the memoirs on Scotland and the Alps, their notebooks, field maps, and separate publications readily reveal their individual scientific styles. In a visual science like geology, the analogy with art history is remarkably close.[40] It is particularly important to recognize that the character of scientific style is best evidenced, not in overarching theoretical concepts, but in material documents deriving from day-to-day practice. These exhibit a surprising diversity within the geological community and offer a concrete method for relating scientific practice to institutionalized traditions of research.

Sedgwick's geological fame rested above all on his insight into structures, an ability to visualize rock masses in three dimensions and interpret their interrelationships after only a few traverses. As

[40] E.g. Gombrich 1977, Baxandall 1972, 1980. Rudwick 1982a characterizes four basic geological styles based on differing conceptions of earth *history* and relates these to social roles *within* the scientific community. My own categorization gives greater emphasis to concrete practice, and relates differences in style to more general contrasts in circumstances and background.

his scientific "son" and student at Cambridge J. Beete Jukes put it, "no man had a quicker eye for a fault or a tighter grasp for the bowels of a country" than "sharp eyed old Sedgwick."[41] Even the language of structural geology bore the traces of his hand. He introduced the word "strike" into English from the German "streichen" and made the first published use of "synclinal" to refer to the well-known phenomena of rock beds bent in a trough. In an important paper of 1835 Sedgwick made an explicit distinction between original bedding and slaty cleavage (Fig. 2.4) that was of fundamental use in interpreting the structural position of highly metamorphosed rocks.[42] His remarkable facility for untangling the interrelationships of complicated rock masses was perfectly adapted for work in a region like North Wales, which provided few of the palaeontological characters or gently dipping strata favored by most contemporaries. Among the twists and turns of the Welsh rocks, geological structures could be interpreted on the grand scale, and with a few days fieldwork Sedgwick could bring order and simplicity to thousands of feet of seemingly chaotic slate and grauwacke.

Murchison, on the other hand, was above all a maker of geological maps. Known for the rapidity with which he grasped the fundamental features of a district, he excelled in inferring the distribution of the underlying formations from a quick survey of the surface. His field jottings from the 1830s reveal a scientific style considerably more deductive than that of Sedgwick, whose notebooks are crammed with thousands of discrete observations. From the beginning Murchison constructed classifications, drew tentative sections, and made correlations. Although this sometimes led to a certain carelessness in his overall structural interpretations, he pressed on and became the most productive Victorian geologist, single-handedly putting a large portion of the earth's crust into order. He pursued geology with much of the same restless vigor that had led to startling feats of pedestrianism in his earliest years, such as walking 452 miles in fourteen days in an Alpine tour of 1816. In many ways his field style epitomizes the emphasis of practical stratigraphy on physical rather than mental activity, on doing geology with the feet and the hammer with the goal of producing maps and classifications.[43]

[41] Jukes to Ramsay, 28 Apr. 1847, ICL(R).

[42] Sedgwick 1835a; Sedgwick to Jukes, [pmk. 15 Dec. 1855], CUL: Add. ms 7652IIIE5; Fitton 1836: 125; C. Lyell 1830-1833, 3: 293. Further references available in Challinor 1978: 295-296, 304. As Phillips (1857: 370-373) emphasizes, geologists in the 1820s often distinguished informally between bedding and cleavage.

[43] To the end of Murchison's life *"omnia vincit labor"* was his avowed motto; Geikie

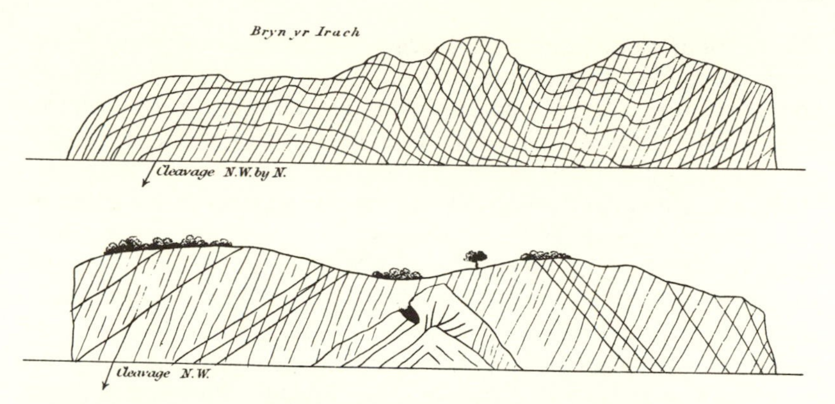

Fig. 2.4. A pair of traverse sections by Sedgwick illustrating the difference between original stratification (here shown by arched lines) and the subsequent effects of slaty cleavage (suggested by the steeply dipping parallel lines).

As a means for long-distance correlation, fossils were ideally suited to this program of research. In a typical early paper, Murchison paralleled the Brora coalfield in northeast Scotland with the Oolitic beds of the Yorkshire coast—an explicit test of the utility of fossils over long distances.[44] In turning to the older strata after 1831, he continued the emphasis on fossil evidence that had proved so fruitful in studying the Tertiary and Secondary rocks. As Whewell said in his 1839 presidential address to the Geological Society, "He has applied, for the first time, to the rocks below the Old Red Sandstone, the method of classification previously employed with such success for the Oolites." In the less flattering words of John MacCulloch, an extreme opponent of the use of fossils, Murchison must have seemed one of the worst of the "namby pamby cockleologists and formation men."[45]

Sedgwick, although interested in the newer palaeontological methods, always stressed the physical characteristics and overall disposition of the rocks as primary indicators of appropriate classifications. Unlike Murchison, Buckland, Conybeare, or John Phillips, he never published any papers on fossils and possessed even less technical familiarity with "organic remains" than most geolo-

---

1875, *1*: 192. His walking prowess is mentioned on pp. 175-176. Porter 1978: 818-821 discusses geology as a physical activity.

[44] Murchison 1829a; Waterston 1982; see also Balan 1979: 385.

[45] Whewell 1839: 79-80; MacCulloch to L. Horner, 2 June 1820, in K. M. Lyell 1890, *1*: 173-175, at p. 174. The phrase was italicized in the original.

gists of the period. But the methods he employed only put him towards one end of a spectrum. Darwin remembered his eagerness to find fossils during their joint excursion in Wales, and Sedgwick frequently referred to their importance in his lectures and addresses.[46] Indeed, it might even be argued that the differences in fieldwork between Murchison and Sedgwick are simply artifacts of the strata in the areas they chose for study. Fossils were scarce among the metamorphosed slates of North Wales and relatively common in the undisturbed sequences of the Welsh Borders. What would have happened if they had changed places, Murchison beginning from the north and Sedgwick from the south? Such a supposition, however, ignores the fact that the two men chose their respective research areas largely because of the particular problems they afforded. With his interest in palaeontological stratigraphy Murchison would never have returned to Wales in 1832 had he not been impressed by the fossiliferous character of the succession below the Old Red. In contrast, the slates and schists of the north offered unusually wide scope for Sedgwick's powers in structural interpretation.

How, then, are the differences in approach to be explained? The key lies in the very different contexts in which Murchison and Sedgwick practiced geology. Because of the lack of a standard pattern for recruitment into the science at this time, these contexts were inevitably peculiar to them as individuals, at least in part. Unlike French savants, British naturalists were not paid professionals divided into sharply bounded "schools" directed by powerful patrons. It is possible, however, to use Murchison's and Sedgwick's distinctive field styles to point to the presence of loosely defined research centers within the wider social milieu of British geology. One of these, located at Oxford and led by William Buckland, has been characterized in a recent book by Nicholaas Rupke. The notion of an Oxford style of geology, located in the University environment and centering on the use of fossils in the study of the Secondary strata, has much to recommend it. But Rupke also makes the much less plausible argument that Oxford became the nucleus for "English geology" as a whole.[47] This claim greatly underestimates the pivotal role of the metropolitan Geological Society; it also fails to recognize the diversity characteristic of geological practice during this period. The

[46] Clark and Hughes, 1890, *1*: 381; also Sedgwick 1830: 204, and Sedgwick to Darwin, 4 Sept. 1831, in Barrett 1974: 164.

[47] Rupke 1983. Morrell 1984 gives a review emphasizing other distinctive aspects of Cambridge geology. See also Secord 1983.

stratigraphical enterprise was pursued in a wide range of contexts. As will shortly be evident, there were crucial differences even between Oxford and Cambridge. These particularities of local context must be explored in detail if scientific practice is to be effectively related to its social circumstances.

In part, Murchison's forceful approach to geology can only be understood in connection with his early military career. The links between science and militarism during the post–Napoleonic era have been little studied, but their importance is undeniable. Topographic mapping had always been associated with the military, as evidenced by the origins of the Ordnance Survey. Early government geologizing continued this tradition. Many leading stratigraphers—Joseph Portlock, William Lonsdale, Henry De la Beche, Henry James—had pursued military careers of various kinds and found their training directly useful in the field. At Great Marlow, Murchison had taken lessons in draftsmanship, a skill that he put to good use in his notebooks and publications. As De la Beche (who was later expelled from the same school) wrote, "What is commonly termed military drawing will be found of the most essential service; indeed without a knowledge of it, the geologist will often find himself much embarrassed, and be unable to record that which he has observed in nature."[48] (Sedgwick, who lacked this kind of training, constantly suffered from an inability to communicate his profound structural insights in visual terms. His notebooks are almost devoid of illustrations.) Murchison's experience as a soldier, beyond offering these practical advantages, also led him to view the rocks through eyes trained for a "campaign." His correspondence is full of "forced marches," "battles," "salvoes," and "counterattacks." His carefully calculated traverses, his organization of amateur "aides-de-camp" at key localities, and his emphasis on covering territory all bear the stamp of a man who had initially aspired to the highest level of military command.[49]

"Fox-hunting," wrote Peter Beckford in 1781, "is a kind of warfare." In Murchison's case the lessons of both subjects seem equally impressed upon his geological style. Contemporaries commonly spoke of Murchison's "eye for country," a reference specifically to the foxhunter's ability to recognize the habitat of his quarry, and more generally to the skill of reading the surface forms of the landscape. In geology, Murchison turned this ability to the discovery of

[48] De la Beche 1833: 600; for De la Beche's abortive military career, see McCartney 1977: 2-3. Murchison's drawing lessons are mentioned in Geikie 1875, *1*: 17.

[49] Further discussion in Secord 1982: 419-421.

revealing rock outcrops and suitable terrain for traverses, in this respect emulating an earlier geologist who claimed to "follow the rocks like the hunter follows a fox." Given the preponderance of independently wealthy gentlemen in the specialist circles of natural history, it is scarcely surprising that skills developed for field sports were fruitfully applied to field science. Moreover, just as foxhunting could be used to bring together different segments of rural society in a celebration of social solidarity, so geology became in Murchison's hands a communal enterprise involving all classes of people. Quarrymen saved unusual fossils for his collections; lawyers, doctors, and clergymen examined particular sections and helped with the map; and the local gentry and aristocracy sketched illustrations, gave the Murchisons free accommodations, and patronized lavish publications.[50]

Murchison's field methods testify to gentlemanly and military orientations within the English geological community. They also show him adopting particular techniques—most notably the use of small numbers of fossils for correlation—as the best means of achieving his aims. In this connection, it is especially noteworthy that his earliest scientific training was not in natural history, but in the physical sciences. Soon after moving to London in 1824, he attended Thomas Brande's chemistry lectures at the Royal Institution. But this approach had little appeal even from the start. Far more enlightening for his interests were the "catechismal lessons" on fossils by William Fitton, apparently part of a program at the Geological Society to teach new recruits the principles of palaeontological stratigraphy. During the previous decade, the effectiveness of fossils in long-distance correlation had been thoroughly demonstrated, and at any rate the possibilities they presented galvanized Murchison into action. Field lessons that summer confirmed his vocation. Visiting Oxford for a few days, he watched Buckland dissect the landscape of the strata surrounding Shotover Hill and immediately embraced a combination of palaeontology and field geology as the perfect outlet for his abilities and energy.[51]

Although based in the metropolis, Murchison commenced with Buckland as his "idol" and in some respects confirms the concept of an Oxford-based center of geological research. But Sedgwick's

[50] [Beckford] 1781: 205; J. MacCulloch to David Brewster, 15 July 1819, EUL: Gen. 129/158. Itzkowitz 1977 provides a social history of foxhunting; see Torrens 1982 and J. C. Thackray 1977 for examples of Murchison's research network in the Welsh Borders.

[51] Geikie 1875, *1*: 117-118, 124-126; Murchison to De la Beche, 9 Feb. 1847, NMW.

structural approach had distinctly different sources. As Rupke notes, academic studies at Oxford emphasized classical and theological studies to the near exclusion of anything else, and Buckland and Conybeare presented geology as the handmaiden of history and biblical chronology.[52] At Cambridge, however, the importance accorded to mathematics in the examination system provided a very different institutional environment for the sciences. The importance of Cambridge's unique emphasis on mathematics for the practice of geology has gone unrecognized in the historical literature. Jack Morrell and Arnold Thackray, while stressing the prestige enjoyed by mathematical physics at the university, assume that early Victorian natural history was "innocent of mathematical connotations."[53] But Sedgwick's field practice, like that of his Cambridge contemporaries, shows that this view requires considerable qualification. Sedgwick, Whewell, Henslow, William Hopkins, and John Herschel were all determined to bring geology's innocence to an end.

At the simplest level, of course, mathematical training enabled geologists to make the geometrical constructions needed to convert field measurements into strike and dip.[54] But the traditional Newtonian methods that Sedgwick had learned and taught in his early years at Cambridge were relevant in more fundamental ways as well. As one of the twin pillars of a liberal education, mathematics was thought to inculcate an ability for abstract thought; proving theorems provided practical experience in logical methods and models for reasoning in other areas of life. In Sedgwick's eyes, an emphasis on physical geology offered almost identical pedagogical advantages. It developed a facility for visualizing structures in three dimensions, for juggling the positions of obscurely exposed rock masses until they meshed into a coherent whole. By basing itself on the evidence of visible sections in the field, geology could exercise the same powers of reasoning developed through the close study of Newton and Euclid.[55]

Related to the preponderant "geometrical" bias at Cambridge was a strongly developed mineralogical and chemical tradition already in existence when Sedgwick assumed his chair. Brande's lec-

[52] Rupke 1983, esp. pp. 51-63.

[53] Morrell and Thackray 1981: 491.

[54] As in Schwartz 1980, where Hopkins helps Darwin solve a problem relating to South American geology.

[55] For a discussion of these ideas, see Whewell 1836, Sedgwick 1843a: 241-245, Becher 1980, and Garland 1980.

tures may not have made much of an impression on Murchison, but an analogous set at Cambridge by Edward Clarke proved of immense service to his future collaborator, "and how." "He gave a start," Sedgwick later remembered, "he kept us awake."[56] Clarke also provided a model for Sedgwick's first lecture course and inspired his earliest papers. However, Clarke's formative influence has never been sufficiently recognized. The early Sedgwick has always been treated as a purely Wernerian geologist, following directly in a tradition derived from the teaching of Abraham Werner at the mining academy in Freiburg, Saxony. As Sedgwick himself recalled (albeit at a later date), he had initially been "eaten up with the Wernerian notions—ready to sacrifice my senses to that creed—a Wernerian slave."[57] And certainly his first paper—read to the Cambridge Philosophical Society in 1820—described the structure of Cornwall and Devon in terms that would seem to demonstrate his adherence to certain elements of the Wernerian system. It even argued for a controversial "neptunist" interpretation of mineral veins, claiming that they are filled in from above as precipitates from aqueous solution. The criterion distinguishing the Wernerian approach from all others, however, was the identification of minerals through the use of external, visible characters; and here Sedgwick took a decisively anti-Wernerian stance. Like his teacher Clarke (who had polemicized repeatedly against Werner), he analyzed his specimens *chemically* with the blowpipe.[58] If Sedgwick was ever a "slave" of the master of Freiburg, it was the very loose bondage characteristic of English geologists at the time he became Woodwardian professor. As Mott Greene has pointed out, Werner was praised by his contemporaries as one of the founders of the stratigraphical enterprise, for his field methods rather than his speculative theories.[59] In this methodological sense, anyone concerned with structural geology in 1818 could scarcely have avoided a debt to the Freiburg school.

Sedgwick was never a strict Wernerian. But during the 1830s he did adopt an approach that perfectly matched his structural practice. More than anyone else in England, he supported the controversial ideas on mountain building developed by the leading French

[56] Clark and Hughes 1890, *2*: 349.

[57] Clark and Hughes 1890, *1*: 251. Also Lyell to Fleming, 11 Oct. 1829, in K. M. Lyell 1881, *1*: 255-256.

[58] Sedgwick 1820. For the blowpipe, see Clarke 1818 and Oldroyd 1972b.

[59] M. Greene 1982: 31-68. See also A. M. Ospovat 1969; helpful articles on Werner (by Ospovat) and on Leopold von Buch (by W. Nieuwenkamp) are in the *DSB*.

geologist Léonce Élie de Beaumont. Because of their importance for Sedgwick's geology (and the later controversy), these concepts deserve to be outlined in detail. According to Élie de Beaumont, the shrinkage of the earth's crust as it cooled produced sudden upheavals of mountain chains along particular "axes of elevation." These catastrophic events were represented in the succession of strata by unconformities and in the fossil record by massive extinctions. A memoir of 1829 identified twelve "systems" of parallel mountain chains, each formed in a virtual instant and characterized by a unique direction. Although Sedgwick doubted that so simple a correlation between tectonic and faunal discontinuities could be maintained, he eulogized the theory at length from the presidential chair of the Geological Society in 1831 and incorporated it into a new syllabus of his lectures published in the following year. He found the Frenchman's version of earth history far more satisfactory than Lyell's. In Sedgwick's eyes a new epoch in geology had just begun, but the revolution was certainly not to be found in the recently published *Principles*.[60]

Sedgwick's enthusiasm for Élie de Beaumont has been recognized ever since his 1831 address, but the utility of the new theory for his scientific practice has never been appreciated. In correlating strata unconnected by sections, Sedgwick had hitherto used only fossil characters and lithology. But in Wales organic remains were scarce and lithological types persisted with monotonous slatiness. Élie de Beaumont's theory of parallel mountain chains opened up an entirely new means of correlation, one particularly well suited to the older strata. According to the theory, mountain chains with parallel strikes had been thrown up at the same time, whereas those with diverging strikes were produced by differing periods of elevation. To the field geologist this meant that strata upheaved by particular episodes of uplift could be correlated by their strike, because a constant compass direction uniquely characterized each period of mountain building.

Months before leaving for Wales in 1831 Sedgwick hailed the transformation that Élie de Beaumont's discovery of "a new faculty of induction" had worked in his geological vision. Henceforth he would take to the field with different eyes, and his notebooks from the 1830s are filled with information on strike, dip, and structure—precisely the information requisite for dating by the principal epi-

[60] Élie de Beaumont 1829-1830, 1831; Sedgwick 1831: 300-312; Sedgwick 1832: 56-57; also comments on Sedgwick's practice in Whewell 1839: 79. For the reaction to Élie de Beaumont, see Lawrence 1978, Rudwick 1971, and M. Greene 1982: 69-121.

sodes of mountain uplift. Of course, other geologists recorded this kind of data, but Sedgwick did so to an unusual degree. For example, his manuscript field maps of Wales (Fig. 2.5) are covered with inked-in strike and dip symbols and only occasionally have the colored washes characteristic of most geological maps of this period. As far as I am aware, Sedgwick was the only member of the Geological Society circle to incorporate the theory of parallelism into his field practice during the 1830s. Conybeare, Buckland, De la Beche, and other geologists showed interest in the theory but pointed to exceptions to the rule of parallelism. Murchison dismissed it outright as "French puff."[61] Sedgwick was aware of the discrepancies, but maintained that the theory held remarkably well for the ancient rocks he was investigating.

Sedgwick's acceptance of this grand synthesis is yet another manifestation of the underlying structural emphasis of his field practice. Although never embracing the later and more extreme versions of the theory of parallelism such as the infamous *réseau pentagonal,* he shared the French geologist's fascination with regularities in nature, the linearities and sinuosities of the strata considered as three-dimensional objects. When awarding the aging Sedgwick the Royal Society's Copley Medal in 1863, Colonel Edward Sabine emphasized his insight into "mountain geometry," which he defined as "that geometry by which we unite in imagination lines and surfaces observed in one part of a complicated mountain or district with those in another, so as to form a distinct geometrical conception of the arrangement of the intervening masses." "This is not an ordinary power," he continued, "but Mr. Sedgwick's early mathematical education was favourable to the cultivation of it." For Sedgwick, determining the structure of the older rocks resembled an exercise in solid geometry, and it was with considerable justice that Murchison proclaimed him in 1828 "our only *mathematical* champion."[62]

As the years passed, both Sedgwick and Murchison altered their scientific practices in ways conditioned by wider changes in geology as a whole. Murchison moved away from his association with Lyell's doctrines, while Sedgwick adopted the tectonic theories of

[61] Murchison to Harcourt, 5 Dec. 1831, in J. Morrell and A. Thackray 1984: 114. Other comments are in De la Beche 1833: 489-491; Murchison 1839: 568-572; Conybeare 1832, 1833b (also contains discussion by Sedgwick).

[62] Murchison to Sedgwick, 18 Aug. 1828, in Clark and Hughes 1890, *1*: 340; Sabine 1863: 35. Élie de Beaumont's later work is well described in M. Greene 1982: 113-119. See also Élie de Beaumont 1852.

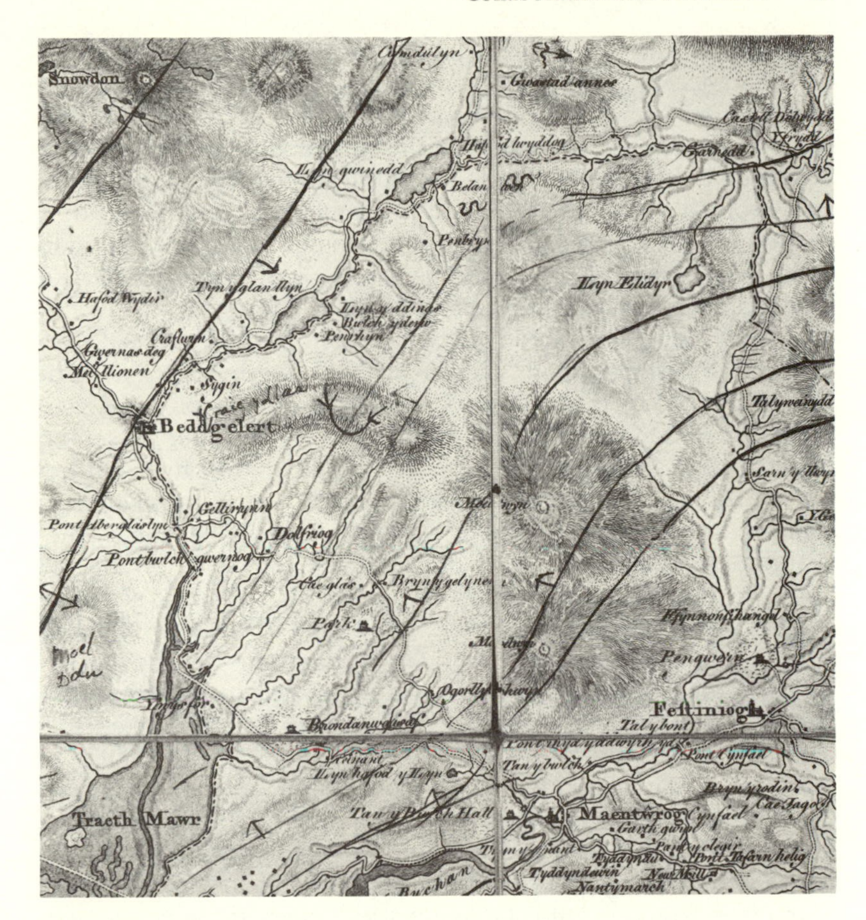

Fig. 2.5. Map of the region southeast of Snowdon, annotated by Sedgwick in the field during the early 1830s. Note the anticlinal and synclinal lines.

Élie de Beaumont. Perhaps most significantly, the decade after 1820 witnessed a continuing shift in the relative importance of different methods of correlation, as the palaeontological techniques pioneered by Smith, Cuvier, and Brongniart gained further ground over the mineralogical emphasis prevalent early in the century.[63]

[63] The general shift from mineral to fossil characters is well characterized in [Laudan] 1974 and Rudwick 1976a, esp. pp. 142-150. Appropriate cautions against underestimating the work of earlier researchers in this respect are in Torrens 1983. Among other works, Allen 1979 and Secord 1985a contain reflections on the disciplinary position of geology.

Certainly by the time Murchison published his first paper, Sedgwick had accepted the importance of fossils for comparisons over long distances and as elements in any finished classification. His later writings make no mention of the blowpipe. Broadly speaking, both men thus supported the aims of the stratigraphical enterprise as it evolved during the 1820s. At the same time, Sedgwick never abandoned his emphasis on structural geology and never ceased to insist that classifications should be based on the superposition of physical masses as determined in the field.

In short, Sedgwick should not be considered either a lapsed Wernerian or a member of an "English school" directed from Oxford. His scientific work is best viewed as an extension of a physical, geometrical approach peculiarly appropriate to the institutional setting of Cambridge. From his first paper he focused on the physical interrelationships of rock masses and strata at the base of the geological column. His language is that of "powers of crystallization," "mechanical aggregation," "great parallel fissures," and "contractile powers." Like others at Cambridge, he actively worked to associate geology with the exact sciences. Crosbie Smith has shown that this approach later served as the foundation for studies in physical geology by the Cambridge mathematics tutor William Hopkins. Ultimately this led to important work by George Gabriel Stokes, Osmond Fisher, George Darwin, and others late in the century.[64] Murchison, in contrast, pursued a streamlined, geographically oriented version of the palaeontological stratigraphy developed in Oxford and London. But where Buckland, Conybeare, and their fellow clerics had used fossils primarily as supports for natural theology, Murchison would eventually demonstrate their effectiveness in an expansionist program of global correlation.[65]

Obviously the celebrated partnership of Sedgwick and Murchison would have been impossible without an overall correspondence of their views. In this sense, Cambridge and London were not so far apart after all. But there can be little doubt that the real secret of their success rested in differences in training and perception, the source of creative friction first in collaboration and ultimately in controversy.

[64] Sedgwick 1820, *passim;* Smith 1985. I am grateful to Crosbie Smith for helpful conversations on the subject of the preceding section.

[65] Murchison 1829a.

CHAPTER THREE

Cambria and Siluria Established

ONCE Murchison began hammering the rocks of Wales and the Welsh Borders in earnest, the problem of relating his results to those of Sedgwick arose almost immediately. They readily agreed on the limits of their fields of research: Sedgwick's territory would extend from the Berwyn range to the higher mountains to the west in Caernarvonshire, while Murchison's would center on Shropshire and the Transition rocks to the southwest. This division of labor, and hence the boundary itself, marked contrasts in the research aims, personalities, and scientific styles of the two men. But at the same time the very existence of separate territories was itself a sign of collaboration, for it enabled them to produce a unified picture of Welsh geology with a minimum of direct competition. Like two men drilling an underground tunnel from opposite sides of a river, they were acutely aware of the need to coordinate their researches. A delicate process of dovetailing was accomplished during the next few years as Murchison and Sedgwick established the boundary between two separate geological "systems." Given the complexity of the sections, the older rocks of Wales could conceivably have been classified in several different ways; the boundary they finally selected satisfied the demands of friendship just as well as it met technical criteria for classification. The interests of the two collaborators and the geological community at large went hand in hand. But in this case, peace was temporary and unwittingly bought at a price.

PRELIMINARY DOVETAILING

Throughout the 1830s Sedgwick took chief responsibility for creating a unified picture of Welsh geology. His book required Murchison's results not just for the sake of completeness; it was only through the missing links in the upper Grauwacke that Sedgwick's sections could be connected with the rest of the British succession.[1] Murchison's own need to relate his findings to those of Sedgwick

[1] Murchison to Sedgwick, 19 Aug. [1832], CUL: Add. ms 7652IA54c.

was much less acute, for he gained nothing more than a base line at the bottom of his sections, an important but not an essential matter. As the summer of 1832 drew to a close, both men felt ready to link their work into a coherent whole. Notebooks, letters, and published progress reports allow the fieldwork of these early years to be traced in great detail—the routes taken and the reasons for them, the provisional groupings, the rejected classifications, the help received from local amateurs and expert palaeontologists—revealing the gradual creation of a scientific picture of a relatively unknown region.

After his first summer in North Wales, Sedgwick had resolved to complete his mapping there in one additional field season. In order to do so he even declined a clerical living at East Farleigh in Kent, which would have opened up the possibility of marriage, prohibited by the terms of the Woodwardian foundation.[2] His choice of geology over a church living now virtually irrevocable, he determined to finish the *Outlines* as rapidly as possible and spent more than four months in 1832 working his way across Wales, checking earlier work in Caernarvonshire and extending sections to the south and east. By October his general views of the succession had assumed the form they would hold for almost two decades. The key to his interpretation, and the linchpin of the later debate, was a dark-colored fossiliferous limestone on the western slopes of the Berwyn range in Merionethshire. This formation, which Sedgwick soon called the "Bala Limestone" after Bala Lake to the west, was about twenty to thirty feet thick, a mere ribbon among the accumulated miles of slate and porphyry in the wilds of North Wales. In 1832 Sedgwick traced it in outcrops for about thirty miles, from Glyn Diffwys in the north to Dinas Mawddwy in the south. The significance of the Bala Limestone was twofold. First, as one of the few easily recognizable fossiliferous bands in North Wales, it gave Sedgwick an excellent marker horizon for tying together his traverse sections. Although, in the absence of a passage series, the relation of these sections to the overlying succession inevitably remained unclear, careful mapping of the Bala Limestone at least allowed the construction of an internally consistent picture of the structure of North Wales. His previous starting point in the Primary rocks had been far less secure. A second advantage derived from the relative palaeontological richness of the Bala Limestone. A few fossiliferous

[2] Clark and Hughes 1890, *1*: 383-386, also K. M. Lyell 1881, *1*: 374-375. It should be noted that Sedgwick received a fellowship stipend of over four hundred pounds each year in addition to his Woodwardian income, so the financial benefits of accepting the living were minimal.

bands were found elsewhere in North Wales, but these could be traced only for short distances and were thus unsuited for use as a base line. Moreover, the presence of twenty or so fossil species in the Bala Limestone—together with its lithological distinctiveness and clear position in a sequence—meant that this stratum could be used to correlate Wales with other regions in Britain that Sedgwick had studied.[3]

Using his new-found key and uniting the work of two summers, Sedgwick outlined his views to Murchison in the autumn of 1832. Lectures to the Cambridge Philosophical Society and the British Association meeting in 1833 carried his findings to a wider audience.[4] In outlining the structure of Wales, he characterized four principal groups. First, above the Bala Limestone were slates occupying the higher reaches of the Berwyn range and much of South Wales. This group, of central importance to the later controversy, was poor in fossils and disturbed by cleavage. Second in the sequence, and below the Bala Limestone, was a calcareous series to the west of Bala Lake. The "Snowdonian slates" came third in the descending succession. A huge volume of igneous strata swelled this group, which occupied much of the spectacular mountain districts of northwest Wales. Searches by previous investigators, notably William Phillips and Samuel Woods, had produced a handful of bivalves, brachiopods, corals, and crinoids, particularly in a limestone on the summit of Snowdon which represented the lowest known fossiliferous horizon. Finally, at the base of the geological column in Britain were the "Primary slates" of Anglesey and Caernarvonshire. Above the highest of these four groups, Sedgwick entered what he termed "the upper calcareous greywacke," the thick series of passage beds being studied concurrently by Murchison.

By the end of the 1832 season, Murchison had filled notebook after notebook with details of the rocks of South Wales and the Welsh Borders. His uppermost groups, in 1835 christened the Ludlow and Wenlock formations (see Fig. 3.1), were mired in a confusion that he publicly clarified only in January 1834.[5] These were

[3] Sedgwick to Murchison, 23 July 1832, in Clark and Hughes 1890, *1*: 391-394.

[4] The reports of both accounts occupy only a sentence. See Clark and Hughes 1890, *2*: 594; *1833 British Association Report*, p. xxxiii; and *Philosophical Magazine*, 1833, 3d ser., *2*: 381. A more complete view can be found in Sedgwick to Murchison, [pmk. 19 Nov. 1832], CUL: Add. ms 7652IC11; copy at IIIG7. D. A. Bassett 1969b also discusses Sedgwick's picture of the structure of Wales.

[5] The errors in Murchison's first classification (1833b) are listed in Clark and Hughes 1890, *2*: 524-525; Geikie 1875, *1*: 217; and J. C. Thackray 1977: 190. Torrens (1983: 127) suggests that the error may be traced to a similar one by Arthur Aikin. The revised classification is presented in Murchison 1834a.

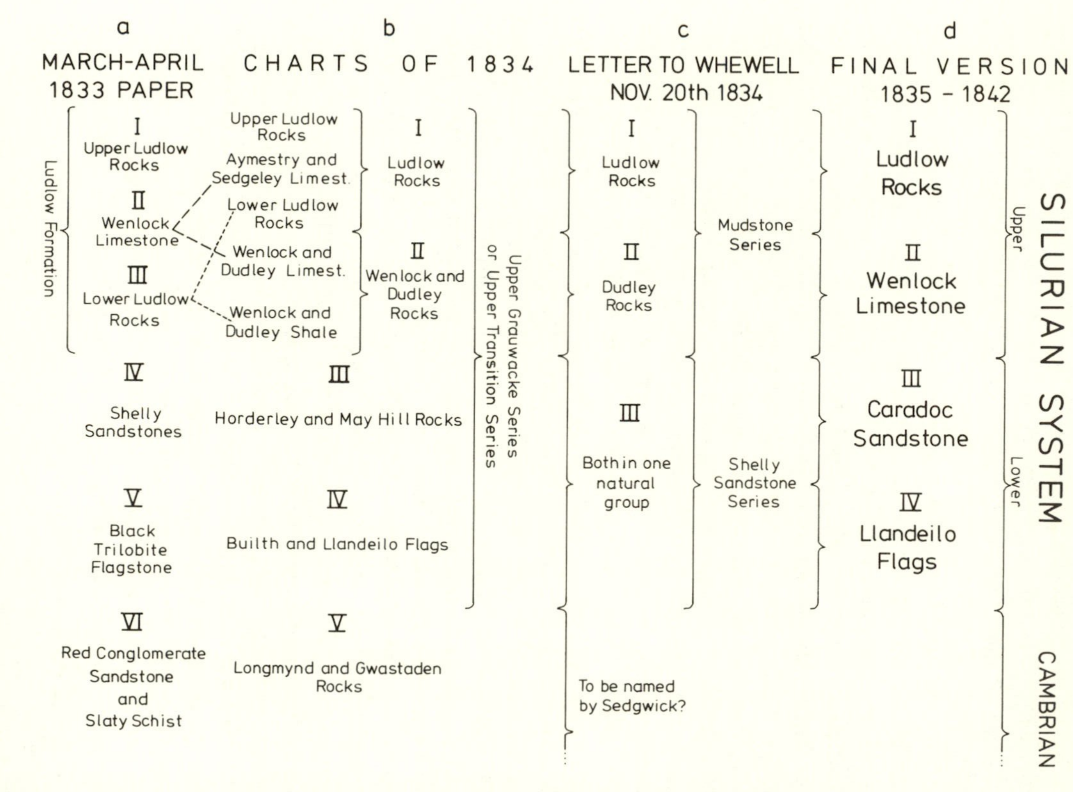

Fig. 3.1. Changes in Murchison's classification of the Transition strata, 1833–1842.

highly fossiliferous shales and limestones—the Wenlock often rich in corals—and constituted his original passage beds below the Old Red Sandstone. For our purposes, however, the underlying formations are of much greater importance, for they were directly involved in the dovetailing process. As part of a paper delivered to the Geological Society in March 1833, Murchison summarized his lower divisions as they had stood since the end of the previous summer. Immediately below the Wenlock was his "Caradoc Sandstone" formation. Called "shelly sandstone" in his early notebooks, it consisted of green, purple, and red sandstones and was typified by strata found on the slopes of Caer Caradoc in Shropshire. Its fossils included crinoids, brachiopods, and shells (Fig. 3.2). The "Llandeilo Flag," or "Black Trilobite Flagstone," came next in the succession. It could be recognized in the field as a dark-colored flaggy group characterized by the showy trilobite *Asaphus Buchii* (Fig. 3.3). Below Murchison's fossiliferous formations was a vast series of conglomerates, sandstones, and slaty schists. These were several thousand feet thick and apparently without any traces of organic remains. Further details of all these formations will be found in Figure 3.4, which reprints a chart published after Murchison had revised the sequence of his upper beds. This chart is especially interesting as an indication of the wide, even eclectic, range of characters that could be used to group together individual beds as a "formation." Lithology, thickness, locality, and fossil content all had a part to play. Even for Murchison, palaeontology could not possibly have served as the only approach to the early stages of work in the field.

The problem to be solved in 1832 was straightforward enough. How could the lower strata of Murchison's sections be related to the upper strata of Sedgwick's? From the first both men assumed that they had described some of the same beds under different headings.[6] In Sedgwick's view the main problem for any dovetailing was presented by his uppermost group, the slates making up the summit of the Berwyns and South Wales. This series immediately above the Bala Limestone remained for Sedgwick "a great obscure group of which I don't exactly know the end and side." During four exploratory traverses into Central and South Wales at the end of the 1832 field season Sedgwick had confronted a veritable ocean of slate, the sections contradictory and highly contorted, poorly exposed and almost entirely without fossils. Murchison had found them equally confusing during his own extensive examination

[6] Sedgwick to Murchison, [pmk. 19 Nov. 1832].

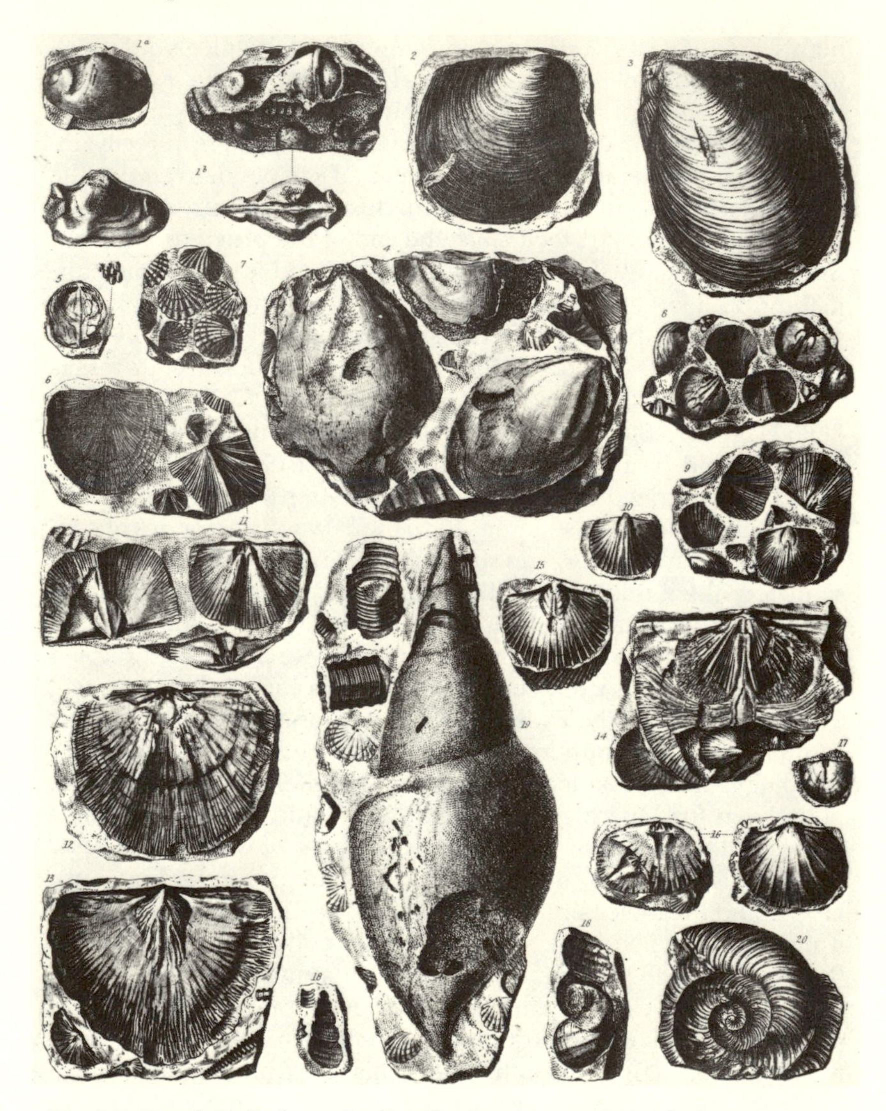

Fig. 3.2. Fossil shells from the Caradoc Sandstone. Several of these were also found in the Bala Limestone, which was assumed to occupy a much lower position in the sequence.

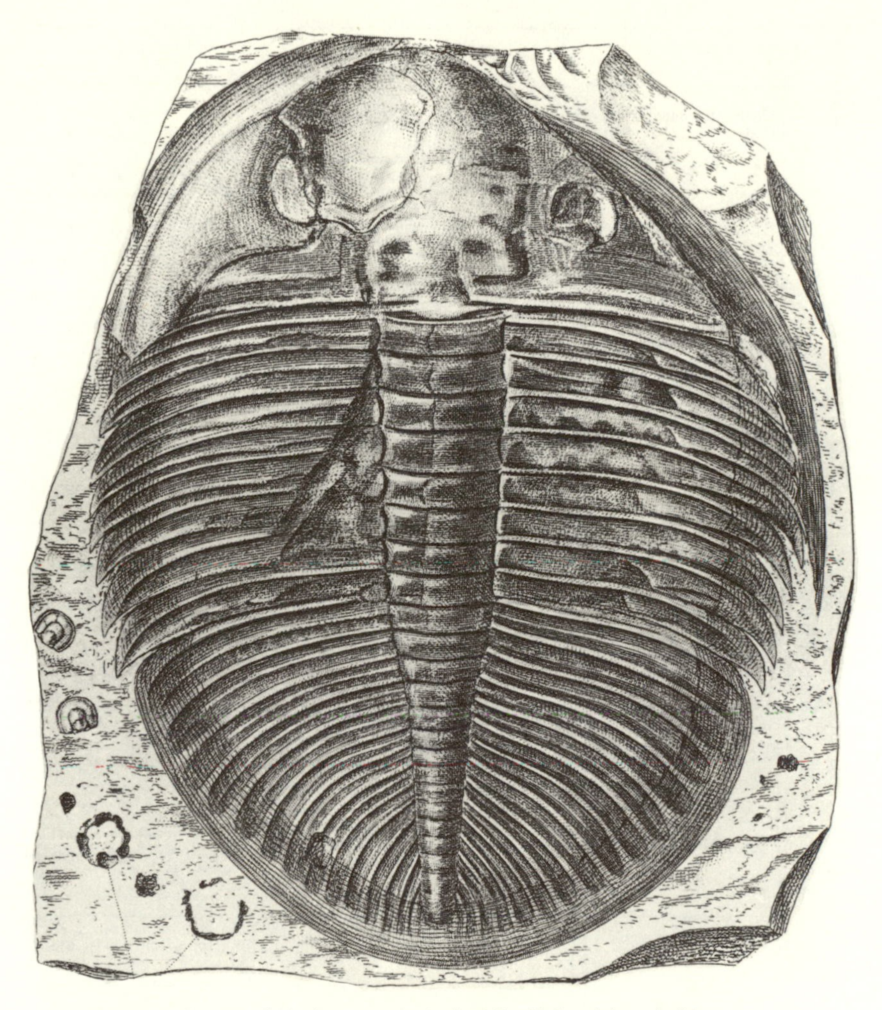

Fig. 3.3. *Asaphus Buchii*, characteristic fossil of Murchison's Llandeilo Flags.

during the same year. Despite their obscurity, these rocks and those of the Berwyns needed in some way to be coordinated with Murchison's lower beds so as to become what Sedgwick called "the connecting link of our work."[7] Many letters passed between London and Cambridge as the two friends welded their investigations together.

[7] Sedgwick to Murchison, [pmk. 26 Oct. 1832], CUL: Add. ms 7652IC8; copy at IIIG4.

	Formations.	Maximum approximate thickness.	Subdivisions.	Lithological Characters.
	Carboniferous Limestone.	Feet. 500 ?	Limestone. Shale.	
	Old Red Sandstone.	10,000.	*a.* Red conglomerate and sandstone. *b.* Cornstone and argillaceous marls. *c.* Tile stones, &c.	*a.* Quartzose conglomerate overlying thick-bedded sandstones. *b.* Red and green, concretionary limestones, with spotted argillaceous marls and beds of sandstone. *c.* Flaggy, highly micaceous, hard, red and green sandstone.
UPPER GREYWACKE SERIES.	I. Ludlow Rocks.	2000.	*d.* Upper Ludlow rock. *e.* Aymestry and Sedgeley limestone. *f.* Lower Ludlow rock.	*d.* Slightly micaceous, grey-coloured, thin-bedded sandstone. *e.* Subcrystalline or grey and blue argillaceous limestone. *f.* Sandy, liver, and dark-coloured shale and flag, with concretions of earthy limestone.
	II. Wenlock and Dudley Rocks.	1800.	*g.* Wenlock and Dudley limestone. *h.* Wenlock and Dudley shale.	*g.* Highly concretionary grey and blue subcrystalline limestone. *h.* Argillaceous shale, liver and dark gray-coloured, rarely micaceous, with nodules of earthy limestone.
	III. Horderley and May Hill Rocks.	2500.	*i.* Flags. *k.* Sandstones, grits, and limestones.	*i.* Thin-bedded, impure, shelly limestone, and finely laminated, slightly micaceous greenish sandstone. *k.** Thin-bedded, red, purple, green, and white freestones. Conglomeritic quartzose grits. Sandy and gritty limestones.
	IV. Builth and Llandeilo flags.	1200.		*l.* Dark-coloured flags, mostly calcareous, with some sandstone and schist.
	V. Longmynd and Gwastaden Rocks.	Many thousand feet.	Comprising all the slaty system of South Wales.	*m.* Hard, close-grained, gray greenish and purple sandstone. Red and gray quartzose conglomerate. Slate-coloured and purple schists. Coarse slates: little or no calcareous matter.

Fig. 3.4 Murchison's classification of the "Upper Greywacke Series," October 1834.

Characteristic Organic Remains.	A few Localities.
Corals differing in species from those of the formations below. Producta hemisphærica. P. Martini. Spirifer triangularis, &c. (Defence and teeth of fishes. Clee Hill, Salop.)	Lilleshall, Steeraways, Orleton, south end of Clee Hills, and Llanymynech, Shropshire. The edge of the South Wales Coal-basin.
a. No organic remains observed.	*a.* Caermarthen and Brecon Fans, SE. part of Black Forest, Brecknockshire; flanks of the Brown Clee Hill, Shropshire.
b. Fishes of undescribed genera.	*b.* Central and north. parts of Herefordshire: eastern part of Brecknockshire: Whitbach near Ludlow, and base of the Clee Hills, Shropshire: Tenbury and Alveley, near Kidderminster, Worcestersh.
c. Avicula, n. s. Pileopsis, n. s. Small Orthocera. Small Ichthyodorulites?	*c.* Pontarlleche, Cwmdwr, Caermarthenshire: Clyro Hills, Brecknocksh.: Tinmill Copse, near Downton Castle, Herefordsh.: Clun Forest, Shropsh.
d. Avicula, n. s. A. retroflexa, Hisinger. Atrypa (Dalman), n. s. Cypricardia, n. s. Homonolotus Knightii, new genus, Konig. Leptæna lata, V. Buch. Orthis, several new species. Orbicula, 2 new species. Orthocera, several new species. Pleurotomaria? 2 new species. Turbo, n. s. Gigantic serpuline bodies, &c. &c.	*d.* Ludlow Castle, Whitcliffe, Munslow, Diddlebury, Larden, Shropshire: Croft Castle, Mortimer's Cross, Titley, Kington, Fownhope, Stoke Edith, Herefordshire: West flanks of Malvern and Abberley Hills, Worcestershire: West flank of May Hill: Presteigne, Pain's Castle, Radnorshire: Trewerne Hills, Corn-y-fan, Brecon, Usk Castle.
e. Pentamerus Knightii, M. C. Pileopsis vetusta, M. C. Bellerophon, n. s. Lingula, n. s. Atrypa, n. s. Terebratula Wilsoni, M. C. Calamopora fibrosa, Goldf., and a few other corals.	*e.* Aymestry, Croft Ambry, Gatley, Brindgwood Chase, Downton on the Rock, Herefordshire: Yeo Edge, Shelderton, Norton Camp, Dinchope, Caynham Camp, Shropshire; Sedgeley, Staffordshire.
f. Phragmoceras, new genus, Broderip, 3 sp. Asaphus caudatus. Ichthyodorulites? small. "Cardiola," Brod., new gen. 2 sp. Nautilus, n. s. Spirulites, 2 n. s. Pentamerus. Atrypa galeata, Dalm. n. s. Pleurotomaria, n. s. Orthocera pyriformis, n. s.; and several others.	*f.* Escarpments of Mocktree and Brindgwood Chase, Gatley, and valley of Woolhope, Herefordshire: Marrington Dingle, Westhope, Hopedale, and Long-Mountain, Shropshire: west side of Abberley and Malvern Hills: escarpments in Montgomery, Radnor Forest, Brecknock and Caermarthen shs.
g. Corals and Crinoidea in vast abundance. Bellerophon tenuifascia, M. C. Euomphalus rugosus. Eu. discors. Conularia quadrisculata, M. C. Pentamerus, n. s. Natica, n. s. N. spirata, M. C. Leptæna euglypha, Dalman. Spirifer lineatus, M. C. S. n. s. Terebratula cuneata, Dalm. Producta depressa, M. C. Orthocera, sev. sp. Asaphus caudatus. Calymene Blumenbachii. The Bar Trilobite and others.	*g.* Lincoln Hill, Benthall and Wenlock Edge, Shropshire; Burrington, Nether Lye, near Amestry, Nash, near Presteigne, Old Radnor: Pwll-Calch, Caermarthenshire: valley of Woolhope, Ledbury, and west side of Malvern Hills: east side of Abberley Hills, Dudley, Worcestershire: Long Hope, near May Hill, and Tortworth, Gloucestershire: Prescoed and Cil-na-Caya, near Usk.
h. As. caudatus variety, C. Blumenbachii. Lingula, n. s. Orthis, n. s., and others. Cyrtia trapezoidalis, Dalm. Delthyris, n. s. Orthocera, n. s. O. annulata, M. C. Crinoidea, &c.	*h.* Buildwas, Hughley, Wistanstow, and Clungunford, Salop: escarpments in Montgomery, Radnor, Brecknock, and Caermarthen shires: west flank of Malvern Hills, Alfrick, Worcestershire: centre of Wren's Nest, Dudley, &c. &c.
i. Pentamerus lævis, M. C. P. oblongatus, n. s. Leptæna, n. s. Pileopsis, n. s. Orthis Callactis, Dalm., and several new species. Terebratula, n. s. / Tentaculites and Crinoidea, abundant, } Corals rare.	*i.* Banks of the Onny, near Horderley, Acton Burnell, Chatwall: the Hollies near Hope Bowdler, Cheney Longville, Acton Scott: east flank of Wrekin and Caer Caradoc, Salop: Eastnor Park, Obelisk, and centre of Woolhope Valley, Hereforshire: May Hill, and Tortworth, Gloucestersh.
k. Nucula, n. s. Pentamerus, n. s. Trilobites of underscribed species, including the genus Cryptolithus of N. America, and 14 species of the genus Orthis have been found, including O. aperturatus, Dalm., all differing from those of the overlying formations.	*k.* Horderly, Hoar Edge, Long Lane, and Corton, Shropshire: Ankerdine Hill, Old Storridge, Howlers Heath, SW. of Malvern Hills, Worcestersh.: May Hill, Gloucestersh.: and the same localities as *i* in Shropshire: Powis Castle, Guilsfield, and Alt-y-maen, Montgomerysh.: Castell Craig, Noeth Grug, and Llandovery, Caermarthenshire.
l. Asaphus Buchii. Agnostus, Brongn., undescribed Trilobites of three species; differing from those of the overlying formations.	*l.* Rorington and Hope, near Skelve, Shropshire: Llandrindod and Wellfield, near Builth, Radnorshire: Tan-yr-Alt to Llandeilo, Caermarthenshire.
m. Few organic remains have yet been observed in this great system, but it is underlaid by fossiliferous strata and limestones, which will be described by Professor Sedgwick.	*m.* The Longmynd, Linley, Haughmond, Lyth, Pulberbatch Hills, Salop: Gwastaden, east of Rhayader, Radnor, &c. &c.: hills west of Llandovery, Caermarthenshire.

The episode affords a revealing instance of the role of preconceptions in shaping scientific interpretations of the natural world. Although the geology of Wales was relatively unexplored, Sedgwick and Murchison were by no means the first investigators in the region. The work of predecessors such as Aikin, Buckland, and Greenough inevitably served as the basis for later interpretations. The pioneering studies by these men strongly indicated that the slates of South Wales should be placed below the strata now being studied by Murchison. Thus Greenough's 1820 map, although strictly intended to display the distribution of formations rather than their succession, did imply that the Welsh strata became progressively older towards the northwest (Fig. 2.3). Important topographical and lithological considerations supported this view. The core of Wales was a mountainous district made of altered slates; in contrast, Murchison was mapping a hilly region composed of clearly stratified sedimentary rocks. Typically, undisturbed strata occupied a higher position in a given geological sequence than those upturned and cut by slaty cleavage. Assumptions of this sort provided useful first-order guides for proceeding in the field.

Both Sedgwick and Murchison also looked to their own recent experience on the Continent for a potential model for the succession in South Wales. Their memoir on the eastern Alps had described a central mountainous core of older rocks surrounded by a single band of chloritic black slate.[8] Similar black slates were often found in South Wales at the base of the Llandeilo Flags. A simple solution to the problem of dovetailing, and one that completed the Alpine analogy, was to place all these black slates on a single stratigraphical parallel, as shown in Figure 3.5 (top). In a short-lived attempt to accomplish this end within the Welsh Borders, Murchison had correlated the Longmynd strata with the Llandeilo Flags in 1832, and he now assumed that an analogous situation would prevail in the succession of South Wales already mapped as far to the southeast as Carmarthen.[9] Sedgwick, his expectations also primed by the Alpine case, similarly anticipated a single zone of black slate between the Llandeilo Flags and the underlying blue slates. During his rapid

[8] Sedgwick and Murchison 1830: 86-88, where these beds are referred to as "chloritic schist with thin bands of limestone."

[9] For Murchison's Longmynd correlation, Murchison 1833b: 476; for its rejection, Murchison 1834a: 14. Sedgwick initially approved of the link; see Sedgwick to Murchison, [pmk. 26 Oct. 1832]. Sarjeant and Harvey 1979 provide a useful history of Longmynd geology, although they mistakenly state that Murchison placed the Longmynd in his lowest group from the beginning (pp. 191-192).

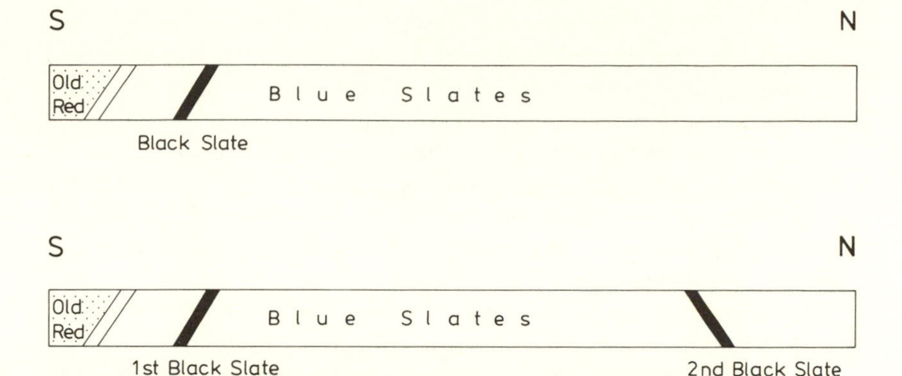

Fig. 3.5. Schematic traverse sections showing alternative views of the geological succession in South Wales. In both the strata become progressively older to the north: (top) expected on basis of Alpine case and Murchison's earlier work; (bottom) Sedgwick's interpretation in 1832 after finding anomalously dipping slates.

traverses at the end of the 1832 season he found at least one section, from Lampeter to Carmarthen, where the expected order seemed to be upheld. Such an arrangement united the work of the two geologists with the least possible difficulty, for it not only matched the Alpine case and Greenough's map, but also kept the amount of stratigraphical overlap to a minimum.[10]

But Sedgwick rejected this view even on the basis of his brief tour. Although eager to put the black slates in a single group intermediate between the Llandeilo Flags and the blue slate zone of South Wales, he concluded that the situation could not be so simple. North of the town of Llandovery, the black slates appeared to be below the blue slates rather than above as the two men had hoped. "I *was most anxious* to pack [the black slate zone] comfortably *under*—the trilobite system & over the fine blue slate zone," Sedgwick told Murchison in November 1832. "But the dips are most perverse. . . . I tried hard to explain this away by interpolating faults but all in vain."[11] As he commented a few days later, "I had my eyes quite open to the Alpine case, & I was as I said before most anxious to pack all the black slates on the outside; but it would not do—nature has not placed them there, at least not all of them."[12] Needless to

[10] Sedgwick to Murchison, [7 Nov. 1832], CUL: Add. ms 7652IC9; copy at 7652IIG5.

[11] Sedgwick to Murchison, [7 Nov. 1832].

[12] Sedgwick to Murchison, [pmk. 19 Nov. 1832].

say, "nature" did not naysay Sedgwick's anticipations directly. But when he approached the Welsh strata according to the basic ground rules of geology, an inescapable series of facts came into being, confounding the expectations so strongly engendered by Greenough's map and his own experience in the Alps. Sedgwick felt that there was little he could do besides creating a second band of black slates at the base of the succession. Out of the two dovetailings considered in the autumn of 1832, he reluctantly chose the more complex one, illustrated in Figure 3.5 (bottom).

In preparing for his Geological Society presidential address in 1833, Murchison asked Sedgwick to contribute a passage on the relation between their work. Sedgwick equivocated on particulars but spoke confidently of the general situation. "In regard to Wales I hardly know what to say," he replied, "the upper system of deposits with its subdivisions is as plain as day light and entirely under your set—so also are the chains of igneous rocks on the outskirts, at least so they seemed from your descriptions: but as for the dovetailing of our work together, I am not so confident that I should like to commit myself in any formal precis."[13] Important details remained to be ironed out, but neither Murchison nor Sedgwick ever seriously questioned the overall succession in Wales until the Geological Survey began mapping the area in the next decade. From their first exposure to the older rocks, the two friends assumed they were studying strata on separate horizons.

A Collaborative Tour and its Consequences

At the beginning of the long vacation in June 1834 Sedgwick left Cambridge to meet Murchison at the spa resort of Great Malvern, where the two men began four weeks of touring. This was the first time they hammered jointly in Wales and the only time they examined their mutual boundary together. This tour deserves a detailed discussion, especially because it illustrates the depth of their assumption that they had been describing rocks of different ages. In linking their work both Sedgwick and Murchison believed they were removing the last vestiges of overlap and confusion; neither suspected that any major structural revisions would be required. By the end of the tour all remaining difficulties had been swept away in a collaborative boundary acceptable to both men.

The dovetailing excursion began in June with a tour of the best

[13] Sedgwick to Murchison, [pmk. 4 Feb. 1833], GSL: M/S11/73.

sections within Murchison's scientific territory (for localities, see Fig. 3.9). On a day excursion to Noeth Grüg, northeast of Llandovery, they examined a "fine development" of the Caradoc Sandstone but did not trace the passage downwards into the older strata (Fig. 3.6).[14] Despite the confusion of the rocks in this mountainous district, the evidence that the section eventually passed conformably into underlying "Sedgwickian" strata seemed unproblematic, a conclusion so self-evident that it did not require checking. The real work of dovetailing lay not in South Wales, but rather to the northeast in the Berwyn mountains. Murchison's typical sections were along the Welsh Border, especially in Shropshire; Sedgwick's were in the Bala region. The critical section connected these areas in a long traverse extending from Welshpool in the southeast to Bala Lake in the northwest, crossing the Berwyns en route. In retrospect the Berwyn section became the key to the Cambrian-Silurian controversy; for Sedgwick, the entire question of historical justice involved in the later phases of the debate hinged upon this single traverse. Here, from the perspective of later events, the boundary was finally set: here the fatal error was made.[15]

The two geologists walked along the line of their section towards the northwest and interpreted the rocks as successively lower elements in a descending sequence. The first important decision concerned the geological horizon of the limestone at a quarry near Meifod, a town in Montgomeryshire. Murchison had seen the Meifod Limestone on an earlier tour and he now placed it in his Caradoc formation for several reasons—its similarity to other limestones in the Caradoc, its general position in the sequence, and the general aspect of its fossils.[16] Judging from his notebook Sedgwick accepted this correlation without question, and he carefully recorded the dip, its angle, and the mineral character of the beds. Continuing to the northwest along the line of their traverse, they soon found outcrops of a fine calcareous slate filled with fossils characteristic of the Llandeilo Flag series. "Among 'em," as Sedgwick wrote in his notebook, "the Asaphus Buchii," the trilobite taken as a certain indicator of this formation.[17] So far all was well, for the rocks gave every appearance of increasing age.

[14] Sedgwick, Notebook "1834 XXVII," p. 12, entry for 22 June 1834, SM.

[15] See Chapter Seven.

[16] Murchison, Notebook "Vol. 13, June 1834," p. 77, GSL: M/N69. Several fossils are listed in the notebook, and in Murchison 1839: 222-224 they are included in the Caradoc formation.

[17] Sedgwick, Notebook "1834 XXVII," pp. 14-15, entry for 27-28 June, SM.

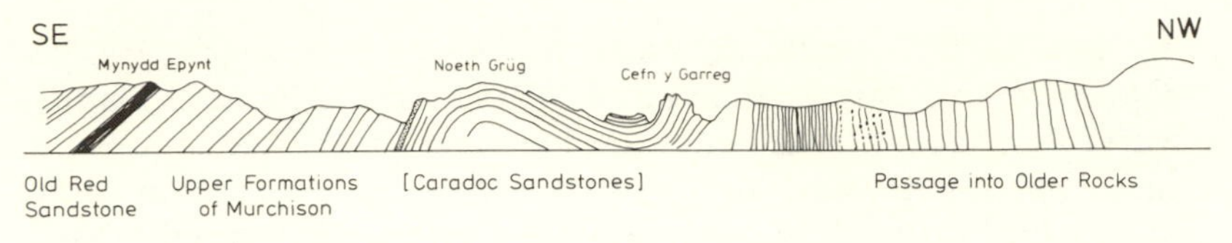

Fig. 3.6. Murchison's traverse section across Noeth Grüg.

On the following day Sedgwick and Murchison ascended the Berwyns from the east. The weather (for once) was perfect, and from their lofty vantage point the view was clear for many miles to the south and west. Sedgwick first explained the structure of the surrounding countryside. According to his previous fieldwork, they stood near the top of a sequence of slate immediately above the Bala Limestone. He then showed how these slates extended southwards in a series of great flexures. In standing on the crest of the Berwyns the two companions thus believed themselves to be on a stratigraphical parallel with most of the slaty rocks that spread across South Wales. They now faced the principal question underlying the entire tour: What was the position of Sedgwick's base line—the Bala Limestone—with respect to the strata being studied by Murchison?

The problem first demanded positioning the Bala Limestone with respect to the Meifod Limestone, which Murchison had just located in his Caradoc formation. Two years earlier Sedgwick had proposed a tentative solution. In a section drawn in 1832 (Fig. 3.7) and sent with a letter to Murchison, he had shown Meifod as a folded-over repetition of Bala.[18] In light of later research by the Geological Survey this linkage proved to be "correct," and during the 1850s Sedgwick claimed that he had abandoned it only through misplaced faith in Murchison's lower groups. If only he had followed his geological instincts, Sedgwick argued, the immense overlap between their classifications would have been recognized from the start. This conclusion totally ignored, however, the conjectural nature of linking Bala and Meifod in 1832. Until these beds were comprehensively related to the overlying sequence, any isolated correlations between them meant almost nothing and could not be relied upon. In 1832 Sedgwick had clearly recognized that the sections in North Wales "would lose 9/10ths of their interest" without Murchison's connecting strata, "as in that case, notwithstanding their complexity & interest they would have neither top nor bottom—."[19]

[18] Sedgwick to Murchison, 23 July 1832, in Clark and Hughes 1890, *1*: 394. Hughes's explanation of this section (2: 518-528) is wildly misleading.

[19] Sedgwick to Murchison, [7 Nov. 1832].

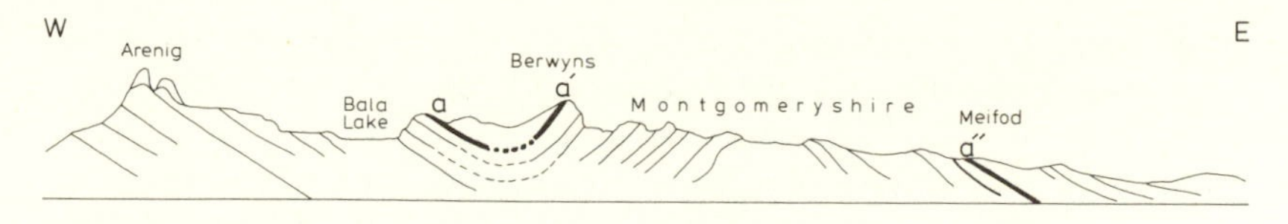

Fig. 3.7. Sedgwick's traverse section across the Berwyns in 1832, illustrating his tentative correlation between the Bala Limestone (a) and the beds near Meifod (a″).

Thus Sedgwick necessarily approached the issue in 1834 as an open one. To be sure, two good pieces of evidence still pointed to a Bala-Meifod link. The first involved geological structure; as Figure 3.7 makes clear, Sedgwick's Berwyn section showed the limestones dipping in a manner that suggested a connection of some kind. A second category of evidence was palaeontological. After surveying Sedgwick's domains from the height of the Berwyns, the two companions descended along the road to the Bala Limestone quarries and collected fossils. At a glance these were almost identical to the Meifod specimens, and in his field notebook, Murchison remarked upon their similarity to the fossils of his Caradoc formation.[20]

But such considerations, however important, were in the end vastly outweighed by the basic premise that Sedgwick's strata were almost entirely below Murchison's. From all contemporary evidence it seems clear that neither man ever contemplated giving the Bala Limestone a place in the sequence so high as the Caradoc. To have united Bala and Meifod on a parallel with the Caradoc *in 1834* would have required very large revisions of the geology of the entire Principality; the overlap between Sedgwick's and Murchison's classifications would have become painfully large, the territorial boundary provisionally established in 1832 would have lost its geological meaning, and the two men would have been describing strata of identical age. Once the Meifod Limestone had been given a place in Murchison's succession, its correlation with Bala had consequences undreamed of by Sedgwick two years earlier. Neither geologist would have contemplated such a major conflation of the sequence without compelling evidence either from fossils or sections. In this situation the palaeontological hints of a link between Bala and the Caradoc formation were weak enough to be discounted,

[20] Murchison, Notebook "Vol. 13, June 1834," pp. 83, 89. For a possible Bala-Meifod linkage, see the letter cited in n. 18; a later and more explicit discussion written soon after Sedgwick's next Welsh excursion is in Sedgwick to Murchison, [Nov. 1842], GSL: M/S11/200.

and the evidence from sections remained ambiguous, for in a traverse lasting two days, only a few critical exposures could be examined. Any reading of a particular section was thus highly conjectural, based on a few outcrops seen in the light of a structural interpretation of the entire district. Under these conditions considerable advantages were presented by a second alternative: putting the Bala Limestone well below the Meifod beds and the rest of the Caradoc and Llandeilo formations, one step lower in the descending section they had traced from Meifod (as illustrated diagrammatically in Fig. 3.8). Apparently the separation of the two beds in this way was largely Murchison's responsibility, although Sedgwick could have said little against it, as virtually all his previous studies pointed in the same direction.[21]

The critical part of the joint excursion was now over and potential territorial difficulties had been averted. The two men spent a final week driving a section east into the heart of Murchison's typical region, travelling from Powis Castle near Welshpool to the Stiperstones, over the Longmynd to Caer Caradoc, and ending at Ludlow on the tenth of July.[22] In a letter to Whewell, Murchison waxed enthusiastic about the tour.

> The Professor & myself having parted company on the most *friendly* terms, & having dovetailed our respective upper and under works most *satisfactorily* to both of us, I hastened back to join my wife. . . . I am not a little proud of having such a pupil in my own region & altho' I think & hope he endeavoured to pick every hole he could in my arrangement, he has confirmed all my views, some of which from the difficulties which environed me, I was very nervous about until I had such a *backer*. But I will say no more of number one than to assure you that we had a most delightful & profitable tour in every way & that our Section across the Berwyns in which the Professor became my instructor, was of infinite use to me. Such are the foldings & repetitions that my "Black flags" of Llandeilo (Asaphus Buchii) are reproduced even on the Eastern slope of these mountains, & it is *only* as you get *into* them, that you take final leave of my Upper Groups & get fairly sunk in the old slaty systems of the Professor.[23]

[21] Even in retrospect both agreed that Murchison had put Meifod in the Caradoc and Bala far below; see Murchison 1852c: 175, and Sedgwick 1852d: 152-153.

[22] For this continuation, Sedgwick, Notebook "1834 XXVII," entries for 1-10 July 1834, pp. 21-25. Sedgwick breakfasted on 12 July with Dr. Robert Darwin, father of Charles (p. 25).

[23] Murchison to Whewell, 18 July 1834, TC: Add. ms a.209[96]. Quoted, with changes, in Geikie 1875, *1*: 222-223.

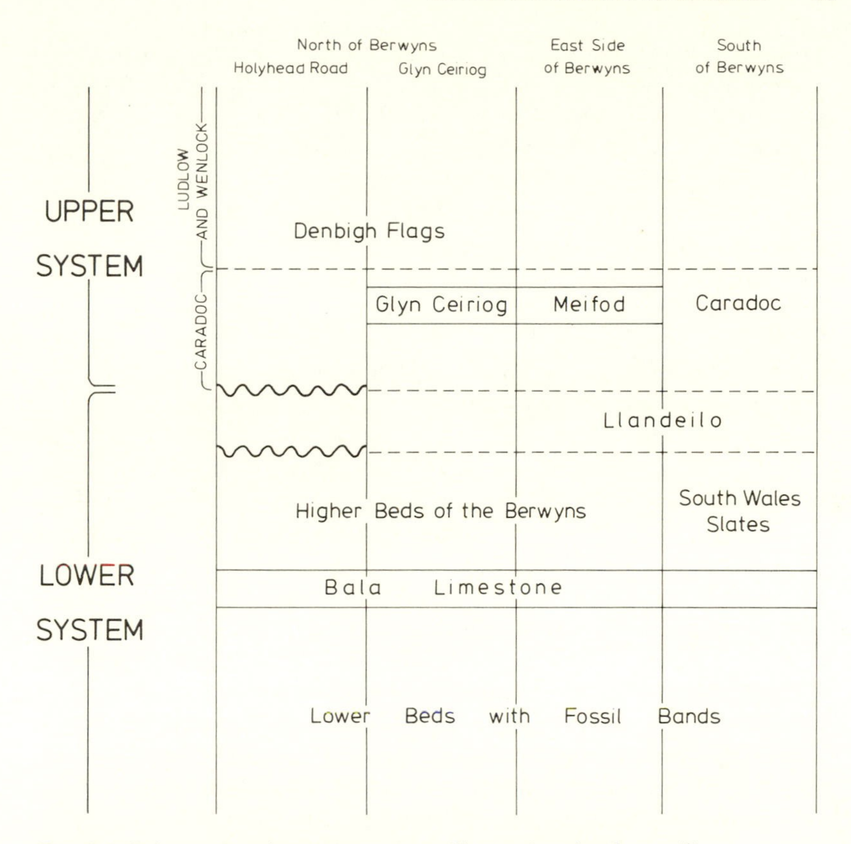

Fig. 3.8. Schematic columnar sections illustrating the dovetailing arrangements during and after the joint tour of 1834. Each vertical column shows an idealized cross section along a single traverse. Formations on the same horizontal level are stratigraphically parallel and hence of the same age, although the lack of illustrations in many of the original sources makes correlations only approximate. Parallel wavy lines indicate the presence of an unconformity, dashed lines show a conformable passage, and unbroken lines (e.g. Glyn Ceiriog and Meifod) represent limestones important in correlation. Vertical dimensions are not drawn to scale.

William Lonsdale, curator of the Geological Society, sent his congratulations and anticipated the completion of Murchison's studies of the Transition strata. For his part, Sedgwick recommended Murchison for a Royal Medal of the Royal Society in the following autumn; he vouched for the accuracy of his friend's sections not on any "implicit faith" as he later claimed, but on the basis of "a personal examination of a considerable part of the country described by Mr. Murchison." A letter written to Murchison in 1836 provides further evidence of Sedgwick's satisfaction with the dovetailing tour. "A section I exhibited, after your sections, at the Cambridge meeting [1833] plainly shewed that I was in *your system*," he wrote, "but exactly *where* was the question—our *joint journey* next year was to answer this question: and I think we have now answered it."[24]

Sedgwick and Murchison had successfully established an uninterrupted sequence from the Old Red Sandstone to the base of the Primary rocks. One task remained: The results of the Berwyn traverse had to be extended so that Murchison's lower formations and the upper ones of Sedgwick would be placed in similarly satisfactory structural relationships throughout the rest of Wales. In most cases this involved little or no change, in others a fairly substantial readjustment. Murchison and Sedgwick often spoke of this task in terms of distinguishing an upper from a lower "system." It should be emphasized, however, that the word "system" as used in this context still referred loosely to any large assemblage of formations and not to clearly defined groupings such as "Cambrian system" or "Silurian system." In this general sense "system" was usually interchangeable with other taxonomic categories like "series" or "group." Despite Murchison's later claims to the contrary, both men believed that the boundary they were tracing was an important one in the succession of the older rocks—not only a territorial limit, but also one possessing a real existence in nature. Within a year this boundary would in fact define the separate Cambrian and Silurian systems, but in the summer of 1834 it served as a convenient marker between two sets of formations.

Several criteria signalled the boundary. In most cases it was marked by an unconformable gap in the sequence, and any such break deep within Murchison's series was usually taken as the dividing line. In the presence of a conformable transition, the change

[24] Lonsdale to Murchison, 11 July 1834, GSL: M/L13/9; Sedgwick to Greenough, 25 Nov. [1834], CUL: Add. ms 7918, box 7; Sedgwick to Murchison, [?25 Jan. 1836], GSL: M/S11/96a & b. MacLeod 1971 discusses the importance of the Royal Medals in Victorian science.

would often be marked by a crumbling black slate or shale at the base of the Llandeilo Flags, the characteristic stratum on which the dovetailing of 1832 had focused.[25] In addition, faulting and numerous intrusive veins were frequently found near the transitional zone. "The *lead veins* seem *generally to mark the newer system* at its junct. with the older rock," Sedgwick wrote in his notebook.[26] Besides these telltale signs at the actual boundary, each set of strata possessed certain distinctive features. Fossils became much rarer or even disappeared altogether in the lower system; similarly, "serrated fucoids" or graptolites (generally then thought to be primitive plants) were for Sedgwick a sure sign of the newer strata. Often the absence of fossils resulted from slaty cleavage, which in itself often provided a practical means of distinguishing between the two sets of formations. He also found that quartz veins typified the older system, while calcareous veins marked the newer.[27] In practice Sedgwick and Murchison used flexible yet readily definable criteria to distinguish their respective sets of formations, thereby maintaining clearly demarcated geological and geographical spheres of activity.

During their joint tour Murchison and Sedgwick divided the task of locating the boundary, with Sedgwick assuming responsibility for its course to the north of Meifod, and Murchison to the south. Once their respective research areas were precisely delineated, their studies of Wales could be completed individually. In previous summers and especially in 1833, Murchison had already spent much time mapping the black slaty band at the base of the Llandeilo Flags. He had generally avoided the underlying rocks to the north, but along a hundred mile frontier in South Wales he had located what both men saw as the lower limit of Murchison's formations.[28]

The completion of the Berwyn section opened the way for Sedgwick to finish his share of the boundary. After parting with Murchison at Ludlow, he reentered North Wales to bring his sections into agreement with the joint determinations. In particular, the placement of the Meifod beds in the Caradoc formation high above the Bala Limestone required certain rearrangements in the rest of Sedgwick's uppermost sections. The most important of these concerned a pair of fossiliferous limestones near Llansaintffraid Glyn

[25] Murchison 1839: 357.

[26] Sedgwick, Notebook "1834 XXVII," entry for 11 Aug. 1834, p. 83, SM.

[27] Sedgwick, Notebook "1834 XXVII," entries for 11-13 Aug. 1834, pp.79-85 and passim, SM. Murchison 1839: 360.

[28] See Murchison's notebooks for 1833, GSL: M/N64-68.

Ceiriog (hereafter Glyn Ceiriog). According to Sedgwick's fieldwork of 1832 the Glyn Ceiriog beds occupied a horizon almost identical to that of the Meifod Limestone. In consequence, if the Meifod Limestone belonged to Murchison's Caradoc formation, then these calcareous strata near Glyn Ceiriog probably did so as well. But an even more important factor in this positioning was the apparent conformability of the series at Glyn Ceiriog to the overlying "Denbigh Flags," a poorly fossiliferous slaty group occupying much of Denbighshire, which Sedgwick had placed in the upper half of Murchison's formations (Fig. 3.8).[29] As we shall see in later chapters, Sedgwick maintained a high position for the Glyn Ceiriog beds throughout the 1840s, even *after* he moved the Meifod Limestone down to the level of Bala. The separation between Glyn Ceiriog and Bala became one of the fundamental points in his view of all the older rocks throughout Britain, in many ways the chief means for avoiding an overlap between Murchison's formations and his own.

In putting Glyn Ceiriog and Bala on separate horizons, Sedgwick adopted a view that he abandoned only after full publication of the official Geological Survey maps in the early 1850s. But he also went a step further in his hammering along the base of the Denbigh Flags and made a correlation that he withdrew on his next visit to Wales. Along the northern end of the Berwyns and continuing due west to the limestone outcrops at Penmacho and Cader Dinmael, Sedgwick found a group of northward dipping strata with an east-west strike.[30] Like the limestones at Glyn Ceiriog these formed part of what he believed to be a band of Caradoc rocks passing conformably into the overlying Denbigh Flags (Fig. 3.8).

The creation of this band of Caradoc strata underlines the importance of Élie de Beaumont's parallelism theory for Sedgwick's field practice. By assuming that strata with the same strike were probably of the same age, Sedgwick could separate "upper" and "lower" systems at the northern extremity of the Berwyn range. The main body of the chain had the north-south direction of strike generally characteristic of the older system. In contrast, the rocks at the northern end were almost perpendicular to this direction, with the strata striking east-west. "In *passing over to Bala,*" Sedgwick wrote in his notebook, "remark a complete change in the strike and character of the stones—But the Berwyns above Corwen are obviously the up-

[29] For Sedgwick's reasoning, I rely partly on his 1834 notebook and also on his actions during his next field tour in Wales in 1842.

[30] Sedgwick, Notebook "1834 XXVII," passim, SM. The limestones at Penmacho and Cader Dinmael had been shown on Greenough's 1820 map.

per system both in dip & strike—."[31] Rather than interpreting this change in strike as the result of a fault or a sharp bend in the strata (as he did upon returning to Wales in 1842), Sedgwick saw it as marking the dividing line between his set of strata and Murchison's, as an unconformity separating masses of rocks uplifted in two separate elevatory movements. Although the name Élie de Beaumont appears nowhere in the notebook, the pervasive influence of his theory could scarcely be more evident.

By the second of August, Sedgwick could survey his boundary drawing with considerable satisfaction. The line between the upper and the lower systems now cut across the northern limits of the Berwyns, crossed the valley of the Dee, and then finally swung to the northwest. During the days that followed, he traced this boundary all the way north to Conway Bay.[32] In the later controversy Sedgwick would maintain that this stratigraphical ordering north of the Berwyns had been accomplished with great difficulty. His notebooks show in contrast that the separation between "old" and "new" was made with considerable ease. For example, near Llanrwst on the road north to Conway Bay, he noted that "slate is black, pyritous, stains the fingers, hard, quartz veins, no serrated fucoids—from all which, as well as from the strike & physical features conclude it to be the *old system*." Often he referred to previously seen rocks in making his correlations. At one outcrop, after measuring the strike and dip, he noted: "Sh[ale] is black, pyritous, & weathers rusty brown . . . soils the fingers, many white calc[areous] veins (like Llansaintffraid Glyn Ceiriog)—*fucoids* ∴ the new system." Here Sedgwick was locating another point within the band of Caradoc strata just below the Denbigh Flags. Elsewhere in North Wales, the sections harmonized with the joint determinations without undue difficulty, although on one or two occasions he had to assume that the rocks "thro' a considerable distance are *turned over*" and placed in an inverted position.[33] In more instances the evidence was merely ambiguous. Igneous disturbances, heavy ground cover, or confusing geological structures sometimes made it "impossible to separate the systems," as the frustrated geologist once complained.[34] With only a few minor exceptions, then, Sedg-

[31] Sedgwick, Notebook "1834 XXVII," entry for 22 July 1834, p. 34, SM.

[32] Sedgwick, Notebook "1834 XXVII," entry for 2 Aug. 1834, pp. 53-56, SM.

[33] Sedgwick to Murchison, 27 July 1834, GSL: M/S11/86; Sedgwick, Notebook, "1834 XXVII," entries for 23 July 1834, p. 35 [*sic*], and for 11-12 August 1834, pp. 79-85, SM.

[34] Sedgwick, Notebook "1834 XXVII," entry for 11 Aug. 1834, pp. 81-82, SM.

wick found that arranging his upper groups in accordance with the results of the joint excursion was relatively simple. Like Murchison, he participated fully in the assumptions underlying their tour. Sedgwick would claim at the height of the later controversy that the Berwyn traverse had forced him to twist sections and discard a correct interpretation elaborated in 1832. But far from enveloping him in "the most perplexing difficulties" as he later remembered,[35] the boundary hunting in 1834 proceeded with increasing confidence.

The collaborative boundary, illustrated here in Figure 3.9, clearly recognized both a geological division and the closeness of a friendship. Before hearing of the success of the joint tour, Whewell jocularly warned Charlotte Murchison of the consequences that would result if Sedgwick and her husband should fail to find a suitable boundary between their formations. "I hope you fall in with them in time to prevent their turning their fratricidal hammers on one another," he cautioned, "which I feared would be the result if they could not agree about the dovetailing of the two portions into which they have partitioned the unhappy principality."[36] But as Murchison told Whewell a week later, he and Sedgwick had "made our formations embrace each other in a manner so true & therefore so affectionate, that the evidence thereof would even melt the heart if it did not convince the severe judgment of some Cantab. Mathematic[os] of my acquaintance."[37] Given the brevity of the tour, the difficulty of the sections, and the meagre amount of secure information at their disposal, the boundary chosen in 1834 had many technical features to recommend it: changes in lithology, number of fossils, amount of slaty cleavage, and direction of strike. Using such criteria and faced by frequent ambiguities, the two friends had welded their formations together in the manner least disruptive to their previously established scientific territories.

The Birth of the Silurian and Cambrian Systems

Murchison celebrated the amicable adjudication of boundaries by issuing an improved strata table (Fig. 3.4) at the end of the summer. In this chart he grouped his four principal formations (equivalent to the Ludlow, Wenlock, Caradoc, and Llandeilo divisions of 1835) into an "Upper Greywacke Series" separated from the slaty rocks of

[35] Sedgwick 1852d: 152.

[36] Whewell to C. Murchison, 11 July 1834, GSL: M/W4/10.

[37] Murchison to Whewell, 18 July 1834, TC: Add. ms a.209[94].

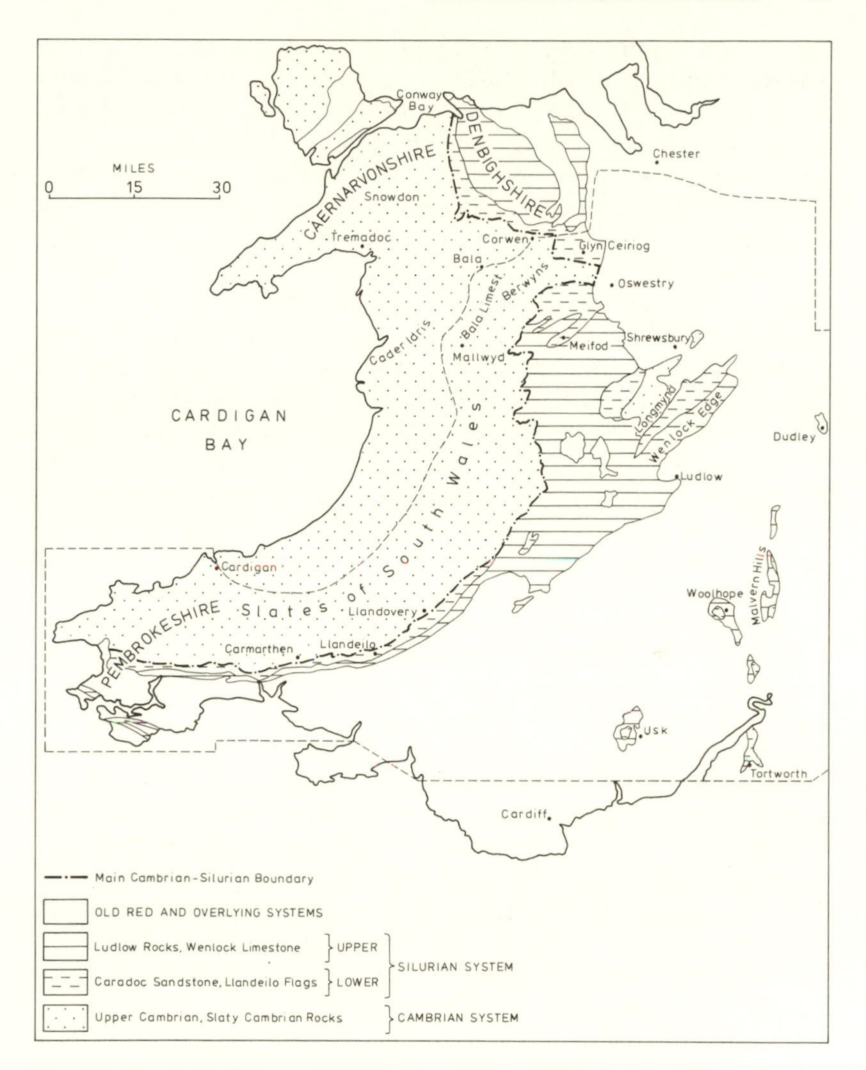

Fig. 3.9. Geological map of Wales, circa 1839, showing the collaborative boundary between Cambria and Siluria. The fine dashed line encloses the area colored by Murchison in his map for the *Silurian System*; north of Meifod this was done with Sedgwick's help.

North and South Wales to be described by Sedgwick.[38] Less than twelve months later, Murchison proclaimed that these four formations constituted a distinctive "Silurian system," and Sedgwick soon christened the underlying groups the "Cambrian system." The naming of these two systems in 1835 might appear to have been a matter of convenience inevitable after the previous summer, with the "new" or "upper" system becoming the Silurian, and the "old" or "lower" system providing the basis for the Cambrian. But the emergence of separate names actually reflected an immensely important shift in the meaning of the word "system" itself. Although initially "system" referred to any convenient and readily identifiable grouping of a large number of strata, with the introduction of the names Cambrian and Silurian the word began to have a much more specific and theoretical meaning in England, one (as I hope to show) that Murchison was particularly eager to exploit in a nascent dispute with fellow geologist Henry De la Beche. In fact, this very notion of system was actively forged in controversy. A system in the new sense marked one of the very largest and most important divisions in the physical and palaeontological record of the strata. Above all, it was a natural period in the history of life, tenanted by organisms largely peculiar to it. Defined in fossil terms, a system could serve as a tool for international correlation; the securely determined sections in a region like Wales typified a sequence of strata with organic remains that would be found elsewhere in Britain and perhaps throughout the world. A system in the new sense was also characterized in structural terms, for changes in the fossil population were often marked in the geological record by unconformities. With the naming of Cambrian and Silurian as two demarcated systems in 1835, the collaborative boundary began to acquire an immense theoretical significance for geology and an importance far beyond the stratigraphy of Great Britain.

Less than a year before the christening of the Silurian, however, Murchison continued to think of his strata primarily in terms of four localized formations, with the larger groupings used to associate them remaining in flux. In November 1834, a few weeks after returning from the field, he asked Whewell (appropriately addressed as "the great geological nomenclator") for help in improving his individual formation names. The proposals outlined in this letter are summarized in Figure 3.1, column c. They demonstrate that Murchison still did not believe that all four formations together possessed any precisely defined unity, for the two uppermost are

[38] Murchison 1834b.

joined in one "series" or "system," and the two underlying in another. Overall this classification would have remained within the traditional framework of the "Upper Greywacke Series." "System" is used in the older, indefinite sense.[39]

Whewell's response to Murchison no longer exists, although judging from later comments, his influence on the final choice of names must have been slight. The really decisive event came in an announcement at a Geological Society meeting at the beginning of December. Henry De la Beche, deprived of his independent income from a Jamaican slave plantation, had from 1832 onwards been paid by the government to geologically color the one-inch-to-the-mile Ordnance Survey maps of southwest England. During his survey De la Beche had discovered plant fossils, precisely like those of the Coal Measures but in what he thought were ancient Grauwacke rocks near Bideford in Devon. Murchison was absolutely incredulous when this finding was brought before the Society. Although he had never studied the succession in Devon and Cornwall, the four summers he spent unravelling the sequence in Wales that separated the Grauwacke from the Carboniferous sufficed to convince him that the observation must be mistaken. As he told De la Beche:

> . . . I know & have *proved*, that *all the classes of organic* remains become more & more divergent in *generic* & *specific* characters from those of the Carboniferous Series in proportion as we descend, and that with the finest possible sections the work of 4 years has not produced a fragment of any vegetable approaching to those of the Carboniferous era; how much more strongly am I led to believe, that the case you have observed has presented it in a *deceitful* form & may have misled *even* so good a Geologist as yourself. If your case be substantiated, away go all rules of mining based upon geological & zoological inductions. Forthwith may the Welchmen whom I have checked in their "decies repetita" coal boring mania, again resume their pick-axes & pierce into the depths of Snowdon in search of the Biddeford coal.[40]

The announcement of De la Beche's anomaly, with all its economic and theoretical implications, sparked Murchison to search for a new name for the older rocks. In a vehement response at the December

[39] Murchison to Whewell, 20 Nov. 1834, TC: Add. ms a.209[95]. Add. ms a.209[133] is a chart which probably accompanied this letter. For later references to this nomenclature, see Phillips to Murchison, 22 Feb. 1835, GSL: M/P14/11, which mentions "Mudstone Series" and "Shelly-sandstone Series" in passing.

[40] Murchison to De la Beche, [Jan. 1835], NMW. Rudwick 1985 gives a more complete account of this meeting and the events surrounding it.

meeting he forswore his own previous use of "Grauwacke," evidently fearing that the word was all too liable to the sort of confusion embodied in the purported discovery.[41]

To counter De la Beche, a name lacking all connection to a single lithological type was clearly required, and immediately after the meeting Murchison opted for "Transition" as an alternative way of referring to the fossiliferous rocks below the Old Red Sandstone. But many contemporaries felt that he was missing an opportunity for reform and urged him to drop what they condemned as a vague and confusing term. "So you will adhere to Transition," lamented Phillips in mid-December. "If you find it convenient *well*, but to me it seems a term of no value except for the Limestones. It is far, far too theoretical & too exclusive. *What is not a Transition series*—What is its *lower* limit." The problem of a lower boundary for Murchison's formations was also raised by Sedgwick a few days later. If Murchison continued to use "that *abominable word*" Transition to refer collectively to his formations, it must at least be qualified by the word "*Upper*," for a vast thickness of older rocks that might also be included within the Transition extended for thousands of feet below any strata studied by Murchison.[42] The potential for territorial encroachment by the use of nomenclature was already becoming obvious. By the spring of 1835, in response to these criticisms, Murchison had decided to group all the rocks between the Old Red Sandstone and the slates of South Wales in a single division called the "Upper Transition."[43] The Carboniferous rocks—with their fossil plants—were pointedly classed with the Secondary strata as an entirely different and overlying set of formations. This scheme, although still well within the standard classifications for these strata, was carefully chosen to underline the distinction between the two groups, to avoid the problems introduced by lithological terms like "Grauwacke," and to combat De la Beche's assertion about Devon.

Murchison finally dropped the traditional nomenclature at the urging of Élie de Beaumont, the most influential theoretical geologist of the early 1830s. Combining Murchison and Sedgwick's work, Thomas Weaver's study of Tortworth, and his own examination of France in company with Ours-Pierre Armand Dufrénoy, Élie de Beaumont emphasized the extent of the unconformity separating

[41] Greenough to De la Beche, [4 Dec. 1834], NMW.

[42] Sedgwick to Murchison, [pmk. 23 Dec. 1834], CUL: Add. ms 7652IC12; Phillips to Murchison, 13 Dec. 1834, GSL: M/P14/10.

[43] This choice of name seems to have been fairly stable by the time of Élie de Beaumont to Murchison, 19 Apr. 1835, GSL: M/E4/5.

the Upper Transition strata from those above and below. In a letter of April 1835, he illustrated his view of the structural relations of Murchison's strata (Fig. 3.10). His diagram separated the Old Red Sandstone, Upper Transition, and underlying slaty Grauwacke from one another by sharp unconformities.[44] Its similarity to the illustrations in his memoir on mountain systems is striking. In the Frenchman's eyes, the four formations of Murchison together constituted a *terrain*, a relatively distinctive unit bounded on either side by unconformities recording the uplift of different *systèmes* of mountain chains in the past. Murchison's work in the older rocks had revealed a grouping of equal importance to the other twelve defined in Élie de Beaumont's scheme of 1829: the Ludlow, Wenlock, Caradoc, and Llandeilo formations made a physical and palaeontological unity that virtually demanded a new name all its own. The French "*terrain*" had a much more definite sense than any comparable English term. But such a usage was particularly appealing to Murchison at this time, so concerned to stress the distinctiveness of his strata. He now began employing the nearest English equivalent, "system," in a sense closely akin to it. Although he made little use of the theory of parallelism in his practical fieldwork, Murchison seems to have turned during the mid-1830s towards a more catastrophic view of earth history and began to oppose the picture of gradualistic change advocated by Lyell.[45] The notion of demarcated systems which he partially adopted from Élie de Beaumont depended (at least in the beginning) on seeing unconformities as records of incredible violence, when mountains popped up like mushrooms across large areas of the earth's crust. According to this view, passage beds like those in the Welsh Borders were the exception rather than the rule. The example of Élie de Beaumont is evident in the idealized section accompanying the July 1835 announcement of the new system; Murchison initially showed his strata lying in violent unconformity to the subadjacent Grauwacke (Fig. 3.11). In the text of this paper and in later versions of the idealized section, he noted the presence of "*transitions* into every formation," but initially

[44] Élie de Beaumont to Murchison, 19 Apr. 1835, GSL: M/E4/5. Also his letter of 14 June 1835, GSL: M/E4/8, which shows that he still had received no response by that date. Perhaps because of this, he also wrote to Sedgwick about the need to replace "Transition": see Élie de Beaumont to Sedgwick, 9 May 1835, CUL: Add. ms 7652IF35.

[45] Page 1976: 158; Bartholomew 1976: 167-168. The latter errs in dating Murchison's apostasy after 1839. The real turning point came in the early 1830s; see Secord (forthcoming).

Fig. 3.10. Élie de Beaumont's sketch of the relations of the rocks below the Carboniferous, April 1835. 1) Ancient grauwacke and schist, 2) Upper transition, 3) Coal measures (including Old Red).

the focus was on the stratigraphical breaks that made his "the first great system below the old red sandstone."[46] Catastrophism in its classic sense was by no means a necessary accompaniment to the new notion of "system," but it certainly was present at its birth.

If terms like "Upper Greywacke Series" or "Upper Transition Rocks" were to be discarded, what was Murchison to put in their place? What name should the new system be given? Contemporary practice sanctioned at least four alternative sources for stratigraphical names. First, the new system could take its name from a prevailing lithological type, thus drawing on the precedents of the "Oolites" and the "Chalk"; Murchison had suggested "Mudstone Series" and "Shelly-sandstone Series" in his letter to Whewell with such an idea in mind. This method of nomenclature, however, suffered from the extreme inconstancy of lithology over long distances, the presence of many lithological types within any large stratigraphical unit, and the repetitions of identical lithologies at numerous points in the geological column. After De la Beche's discovery of Coal Measure plants in rocks supposedly on the horizon of the Grauwacke, such a scheme held scant appeal for Murchison. A second alternative involved taking a palaeontological rather than a lithological name and characterizing the strata by a prevalent fossil type. Murchison's early designation of the Llandeilo formation as the Trilobite Flags was in part based on such a criterion. This alternative suffered from many of the same liabilities as the first, however. In particular, almost all families of fossil species ranged too widely to typify a single stratigraphical group, and even those possessing a range of the required limits might eventually be traced into the adjoining parts of the sequence. Third, Murchison could follow a common contemporary practice by christening his strata with a coinage from the Greek, just as Lyell had done in naming his subdivisions for the Tertiary the "Eocene," "Miocene," and "Pli-

[46] Murchison 1835: 48. For the section, see p. 50. A conformable version appears in Buckland 1837, 2: pl. 66.

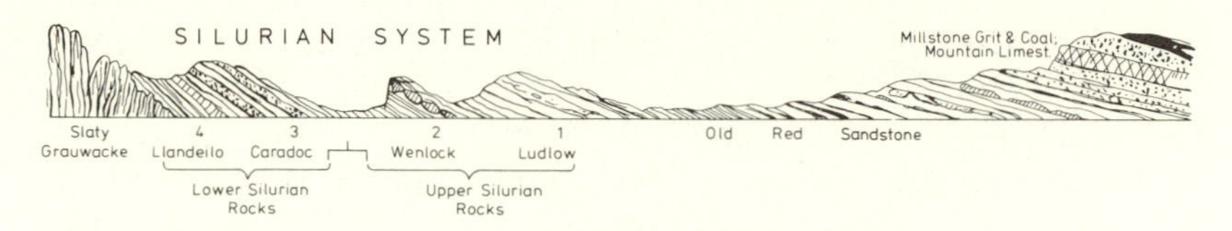

Fig. 3.11. Murchison's idealized section across the Silurian system. The sharp unconformity between the "Slaty Grauwacke" and the Lower Silurian corresponds to that between groups (1) and (2) in Figure 3.10.

ocene."[47] By the use of Greek, Élie de Beaumont thought that Murchison could combat directly the incorrect assumptions engendered by the misleading term "Transition." Several neoclassical alternatives accompanied his suggestion that the Upper Transition be renamed, among them Greek equivalents for "false transition rocks," "hypo-secondary rocks," and "interpolated rocks." Although Lyell's Tertiary nomenclature had been well received, Murchison hesitated to follow this precedent. As he had told Whewell during the previous autumn, "I am *not* anxious to have a fine column of Greek names."[48] Many other British geologists (including several more familiar with classical languages) shared this antipathy to Greek. Classical coinages suffered guilt-by-association with the confused nomenclature of mineralogy; they were too easily invented, were often constructed on false linguistic principles, and had distinctly foreign associations. As Sedgwick told the Geological Society in 1835:

> The pedantry and the gross absurdity of many terms in mineralogy are obvious to every one; and I hope we shall take a lesson from the progress of that kindred science, and not be too hasty in seeking the useless decorations of a too classical nomenclature. Indeed, there already appear in our descriptions some words of doubtful and portentous etymology, with Grecian heads and Gallican tails, which figure but oddly amidst the ordinary staple of our homely pages.[49]

Murchison's eventual choice of "Silurian" offered an innovative variant on a fourth alternative. Geographical place names had

[47] Lyell's Tertiary nomenclature is discussed in Rudwick 1978 and L. G. Wilson 1972: 305-307.

[48] Élie de Beaumont to Murchison, 19 Apr. 1835, GSL: M/E4/5; Murchison to Whewell, 20 Nov. 1834, TC: Add. ms a.209[95].

[49] Sedgwick 1835a: 480.

graced some of the longest-enduring subdivisons of the geological column, particularly in the Secondary rocks that Murchison had previously studied. In England, this principle had been pioneered by Smith with his "Kelloway Rock" and "Portland Beds," while it was well known on the Continent as a result of the "système jurassique" named after the Jura mountains in Switzerland. From the beginning Murchison had planned to name his individual formations after the localities where their place in the succession was best illustrated. No one doubted that terms like Llandeilo Flags, Caradoc Sandstone, Wenlock Limestone, and Ludlow Rock would rest comfortably among those "homely pages" of British scientific literature.

Smith's names derived from local usage and brought the language of quarrymen and miners into the realm of science. In choosing "Silurian" Murchison combined the more genteel tradition of antiquarianism with his own background in the military.[50] According to Murchison, at the time of the Roman occupation the Welsh Borderland had been inhabited by the Silures, a stoutly independent British tribe led by their chief Caractacus. In a far earlier epoch, as he was now demonstrating, the same region had swarmed with trilobites, brachiopods, and other denizens of a shallow sea. By giving these fossilized life forms and their surrounding rocks the name "Silurian," Murchison spoke simultaneously as military historian and geologist: the presumed territory of the Silures became a standard of comparison for other regions containing deposits of the same age, on the same horizon, and with similar fossil species. Of course, the outstanding resistance of the Silures to a foreign power gave their name a special appeal to Murchison, as might have been predicted from the prevalence of military language throughout his writings. The strength of this military connection is demonstrated by his arguing at length for a rather dubious extension of the known territorial limits of the ancient Silurian realm, so as to ensure that Caractacus had actually once occupied the typical area for the Wenlock and Caradoc formations. "British geologists, therefore," Murchison wrote in announcing the new nomenclature, "will not doubt that 'Siluria' is a name entitled to be revived, when they are reminded that these struggles of their ancestors took place upon the very hills which it is proposed to illustrate under the term 'Silurian system.' "[51] And the militaristic overtones of the name are also ap-

[50] Murchison 1835, esp. pp. 49-52; Murchison 1839: xxxi-xxxii.

[51] Murchison 1835: 51. For a more complete discussion of these points, see Secord 1982: 421-423.

propriate in another way, for they recall the historical origins of the Silurian concept as a weapon in the emerging battle with De la Beche over the Devon Coal plants. Notably, when Murchison put forward the new system in a short article in July 1835, he stressed the need "to define the boundaries of groups naturally distinct from each other," particularly with regard to the Grauwacke and the Carboniferous. Although he directed his fire on this point against unspecified "German geologists" (who must have included Heinrich von Dechen among others),[52] any geologist in the know would have recognized that the primary target of the Silurian was in fact De la Beche.

Shortly after receiving a letter from Murchison about his choice of "Silurian," Élie de Beaumont responded enthusiastically to the new coinage. He immediately recognized the general principle of geographical nomenclature that Murchison had followed and hoped that "Silurian" would become as popular as the "système jurassique." Élie de Beaumont then suggested that Murchison name the next lower *système,* the older formations in the mountain districts of Wales.[53] In a few weeks this led to the creation of a separate "Cambrian."

Taking his lead from Murchison's use of a geographical name with antiquarian associations, Élie de Beaumont encouraged the adoption of "Hercynian" for this lower group, a term derived from the "Sylva Hercina" of the Roman historian Tacitus and referring to the Harz mountains in central Germany. Murchison, however, hesitated to choose a name taken at such a distance from proven Silurian sections. Although the rocks of the Harz might in fact be true antecedents of the Silurian, he preferred a name from Wales, where the succession into the older rocks had already been demonstrated beyond a doubt.[54] Moreover, as a British geologist he wanted the rocks to have British names. So too did Buckland. "If you & Sedgwick are not quick to assert your prior right of naming this Group," he told Murchison, "depend on it E. de Beaumont will anticipate

[52] Murchison 1835: 47. Dechen had recently issued an augmented translation of De la Beche's *Geological Manual,* which embodied precisely the "confusion" that Murchison deplored (see De la Beche 1832). Dechen is explicitly mentioned in this context in Élie de Beaumont to Murchison, 19 Apr. 1835, GSL: M/E4/5. For the strong circumstantial evidence concerning De la Beche, compare Murchison to De la Beche, 19 Dec. 1834, [Jan. 1835], NMW, with Murchison 1835, esp. pp. 47-48.

[53] Élie de Beaumont to Murchison, 27 June 1835, GSL: M/E4/9.

[54] For Hercynian, Élie de Beaumont to Murchison, 27 June 1835, GSL: M/E4/9; Élie de Beaumont to Sedgwick, 26 July 1835, CUL: Add. ms 7652IF36. For Murchison's preference, Murchison to Buckland, 2 July 1835, GSL: Buckland/M11.

you with his name Hercynian which is very tempting & classical & wd. no doubt be readily adopted on the Continent."[55]

Murchison described his plans for naming this underlying system to Buckland soon after announcing the Silurian. With Sedgwick's approval he hoped to publish a sequel proposing a "Snowdonian system." This system would be divided into two parts. The first would include the slates of South Wales and the associated strata between Sedgwick's Bala Limestone and Murchison's Llandeilo Flags, while the second would "embrace every thing which Father Adam can descend into," down to the lowest Primary rocks of Anglesey. If Sedgwick objected to "Snowdonian system," Murchison would feel justified in unilaterally naming the first of these two divisions the "Demetian system," after the ancient name for Carmarthenshire in South Wales. Although he planned to consult Sedgwick, he also believed that his own studies of the South Wales slates had given him sufficient right to name at least the top half of the sub-Silurian rocks.[56] In the event, Sedgwick accepted the idea of a single name for the sub-Silurian strata, but rejected the "Snowdonian" suggestion. "We will settle the nomenclature when we meet," he told Murchison in late July 1835. "I don't much like *Snowdonian*. It is a beastly modern word of the Saxon tourists & unknown to the Cumrajs—I like Cambrian or Cumbrian better."[57] In early August they decided on "Cambrian," the ancient Roman name for Wales, presumably because the succession had been proven there rather than in Cumberland in northern England.

This sequence of events demonstrates that the chief impetus for a distinctively named "Cambrian system" came from Murchison. At this stage, Sedgwick had little reason to be concerned with finalizing a nomenclature for the older rocks. That could wait for the composition of the *Outlines*. When writing to Whewell in November Murchison had lamented his friend's inattention to such matters. "These results," he wrote of his formations, "have now become doubly interesting as concerns himself but I can obtain no opinion from him nor will he stick to it without you force him."[58] Classifications served for Sedgwick as heuristic tools, easily erected and as easily abandoned in the face of new evidence. What really mattered was the correct determination of the sections. Thus he never took

[55] Buckland to Murchison, 17 July 1835, NMW.

[56] Murchison to Buckland, [undated P.S., July 1835], GSL: Buckland/M/10.

[57] Sedgwick to Murchison, 25 July 1835, GSL: M/S11/90.

[58] Murchison to Whewell, 20 Nov. 1834, TC: Add ms a.209[95].

the trouble to announce the name "Cambrian" in a rapidly published and widely distributed periodical like the *Philosophical Magazine*, the monthly journal that had heralded the Silurian. In fact, the first use of the new term in a regular research report was made by Murchison in 1836.[59]

Murchison and Sedgwick signalled the founding of the two systems at the August 1835 meeting of the British Association. The structure of their paper, as briefly abstracted in the Association's published proceedings, reflected the separation of their respective investigations in Welsh geology. Murchison started off with a recapitulation of his July paper for the *Philosophical Magazine*: the Ludlow and Wenlock formations were joined in the Upper Silurian, while the Caradoc and Llandeilo together formed the Lower Silurian (Fig. 3.12). For an account of the slaty beds below the Llandeilo, he turned the podium over to the professor, who briefly described his own sections in North Wales. Sedgwick placed the rocks above the Bala Limestone in an Upper Cambrian group, the underlying beds of Merionethshire and Caernarvonshire (including the Snowdon beds) in a Middle Cambrian group, and the chlorite and mica schists in Anglesey and southwest Caernarvonshire in an unfossiliferous Lower Cambrian group. This characterization of the Cambrian system dealt solely with structures and sections, and although the presence of fossils was noted, Sedgwick made no attempt to list them in the brief time at his disposal. He concluded the presentation by "explaining the mode of connecting Mr. Murchison's work with his own," for this had always been his special concern. With the establishment of the Cambrian and Silurian systems, geologists believed that scientific order had been brought to the ancient chaos of Grauwacke and Transition. In the words of a speech given by Sedgwick at another British Association gathering a few hours later, a geological "terra incognita" had become a "known country, which though just discovered, already exhibits every symptom of regularity and improvement."[60]

[59] Murchison 1836. Soon afterwards, however, Élie de Beaumont printed a letter from Sedgwick that provided a more complete description of the Cambrian: see Sedgwick 1836, read to the Société Géologique de France on 21 Mar. 1836. A report of Murchison's work had already been printed; see *Bulletin de la Société Géologique de France*, 1835, *7*: 90-91, 126-132.

[60] Sedgwick and Murchison 1836. Sedgwick's speech is reported by Hardy 1835: 108-109. Hardy also reports an interesting discussion that followed the reading of the original paper.

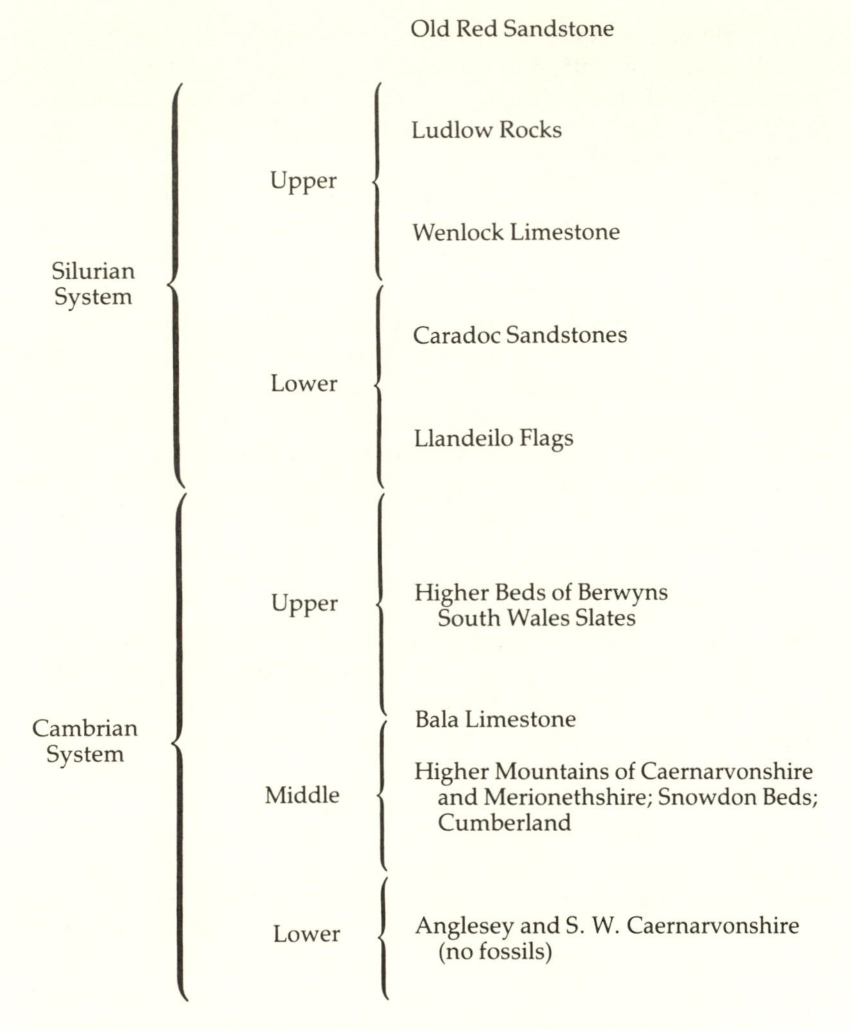

Fig. 3.12. Murchison and Sedgwick's classification of the strata below the Old Red Sandstone as announced at the Dublin meeting of the British Association for the Advancement of Science in 1835.

The Status of the Systems

With the naming of the Cambrian and Silurian as separate systems Sedgwick and Murchison set their seals of approval on an arrangement that lasted until early in the following decade. All that remained was to complete the publication of their respective researches, a big book by Murchison on the geology of the Welsh

Borders and Sedgwick's volume on the older rocks throughout England and Wales. The long-awaited *Silurian System* finally appeared in 1839, and with the reading before the Geological Society of a short "Synopsis" of Sedgwick's work on the older rocks in 1838, the *Outlines* seemed closer to completion than ever before.[61] Taken together these two works make it possible to assess the status of the systems and the boundary between them.

After many delays the two large quartos of Murchison's *Silurian System* were sent to the subscribers and booksellers in early January 1839. With eight hundred pages of text, dozens of sections, numerous sketches of scenery, a large colored geological map, and a price of eight guineas, it was an imposing work, in the eyes of contemporaries a monument to the author's industry and scientific skill. Fitton declared in the *Edinburgh Review* that Murchison's success provided "the young student of geology" with a living example of the virtues of hard work. "It must contribute largely to the advancement of Geology," wrote another reviewer, "and is, indeed, in our judgment, not only a credit to the individual, but to the national character."[62] Murchison dedicated the *Silurian System* to Sedgwick, both in friendship and as a tactful acknowledgement of his early prominence in the study of the ancient strata.[63]

Murchison described at length not only the older rocks of his typical region in the Borderlands, but also the overlying coalfields and Secondary strata found there. The main interest of the work rested in the Silurian system itself, in Murchison's skillful elaboration of his original discoveries (and those of dozens of local minions) after seven summers of hammering, mapping, and fossil collecting. The second volume, which included descriptions of several hundred fossil species, made abundantly clear the importance of the Silurian fauna. Each of the four formations within the system, practically defined in the field on lithological and sectional grounds, was now accompanied by a full complement of distinguishing fossils identified by specialized experts (Figs. 3.2, 3.4).[64] Within these formations, the difference between the fossils of the Upper Silurian (Ludlow and Wenlock) and those of the Lower (Caradoc and Llandeilo) was particularly imposing. Despite these distinctions, however, Murchison maintained the essential unity of his Silurian. Different formations recorded piecemeal faunal change, he believed, but as a system the Silurian possessed relatively sharp palaeontological boundaries. As

[61] Murchison 1839; Sedgwick 1838.

[62] [Fitton] 1841: 16; [anon.] 1839: 113. [63] Murchison 1839: iii.

[64] The preparation and publication of Murchison's book is well described in J. C. Thackray 1978.

he wrote in the concluding pages of his book: "Let it not, however, be imagined that I wish to inculcate the doctrine, of every ancient formation having been tenanted by creatures absolutely peculiar to it. The large natural groups of strata only, or, so to speak, systems, can be thus distinguished."[65] On this basis, Murchison identified eight systems applicable at least to all of Great Britain and very probably further afield. This comprehensive "systematizing" represented a significant extension and codification of the approach initiated with the naming of the Silurian. The four lowest systems (as shown in Fig. 3.13) were the Carboniferous, the Old Red Sandstone, the Silurian, and the Cambrian.[66] The limestones of Devon and Cornwall had not yet been removed from the Cambrian and Silurian into a separate "Devonian system" parallel with the Old Red; as a result, the "Old Red system" was largely unfossiliferous except for a few fish, represented in Figure 3.13 by the letters *m* and *n*. Thus from the Silurian to the Carboniferous, extinction appeared almost universal, with only a few hardy species of shells surviving the vicissitudes of a great planetary change. In this way, Murchison could sharply divide the Silurian from the next overlying fossiliferous system. This massive extinction is illustrated schematically in Figure 3.13, where the letters *a* through *j* represent the fossils of the Silurian, and *p* through *t* stand for those of the Carboniferous.

What of the separation between the Silurian and the underlying Cambrian? Here the situation was less clear, and Murchison adopted a less rigid palaeontological definition of "system," allowing for a relatively large transitional zone between the two. The *Silurian System* explicitly left the characterization of the Cambrian proper to Sedgwick and never described any strata or fossils definitively belonging to the lower system. When an unconformity or faulted zone existed between the two systems, Murchison ended his account there and referred readers to Sedgwick for details of the underlying strata. In a few instances—in the critical sections across Noeth Grüg and the Berwyns, for example—passage beds between the Cambrian and the Silurian were described, marked by the same criteria that had been used in the field. Murchison's descriptions, however, never carried the reader into rocks below these transitional zones into actual Cambrian strata with peculiarly Cambrian fossils.[67] These literary limits simply respected the territorial divisions made jointly during the early 1830s. Again, this is not to imply

[65] Murchison 1839: 582.

[66] See Murchison 1839: xxviii, and the small index map included with the work.

[67] Murchison 1839, index under "Cambrian," esp. pp. 307-310, 352, 357-362, 398, 635.

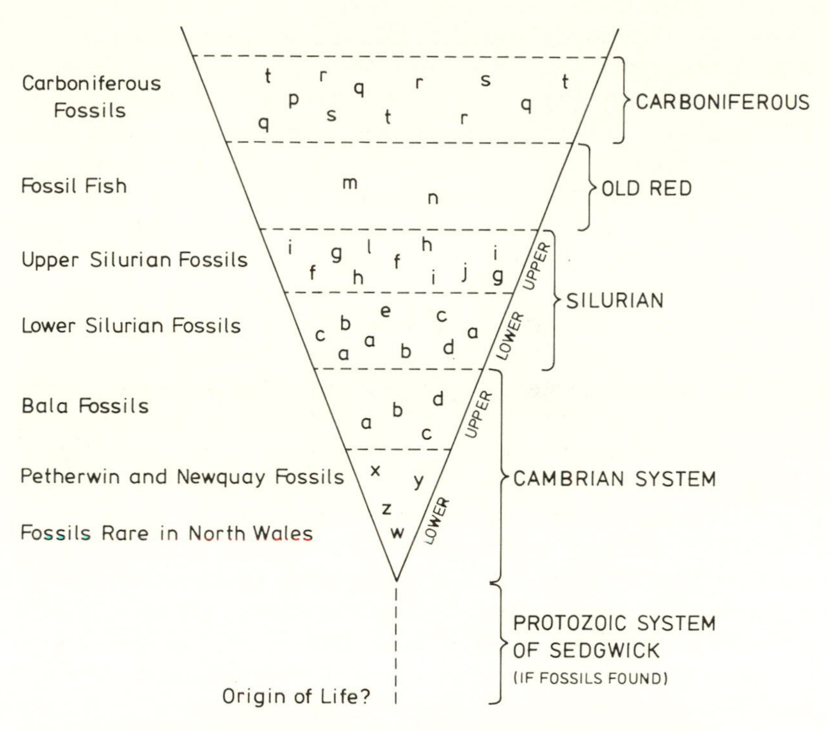

Fig. 3.13. Diagrammatic representation of the faunal evidence for divisions into systems, with the classification presented in Murchison's *Silurian System* (1839) and Sedgwick's "Synopsis" (1838).

that the dividing line between Cambrian and Silurian had *no* scientific significance, as Murchison later claimed. Rather, at this time the geological and the territorial meanings of the collaborative boundary were mutually reinforcing.

Sedgwick's studies of North Wales remained largely unpublished even as the *Silurian System* entered the libraries of geologists all over Europe and America. However, in an important paper delivered before the Geological Society in the spring of 1838, Sedgwick briefly summarized the results of his researches. This lecture, entitled "A Synopsis of the English Series of Stratified Rocks Inferior to the Old Red Sandstone," gave listeners a picture of the Cambrian and Silurian rocks throughout England and Wales and a preview of the second volume of the *Outlines*. A reader turning from the *Silurian System*, curious as to the character of the underlying Cambrian, would have obtained complete satisfaction in several respects from the

"Synopsis." The paper began by outlining the importance of superposition and geological structure in the classification of stratified rocks. By 1838 Sedgwick felt sure of his sections throughout England and Wales, and thus believed that his principal criteria for classification had been met. His arrangement was essentially unchanged from 1835, with a few minor exceptions (compare Fig. 3.13 with Fig. 3.12). Most significantly, he now excluded the unfossiliferous gneisses and schists of Anglesey and Snowdonia from the Cambrian system and placed them instead in a "Primary Class" of unstratified rocks distinct from the overlying Cambrian and Silurian.[68]

As the denial of systematic status to the unfossiliferous Primary Class demonstrates, Sedgwick had by 1838 accepted a definition of "system" dependent at least to some degree on palaeontological evidence. If fossils were found in the Primary Class, it too, like the Cambrian and Silurian, would deserve the name of system, and Sedgwick tentatively suggested "Protozoic system" should such discoveries in fact take place (Fig. 3.13). Further evidence that he relied at least partially on fossils in defining the Cambrian system is provided by an edition of his lecture syllabus issued in 1837, which mentions the probable presence of such strata in distant lands. Such long-distance correlations, as Sedgwick was well aware, could only be made secure through palaeontological comparisons.[69]

In dividing his strata into systems Sedgwick always attempted to coordinate both physical and fossil evidence into a convincing unity. As one would expect from a follower of Élie de Beaumont's theory of parallelism, he discussed the difference in strike between the Cambrian and Silurian systems at some length, explaining deviations from the general rule of parallelism as effects of interference.[70] Along with these similarities in strike Sedgwick also showed how the various regions occupied by each system shared common lithological features and broad similarities in the sequence of strata. For the evidence supporting his classification to be conclusive, however, the additional testimony of organic remains was required. Here the picture was as suggestive as it was incomplete. The "Synopsis" mentioned the presence of fossils in both the Upper and Lower Cambrian without including any extensive lists such as those Murchison had provided for the Silurian several years earlier. According to Sedgwick, the Upper Cambrian system in the county of

[68] Sedgwick 1838, esp. pp. 676-679.
[69] Sedgwick 1838: 684; Sedgwick 1837: 48-54.
[70] Sedgwick 1838: 677.

Westmorland in the Lake District contained fossils, but these were scarce, and his main knowledge of its fauna derived from the Bala Limestone. "Many of the fossils are identical in species with those of the lower division of the *Silurian System,*" he told the Geological Society, "nor have the true distinctive zoological characters of the group been well ascertained."[71] This faunal similarity between the Upper Cambrian and the Lower Silurian in Wales is shown schematically in Figure 3.13, where the letters *a* through *e* represent fossils identified by Murchison as Lower Silurian. Even fewer characteristic species were found in the lower half of the Cambrian, which with one or two exceptions (such as the beds on the summit of Snowdon) was almost entirely barren.

Sedgwick's real hopes for a distinctive Cambrian fauna rested in the region farther to the south, in the limestone beds of Devon and Cornwall. For the preceding three years, a bitter controversy had raged over the rocks of this region, bringing many leading geologists into conflict. Prominent among the protagonists were Sedgwick and Murchison, whose joint field studies of the structure of Devon had by this time convinced even De la Beche of the need to shift his troublesome plant-bearing grauwacke to a higher position in the stratigraphical column.[72] Although the precise dating of these "culm beds" remained a matter for violent disagreement, the focus of the dispute was gradually shifting to the age of the calcareous strata below them. Initially, almost all geologists placed these limestones and their fauna well within the Cambrian. When the "Synopsis" was read in 1838, however, this was no longer the consensus, for palaeontologists like Lonsdale and James de Carle Sowerby were identifying a surprising number of fossils allied to the Carboniferous in these supposedly ancient rocks. Even Sedgwick had begun to suspect that several of the limestones belonged on a much higher stratigraphical level. But he could still maintain with considerable cogency that *some* of these zones would remain within the Cambrian. At Newquay and South Petherwin in Cornwall, these contained corals, encrinites, brachiopods, ammonites, and clymeniids, represented in Figure 3.13 by the letters *w* through *z*. Thus when Lyell requested some peculiarly Cambrian fossils for inclusion in the first edition of his *Elements of Geology* (1838), Sedgwick could suggest the genus *Endosiphonites,* recently named by his student assistant at Cambridge, David T. Ansted. This important fos-

[71] Sedgwick 1838: 679.

[72] See Rudwick 1985, and 1979a: 14-17.

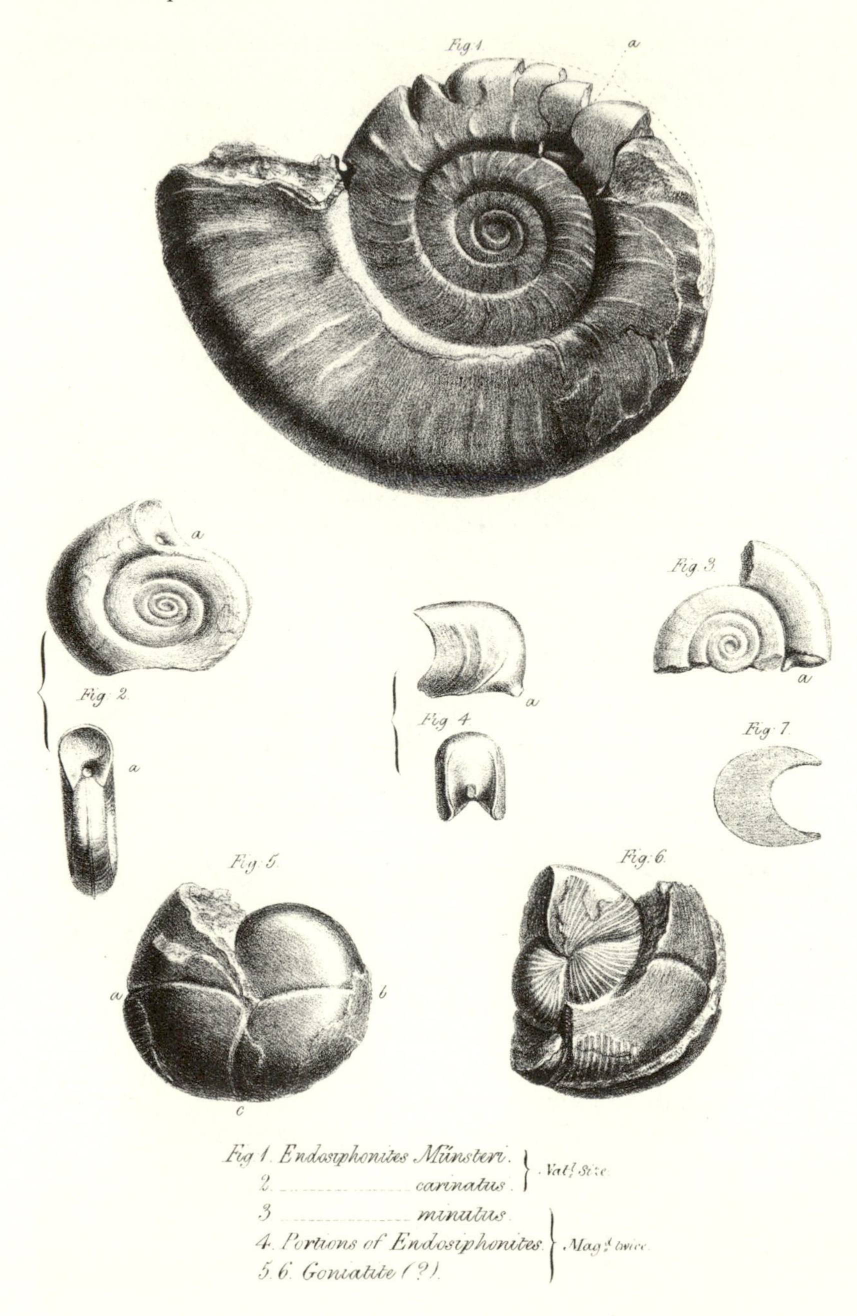

Fig. 3.14. Cambrian fossils of 1838: various species of the genus *Endosiphonites* illustrated by Ansted, including (*1*) *E. Münsteri* and (2) *E. carinatus*.

sil, named and described specifically to help enliven the Cambrian, is illustrated here as Figure 3.14.[73]

In 1838 Sedgwick was certain that a vast sequence of fossiliferous rocks extended for many vertical miles below the lowest beds of the Silurian. Both he and Murchison felt sure that some of these ancient strata must contain fossils distinct from the Silurian, so many thousands of feet higher in the sequence. As firm believers in the doctrine of organic progression, they both thought it extremely unlikely that a single fauna could inhabit the earth for so long a period of time. Some record of the progress of life on earth towards its present state must have been preserved. Thus for Sedgwick, the fauna of the Newquay and South Petherwin limestones—with so many non-Silurian species—pointed to the existence of organic remains as yet undiscovered in Cambrian rocks elsewhere, particularly in Wales itself. Fossil evidence from Cornwall thus joined sectional evidence to point to a conclusion in full accordance with the theory of progression: as Sedgwick said of these specimens in his "Synopsis," "as they occupy a position so much lower, so, as a group, these fossils are distinct from those of the Silurian system."[74]

Because of the potential significance of the Newquay and South Petherwin fossils, Sedgwick had no need to apologize for any lack of a specifically Cambrian fauna. He merely remained temporarily unable to define the smaller subdivisions of his system in palaeontological terms. In any event, his definitions of the Silurian and Cambrian systems depended primarily on sections and only secondarily on fossils. With the physical evidence securely in place for both, the bulk of the older rocks now awaited the palaeontological characterization already possessed by the upper parts of the Silurian. While Sedgwick emphasized the structural underpinnings of the Cambrian, Murchison stressed the faunal riches of his system. But, in spite of this difference, in 1838 the two friends were united in the belief that these criteria for classification would eventually be in full accord, giving a double sanction for their separation of Cambria and Siluria.

[73] Ansted's name, however, lacked priority and was never generally adopted. Sedgwick 1838: 683; Ansted 1838; Lyell to Sedgwick, 20 Jan. 1838, in K. M. Lyell 1881, 2: 35-37, and C. Lyell 1838: 464-466. I am indebted to Martin Rudwick for pointing out the difficulties that Sedgwick faced in using these fossils, although I would not agree in dating widespread disillusionment with the Cambrian so early as 1838; see Rudwick 1985.

[74] Sedgwick 1838: 683.

CHAPTER FOUR

The Spread of Siluria

MEN OF SCIENCE promptly hailed the Cambrian and Silurian systems as major achievements of English geology during the 1830s. "Steps of this kind," Whewell told the Geological Society in 1839, "have formed, and must form, the great epochs in the progress of all sciences of classification, and especially in ours; and I need not remind you how great the importance and the influence of such steps amongst you have been."[1] The magnificent volumes of the *Silurian System* declared the seemingly unquestionable existence of Murchison's system, securing its continued use across the globe. Publication of Sedgwick's researches appeared imminent, and although geologists recognized the difficulties presented by the ancient slaty rocks, most believed that they would yield enough distinctive fossil forms to justify the Cambrian as a separate system.

With the establishment of a generally acceptable classification for the older rocks, the fruitful phase in the collaboration between Sedgwick and Murchison neared its end. In their last important work together, the two geologists created a separate "Devonian system" between the Silurian and the Carboniferous, on a parallel with the largely unfossiliferous Old Red Sandstone. The naming of this new system in April 1839 culminated the violent dispute about the older rocks of Devon that had flared intermittently ever since De la Beche's controversial discovery of Coal plants in the Grauwacke some four years earlier. Initially the evidence for the Devonian was scanty indeed, and Murchison and Sedgwick toured Germany during the summer of 1839 in search of Continental support for their views. This trip and the paper of the following year that grew out of it were the last projects of importance undertaken jointly by the two men.[2] Ironically, although created through close collaboration, the Devonian held consequences that upset the delicate balance between Cambria and Siluria. In the end these proved the undoing of

[1] Whewell 1839: 77.

[2] Sedgwick and Murchison 1840. There was also a last short excursion together in 1850; see Clark and Hughes 1890, 2: 182-185. For the Devonian controversy, see Rudwick 1979a and 1985.

the partnership. Correspondence and friendship continued almost unabated for at least another decade, but after 1840 the paths of Murchison and Sedgwick rarely crossed and differences implicit in their aims from the very beginning came into the open.

The Disappearance of the Cambrian Fossils

After the Continental tour, Sedgwick once again faced the task of completing Conybeare and Phillips's *Outlines.* Quite obviously the book was taking far more time than had been anticipated when he first took on the project. In part the delays resulted from his inherent dislike of sustained literary composition; although he could write passages of great brilliance, a connected argument of substantial length seemed beyond his powers. Sedgwick all too evidently preferred the field, the lecture hall, his museum, or the banks of a fishing stream to the narrow confines of his study. Nevertheless, procrastination was only part of the problem, for the structural relations of the older rocks were proving much more complex than anyone had ever imagined. For example, the issues raised by the confusing sections in Devon and Cornwall had required nearly four full field seasons rather than the brief resurvey suggested by Conybeare in 1828.[3] After the 1839 tour with Murchison, Sedgwick searched for a harmonious arrangement of the older rocks that would enable him to write his great work. For the remaining fifteen years of his active life as a geologist, he concentrated his research on this task and never geologized outside the British Isles again.

In November 1841 Sedgwick came before the Geological Society with a "Supplement" to his 1838 "Synopsis," applying the lessons learned in Devon to the rest of the older stratified rocks in Britain. While it repeated much of his earlier paper verbatim, this reformulated summary of his big book contained critically important modifications. Because of the similiarity to the "Synopsis," the importance of the "Supplement" has previously been ignored, although this paper more than any other spelled the demise of the Cambrian as a separate system and signalled publicly a breakdown in the classification established with Murchison.[4]

The "Synopsis" had emphasized that the Cambrian system, while well defined as a physical unit, showed only promising indi-

[3] Clark and Hughes 1890, *1*: 458-461, 489, 513, 523-526.

[4] Sedgwick 1841.

cations of a distinctive set of fossils (Fig. 3.13). As we saw in the previous chapter, the most likely candidates for such a fauna in 1838 were found in the highly fossiliferous limestones at South Petherwin and Newquay in Cornwall, which Sedgwick (like several other contemporary geologists) placed in the Cambrian. These limestones contained corals, brachiopods, encrinites, orthoceratites, goniatites, and clymeniids, and as might be expected from beds on a horizon so far removed from the Silurian, the various species were almost all distinctive. But less than a year after the reading of the "Synopsis," these limestones—and hence all the remaining "Cambrian" fossil species—were irretrievably removed to a position much higher in the geological column, *above* the Silurian rather than below it. This dramatic shift, which completed a gradual displacement of this fauna to this level, produced the Devonian as a separate fossiliferous system between the Silurian and the Carboniferous (Fig. 4.1). As explained in Murchison and Sedgwick's paper of 1839 which put forth the new system, several considerations pointed to the need to remove the last vestiges of the limestone fauna in Devon and Cornwall from the Cambrian. Chief among them was the testimony from expert palaeontologists that these fossils were intermediate in character between those of the Carboniferous on one hand and the Silurian on the other. Sedgwick hesitated for months to make the change and put his name to the paper creating the new system only after long arguments with Murchison.[5] One reason behind his reluctance to reclassify this fauna should be immediately evident: after the subtraction of these fossils the Cambrian could no longer be accepted as a palaeontologically valid system, whatever its importance as a physical and lithological unit. In short, the Devonian became a system at the expense of the Cambrian.

The difficulties for the Cambrian raised by the outcome of the Devonian controversy were compounded by the unpacking of Sedgwick's accumulated fossil collections in 1841. Owing to cramped conditions in the Woodwardian Museum, most of the fossils gathered during Sedgwick's twenty-year tenure of the chair of geology had remained in their original crates. Spacious new accommodations were completed during the summer before the reading of the "Supplement." With the assistance of David Ansted, an undergrad-

[5] Sedgwick and Murchison 1839. In later years neither Sedgwick nor any of his biographers ever admitted that he had hoped before 1840 for distinct Cambrian fossils, and the effect of the Devon reclassification has in consequence been obscured.

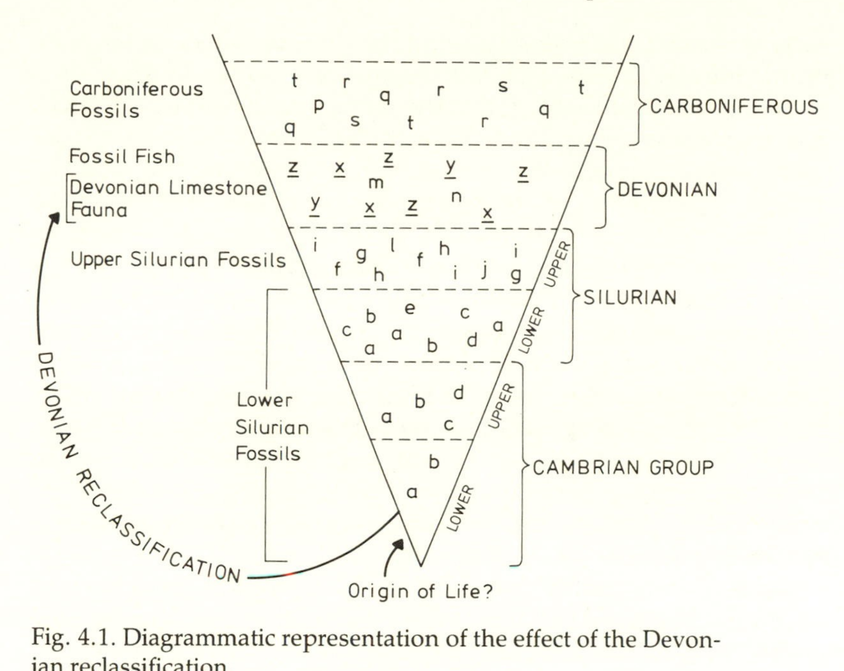

Fig. 4.1. Diagrammatic representation of the effect of the Devonian reclassification.

uate of Jesus College, Sedgwick began unpacking boxes and sending specimens to specialist palaeontologists like James de Carle Sowerby for identification. The first fruits of this examination offered little hope that a distinctive fauna for the Cambrian system would be found. From the Berwyns, the Bala region, Snowdonia, and Cumberland—in other words from all those areas still stratigraphically in the Cambrian after the Devon reclassification—the fossils seemed almost identical to those illustrated in the *Silurian System*. "Below the Caradoc sandstone," Sedgwick announced in his "Supplement," "there seems to have been very few new types of creation, as far at least as we can learn from any *positive* facts in the country here described."[6]

How then did Sedgwick adjust his classification to these surprising discoveries? The changes were small but extremely significant. He now divided the rocks below the Silurian into three major divisions: the sub-Silurian or Upper Cambrian (an intermediate hybrid group which included Murchison's Llandeilo formation); the Lower

[6] Sedgwick 1841: 550. The unpacking of his collections is described in Clark and Hughes 1890, 2: 26; for the identifications, Sedgwick to J. de C. Sowerby, 25 Nov. 1841, BMNH(S), and Sedgwick to Phillips, [1841], ff. 5-8, UMO.

Cambrian; and the unfossiliferous Lower Cumbrian, or Primary group, which included the chlorite slates of Anglesey. In 1838 Sedgwick had confidently referred to the Upper and Lower Cambrian as a system and suggested that its grand thickness and lithological development might eventually be matched by a special fauna. But in 1841 he could no longer make such a claim. Where the "Synopsis" had consistently described a Cambrian "system," the "Supplement" generally reverted to the less theoretical term "groups." On the other hand, even if Sedgwick now denied the Upper and Lower Cambrian the status of a separate system, he still believed that these names continued to mark important physical, structural, and lithological units in the older rocks. In his opinion, vertical miles of slates and schists containing a few unevenly distributed organic remains could not be classified solely on a palaeontological basis, however necessary fossils might be for the creation of systems. When the last hopes for a Cambrian fauna vanished, structure remained as the essential groundwork for a revised classification.[7]

Sedgwick anticipated, however, that his physical groups might yet receive a limited sanction from palaeontology, and in the final pages of the "Supplement" he proposed a highly imaginative method for establishing this. Although ultimately unsuccessful, this method guided much of his work on the older rocks during the next few years and is well worth examining, particularly as it has previously been misunderstood and misrepresented.

At the highest theoretical level, Sedgwick (like Murchison and most other geologists of the period) advocated a directionalist interpretation of earth history, with changes in the environment paralleling the appearance of organisms uniquely adapted by the creator to their surroundings. This divinely ordained matching between the history of life and the history of the earth resulted in a progressive—though not evolutionary—fossil record. By the 1840s the evidence for this record extended from the presumed lower limits of animal life, through the invertebrate classes, to fish, reptiles, mammals, and finally man. Peter Bowler has minimized Sedgwick's interest in any inherently progressive tendencies in the history of life, preferring to see him as an advocate of the view that such "advances" were incidental byproducts of the physical history of the earth. However, Sedgwick almost certainly believed that the earth

[7] Sedgwick 1841: 550; 1838: 684. The alteration from "system" to "group" is evident when Sedgwick 1838: 679 is compared with Sedgwick 1841: 548-549. But the change is not always made, perhaps a sign that he harkened back to the traditional meaning of "system."

and its living beings developed in tandem according to a comprehensive divine plan. This is especially clear in relation to his local stratigraphical studies, for the specific environmental conditions in an individual locality could scarcely be responsible for uniquely determining the forms of life found there. In his eyes this would have been dangerously near to certain doctrines of evolutionary materialism.[8]

In the "Supplement" of 1841, Sedgwick hinted at a possible means for using the theory of progression to add a palaeontological dimension to his sub-Silurian physical groups. In his interpretation of progression (as I have illustrated it in Fig. 4.2) the Lower Silurian fauna had appeared in successive stages, beginning with the initial stirrings of Silurian life in the lowest strata of North Wales (a) and ending with the full range of species (a–g) recorded in the *Silurian System*. He tentatively suggested that the lithological and physical Cambrian groups might obtain a limited palaeontological characterization through a careful charting of the horizons at which the successive elements of the Silurian fauna first appeared. In other words, Sedgwick hoped that separate Cambrian fossil groups would be identified through the gradual unfolding of the same fauna that found full expression in Murchison's formations along the Welsh Borders. By means of what might be called a "progressionist" method of correlation, Silurian fossils could be used to identify Cambrian rocks.[9]

Of course, my diagram is highly idealized; Sedgwick by no means anticipated finding such a perfect record of the early history of creation among the contorted limestones and slates of the ancient rocks. It is not even clear from his initial proposal whether Sedgwick expected life forms of increasing complexity and degree of organization at higher levels in the geological column, although slightly later statements make it almost certain that he did so. The initial proposal, in fact, was couched in purely stratigraphical terms and only implicitly involved a hypothesis about the development of life. Eventually the plan became more explicit.[10] In any case Sedgwick's scheme for characterizing his sub-Silurian groups promised to coordinate fossil and physical evidence into a unified, coherent, and "natural" classification of the older rocks of Britain.

For all their inventiveness, the suggestions of the November 1841

[8] For progression, see Bowler 1976 (esp. 36-37); D. Ospovat 1981: 6-38; and Rudwick 1971, 1976a: 164-217.

[9] Sedgwick 1841: 550.

[10] [Sedgwick] 1845c: 31; Sedgwick 1846: 129.

"Supplement" were greeted with scepticism by the Fellows of the Geological Society. After such a poor reception, Sedgwick needed to return to the field as soon as possible to buttress his views, especially as he had not visited Wales since 1834. He had originally planned to tour in the summer of 1842 with John Eddowes Bowman, a retired banker living in Manchester who had actively pursued scientific studies from the early years of the century. During the late 1830s Bowman focused his attention on North Wales, and in 1840 and 1841 he found Lower Silurian fossils at the very base of the Snowdon range, farther to the west of the Berwyns than ever suspected. Even the most typical fossil of the Llandeilo formation, the *Asaphus Buchii*, was present—a finding of considerable relevance to Sedgwick's novel progressionist method for identifying Cambrian groups. These and other discoveries, reported by Bowman in long letters to Sedgwick in October 1841, reinforced the need for revision already necessitated by the loss of a potential Cambrian fauna.[11] Bowman's death at the end of the year cut short their plans for a tour. Sedgwick travelled for a few days in 1842 with the Irish geologist Richard Griffith, but his chief companion was a twenty-one-year-old palaeontologist named John W. Salter. Salter had determined many of Murchison's species for the *Silurian System* while apprenticed to Sowerby; during the winter of 1841 he had greatly impressed Sedgwick, who commented that the young man "seems quite alive among the old fossil species."[12] For the first time in over twenty years of fieldwork Sedgwick felt the need for an assistant skilled in identifying fossils. Salter's expertise was just what he was looking for.

The results of their five-week tour were mixed. Despite the doubts of his audience at Somerset House, Sedgwick successfully reaffirmed his principal section from the Menai to the Berwyns: whatever the precise nature of the organic remains found in North Wales, an immense thickness of strata extended many thousands of feet below the lowest beds at Bala. Altogether he could list eight or nine separate limestones beneath the Upper Silurian Denbigh

[11] An outline of the planned tour is in Sedgwick to Murchison, [early July 1842], GSL: M/S11/195. Information on Bowman is in the *DNB* and in Tayler 1846; his findings are announced in Bowman to Sedgwick, 4 Oct. 1841, 27 Oct. 1841, CUL: Add. mss 7652ID100, 130b, and in a paper to the Glasgow British Association meeting in 1840: see Bowman 1841a (also relevant is Bowman 1841b).

[12] Sedgwick to J. de C. Sowerby, 25 Nov. 1841, BMNH(S). Salter's unfortunate career is described in Secord 1985b; for Griffith's presence, see Clark and Hughes 1890, 2: 48, and Sedgwick to Phillips, 28 Sept. [1842], UMO.

Flags, including the beds on the summit of Snowdon. While having already abandoned all hope of finding enough distinctive species in these zones to characterize a system of life separate from the Silurian, Sedgwick (with Salter's help) made extensive additions to the fossil lists. In the three principal limestones (Glyn Ceiriog, Bala, and Snowdon), Sedgwick and Salter found fifteen or twenty distinctive species. All the rest were Lower Silurian just as they had anticipated.[13]

The two companions had also hoped that the fossils of the limestone bands in the ascending succession would match the expectations raised by the theory of progression (Fig. 4.2). After extensive searches, however, Sedgwick was forced to concede total failure in distinguishing the fossil groups of the lowest beds of his section from those at its top (Fig. 4.3). He expressed his disappointment to Murchison soon after returning from the field:

> In regard to N. Wales, you know my general views—I *stated last year* (see the abstract) that on unpacking my Welsh fossils I could not discover any trace of a lower zoological system than that indicated in your lower Silurian types. I did however *expect* to find certain definite groups indicating a succession in the ascending steps of a vast section (certainly many thousand feet thick). And my hope was last September to prove this point but I failed utterly as I told you before; & at present I really know no such definite groups.[14]

Notably, three years later Sedgwick used this failure to trace the gradual appearance of Silurian life in refuting the theories of the *Vestiges of Creation*, an anonymous transmutationist work which required precisely such an advance from simple to complex in the geological record. "We have spent years of active life among these ancient strata," Sedgwick wrote in the *Edinburgh Review*, "looking for (and we might say longing for) some arrangement of the fossils which might fall in with our preconceived notions of a natural ascending scale. But we looked in vain; and we were weak enough (perhaps our author might tell us) to bow to nature."[15] When he penned these words in 1845 Sedgwick was possibly being disingenuous, as he had by then achieved some success with the progressionist scheme. But immediately after the summer of 1842 the Upper Cambrian groups had lacked even this minimal palaeonto-

[13] Sedgwick to Murchison, 20 Oct. 1842, GSL: M/S11/198a & b.

[14] Sedgwick to Murchison, [Jan. or Feb. 1843], GSL: M/S11/209a & b.

[15] [Sedgwick] 1845c: 31.

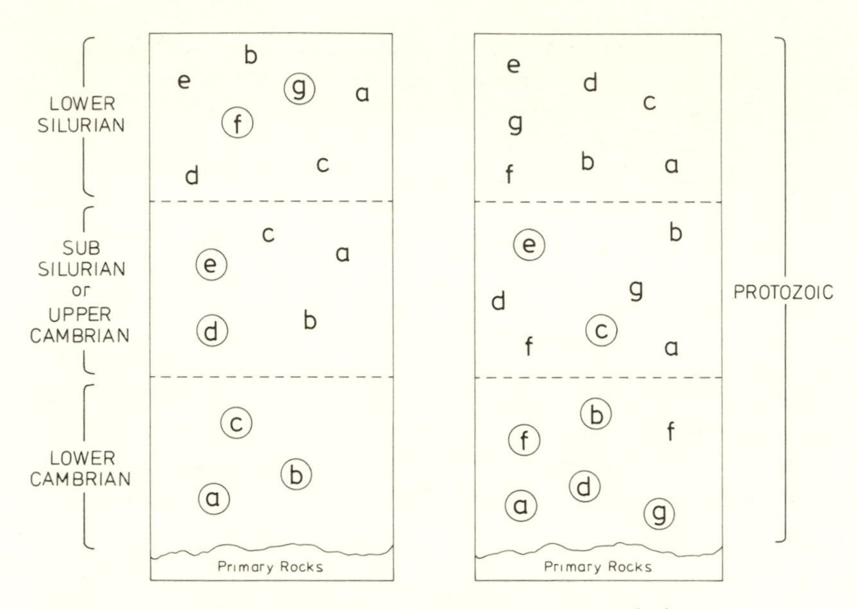

Fig. 4.2. (above, left) Interpretative diagram illustrating Sedgwick's proposed progressionist scheme (1841) for adding a fossil dimension to his sub-Silurian physical groups. All letters (a–g) represent Lower Silurian species; the first appearance of each in the geological record is circled.

Fig. 4.3. (above, right) Initial failure of Sedgwick's progressionist scheme in 1842, with the "Protozoic" classification of the following year.

logical definition. The collaborative boundary had failed an important test, and Sedgwick became increasingly convinced that a major change in classification would be required.

Palaeontology and the Silurian Empire

Unlike Sedgwick, who prudently limited his researches to the British Isles after 1839, Murchison turned his attention to a far wider field and began geologizing on an unprecedented international scale. In the early 1840s he hammered his way across European Russia, carrying English stratigraphical nomenclature to the western edge of Asia. Sedgwick viewed the loss of the anticipated Cambrian fauna from a context of local studies of the older rocks of Great Britain and was concerned primarily with structure and lithology. But in the enterprise of international correlation shared by Murchison and many contemporaries, geological classifications depended on palaeontology alone. In this context the loss of a fauna for the Cambrian automatically necessitated not only its demotion from the rank of system, but also the complete abandonment of the word except as a local term for the slaty rocks of Wales.

Before the Cambrian failed to live up to its early promise, both systems had rapidly spread across the global geological map. Their potential international importance had been recognized at the British Association meeting in 1835 when the joint classification had been announced, with Greenough suggesting that the new arrangement might be extended to Spain, Northern Europe, and America. Within weeks Buckland had applied Cambrian and Silurian outside the type area during a tour of the Ardennes. Similar comparisons had followed in Ireland, Turkey, Scandinavia, the United States, and France, to the extent that the main units of the *Silurian System* were well on their way to being established as global standards of correlation even before the book itself actually appeared.[16] Announcements and endorsements appeared in Buckland's *Bridgewater Treatise*, the textbooks of John Phillips, the early editions of Lyell's *Principles of Geology*, as well as in the annual Geological Society presidential addresses.[17] In America the English diplomat and geologist George W. Featherstonhaugh proved an effective agent, while Ebenezer Emmons and other state survey geologists paralleled their own arrangements with the new classification. Élie de Beaumont, who had done so much to encourage the abandonment of Transition and Grauwacke, was one of several Parisian naturalists who began deploying Murchison and Sedgwick's nomenclature immediately. His great map of France, undertaken with Dufrénoy, regularly referred to Silurian and Cambrian. Together with Édouard de Verneuil and Ami Boué, they made the Société Géologique de France (founded in 1831 on the model of its London counterpart) into a center for the consolidation of the new classification.[18]

Murchison, eager to underline the importance of his Silurian ac-

[16] Hardy 1835: 109 reports Greenough's comments. Some of the early adherents are mentioned in Murchison 1839: 580-586, 701.

[17] Buckland 1837, *1*: 527-528; Phillips 1836: 19-20; C. Lyell 1835, *4*: 305-307, where he accepts only the Silurian; for the Cambrian, see C. Lyell 1838: 464-466. Relevant presidential addresses include Whewell 1839: 77.

[18] See various articles in the *Bulletin de la Société Géologique de France;* also Dufrénoy and Élie de Beaumont 1841 and Boué 1840. Verneuil's work in Spain, Germany, and America was of great importance; e.g. Verneuil 1847, a major comparison of the older rocks of Europe and America. Earlier American work is discussed in L. G. Wilson 1967, Schneer 1969b, and J. M. Eyles 1978: 388.

complishment, frequently gave the impression that these foreign geologists were engaged in applying the new nomenclature directly to their own strata, just as he and Sedgwick had done within the British Isles. But this happened only occasionally. Among the international geological community, the new divisions did serve as standards of comparison; however, locally derived names continued to provide the principal means of labelling strata. This, of course, rendered their conclusions relatively immune from changes in the British sequence; few were ready to have their detailed findings stand or fall on the accuracy of the work of men from a different country. And sometimes, particularly in the United States, this hesitancy reflected fears of foreign domination. Like other American intellectuals, geologists were acutely conscious of the need to establish an independent cultural life. Such sentiments exacerbated worries, widespread even in England, that one small district in the Welsh Borders was being too rapidly raised to a geological model for the world. As an old hand in India asked Sedgwick in 1840, why should views reflecting years of experience among the Himalayan wilds "be changed according to Mr. Murchison's last paper on Shropshire"?[19] Robert A. C. Austen, an independent-minded Devonshire geologist, ridiculed Darwin's 1838 announcement of Lower Silurian fossils in the Falkland Islands as the ultimate *reductio ad absurdum*. "If Darwin was a joker of jokes," he told De la Beche, "which he is not, I should esteem this as one of his best."[20]

Cambrian and Silurian had shared equally in much of the early global "systematizing." But as the 1830s drew to a close Sedgwick's continuing failure to produce a full-fledged fauna began to take its toll: fossiliferous areas previously compared to the Cambrian were ceded instead to Siluria, and Murchison's territory grew ever larger. Murchison's own part in the territorial extensions of the Silurian was a large one. Because of the ill health of his mother-in-law, he had always confined his geologizing to the relatively narrow limits of Western Europe. After her death in 1838, however, he considered himself "a free agent, for the first time these 20 years," finally able to extend his researches to distant lands as yet barely explored by geologists.[21] A letter from Leopold von Buch, one of the first geol-

[19] J. H. Batten to Sedgwick, 20 Jan. 1841, CUL: Add. ms 7652IF80.

[20] Austen to De la Beche, 8 Aug. 1838, NMW. Austen referred to C. Darwin 1839-1843, 3: iii. The part in question had actually appeared in May 1838. See also C. Darwin 1977a and Murchison 1839: 583.

[21] Murchison to Sedgwick, 19 Jan. 1838 [not sent], in Craig 1971: 499. An expurgated text of this letter is given in Geikie 1875, 2: 261-264.

ogists to propose extensions of the Silurian and Cambrian in Europe, pointed out that vast reaches of the Russian empire held gently dipping Transition strata, scarcely disturbed and full of fossils. Here was an ideal opportunity for scientific work on a grand scale, and Murchison grasped it at the first available moment. The Russian strata had been surveyed in outline two decades earlier by the British ambassador in St. Petersburg, William Fox Strangways, and more recently by the Prussian Georg Adolf Erman and the Russian mining engineer Grigory Petrovich Helmerson; in addition, palaeontologists had offered scattered descriptions of Russian fossils. But compared with the state of the science in the rest of Europe, let alone in England, Russian stratigraphy appeared patchy and preliminary, particularly because it had not yet been comprehensively correlated with the standard type sequences.[22] For Murchison, this relative backwardness rendered Russia all the more attractive as a field of research, for the originality and importance of his own work would appear in even sharper relief. In 1840 he sped across the countryside in a *tarantass* drawn by six horses, accompanied by Verneuil and two Russians: the naturalist Alexander von Keyserling and a young officer, Nikola J. Koksharov, later an eminent mineralogist. With the blessings of the Imperial authorities, they covered almost ten thousand miles in their first summer's research and travelled even farther east during a more extended tour of the following year. In the course of these investigations they unearthed the bones of extinct mammoths, laid the foundations for a "Permian system" above the Carboniferous, discovered further proofs of the accuracy of the Devonian, and above all demonstrated the utility of a strongly bounded classification relying on palaeontology.[23] The tour had potential economic consequences of the first order as well. From the beginning of his Welsh Border studies, Murchison had been convinced that the Silurian predated the commencement of land plants; hence his claims that seaches for coal in the early strata of Britain were inevitably a waste of money. During his Russian journeys he concluded that gold only rarely appeared in Secondary or more recent strata. Therefore, although Siluria might not contain

[22] This earlier work is briefly described in Vucinich 1965: 344-346 and J. C. Thackray 1979: 421. Carver 1980 gives a useful corrective to tendencies to underestimate the extent of early scientific culture in Russia. Buch extended the new systems in a publication of 1836; see also the comments in Buch to Murchison, 23 Feb. 1840, GSL: M/B33/1, partly printed in Geikie 1875, *1*: 291-292; and Murchison et al. 1845, *1*: vi-vii.

[23] Murchison et al. 1845, passim. The tours are narrated in Geikie 1875, *1*: 295-302, 315-357, and a brief summary is given in J. C. Thackray 1979.

the coal needed to fire the Industrial Revolution, it seemed the most common repository for a precious metal soon to be very much in the news.[24]

Murchison's desire to give geology a role in economic development, his interest in travelling abroad, and his wish to be known as a wide-ranging explorer in the Humboldtian mold were all strongly linked to his commitment to the use of fossils in stratigraphy. With the advent of the Russian journeys he could employ this method on a vast scale, using Verneuil's palaeontological assistance to color great stretches of the geological map in the shades of the British sequence. At a single bold stroke the greater part of Europe could be brought into stratigraphical order. Murchison looked across the Atlantic to an even wider field in America and made serious preparations for a tour (never undertaken) of the geology of the United States. As he had told Sedgwick upon his return from Russia in 1840, "nothing short of Continental masses will now suit my palate."[25]

In Murchison's view the success of the Silurian gave a secure foundation not only to international correlation but also to his personal reputation. His two presidential addresses to the Geological Society, the first in February 1842 and the second a year later, were marked by the triumphant tone of a man who had just mapped the geology of half a continent. In the 1842 speech he delivered the first of many accounts of his early Silurian work with the intent of establishing his place in the history of science. "The perpetuity of a name affixed to any group of rocks through his original research," Murchison said, "is the highest distinction to which any working geologist can aspire."[26] Speaking from this self-defined pinnacle of success, he could finally forget the failures of his early ventures as a soldier and landed gentleman, for his reputation now rested securely on a grand geological system, a major scientific accomplishment already part of the daily usage of British geologists. Murchison's biographer summed up this possessive attitude towards the Silurian by comparing his subject to an entrepreneur engaged in a successful commercial enterprise. "His domain of 'Siluria,' " Geikie wrote, "became, in his eyes, a kind of personal property, over

[24] Murchison 1850, and 1844: xcix-c.

[25] Murchison to Sedgwick, 1 Sept. 1840, printed in Geikie 1875, *1*: 302-303; copy at CUL: Add. ms 7652IIID32. For the abortive American journey, see Silliman 1841 and Geikie 1875, *1*: 312.

[26] Murchison 1842: 649.

which he watched with solicitude. Or, it might rather perhaps be compared to a vast business which he had established, of every original detail of which he was complete master, and which he laboured to extend into other countries, while he kept up through life a close correspondence with those by whom the foreign extensions were so abundantly and successfully carried out."[27] What contemporary industrialists were doing for the British economy, Murchison wished to accomplish for British science. As he explained to Sir Robert Peel: "Just as our goods are patterns for the world, so may our geological types be recognized in the remotest parts of the world. This is my deepest geological aspiration."[28]

From his central position in the London scientific community, Murchison was able to orchestrate an effective campaign for the new classification. The recent date of his family's rise to fortune—and its shadowy history in his father's dealings on the Indian subcontinent—made the need to establish a secure position through science all the more imperative. Murchison's standing in the social world of the metropolis had been greatly enhanced in 1838 by the purchase (with substantial financial assistance from Charlotte's dying mother) of a new home in fashionable Belgravia. Famous for its grand dinners and soirées, this house became a haunt of London's social and scientific elite, a place where men of science could rub shoulders with politicians, visiting dignitaries, explorers, and literary men. The evident purpose of these soirées was to maintain and improve the social status of scientists, particularly geologists, and especially Murchison himself. The young John Ruskin described the scene at one reception in 1842:

> The old lady at Athlones said that Mr Murchison had told her there would be above 700. I could not count more than 150 in the rooms at a time—but as they kept rushing in and out, there could not have been many fewer passing through during the two hours I staid. . . . I don't know what fortune Murchison has—but this is coming it rather strong—rooms all pale grey & gold—magnificent cornices—with arabesques like those of Pompeii in colour, furniture all dark crimson damask silk & gold—no wood visible—at least four footmen playing shuttlecock with peoples names up the stairs. . . .[29]

[27] Geikie 1875, *1*: 243.

[28] Quoted in Stafford 1984: 13, an article that discusses Murchison, the Geological Survey, and colonial development.

[29] J. Ruskin to John J. Ruskin, 28 Feb. 1842, in Burd 1973, *2*: 709-710. "Athlones"

The glittering magnificence of these affairs led some to deplore the scientific lionizing that they engendered, with all attention seemingly drawn to the latest superficial sensation while men of solid merit labored away unseen.[30] Whatever the truth of such criticisms, there can be little doubt that Murchison's name was becoming a byword for an elitist and authoritarian vision of scientific activity. It was a reputation that the ever-expanding Silurian realm itself inevitably reflected.

From the beginning Murchison intended the Silurian classification to be a carrier of his conservative and militaristic attitudes towards science. In the preceding chapters we have seen that his field style reflected his military upbringing and that the Silurian name itself purposefully recalled the warlike Welsh Border tribe led by Caractacus.[31] Friends and correspondents constantly compared him to Caractacus and, after the journeys to Russia, to his patron the Czar Nicholas I. "Murchison has come back grander by far than ever from his Russian travels," John Lockhart told the publisher John Murray. "I fancy he must now take rank as Grand Duke."[32] But the imperial expansion of the Silurian classification soon became more than just a personal achievement. By the time Murchison's Russian researches were published in England later in the decade, the educated Victorian public acclaimed Siluria as a great national accomplishment. As one commentator wrote in 1847, "such labours by one who after serving his country as a soldier in the field, now adorns her by his discoveries in science, deserve all praise." Active promoter of the Royal Geographical Society, tireless friend of explorers and travellers, Murchison considered his advocacy of the Silurian as part of an integrated plan to foster British national greatness. In his view and that of many contemporaries, palaeontological stratigraphy promised to model the world's geology upon the rocks of the British Isles. As I have emphasized elsewhere, the spread of the Silurian—and the other names of the geological column—must be viewed in connection with the network of British influence and trade being established at this time. Just as Siluria be-

---

refers to the mansion of Ruskin's acquaintance, George Godard Henry de Ginkel, Earl of Athlone. Financial arrangements for the Belgrave Square house are mentioned in Murchison to Sedgwick, 19 Jan. 1838 [not sent], in Craig 1971: 498-499.

[30] J. D. Hooker to T. Anderson, 7 July 1861, in L. Huxley 1918, *1*: 406, and J. S. Henslow to T. H. Huxley, 7 Dec. 1858, ICL(H): *18*, f. 133.

[31] [Anon.] 1854: 815.

[32] Lockhart to Murray, 28 Sept. 1840, in Smiles 1891, 2: 392-393. I owe this reference to Patrick Boylan.

came associated with the centralized scientific world of the metropolis, so also did it become part of an emergent ideology of cultural and economic imperialism.[33]

At the same time that Murchison became convinced of the vast geographical extensions of the Silurian, he also began to hope that his system could be extended stratigraphically, down to the bottom of Sedgwick's so-called Cambrian divisions. A single sentence of the *Silurian System* had stated that the collaborative boundary might need to be shifted to a lower horizon, especially as the Bala Limestone fossils so evidently matched those of the Silurian. "As these shells abound also in the Lower Silurian Rocks," he had written, "it would seem that as yet no defined line of zoological division can be drawn between the Lower Silurian and Upper Cambrian groups, and that as our knowledge extends, we may probably fix the lower limit of the Silurian System beneath the line of demarcation which has for the present been assumed."[34] It should be stressed, however, that this passage referred only to the well- known presence of Lower Silurian fossils in the *Upper* Cambrian system, especially at Bala. When adding these words to his text in 1838, he certainly had not believed that the Silurian fauna would persist to the base of the fossiliferous rocks; the faunal expectations engendered by the Cornish and Devon limestones and the vast thickness of rocks in North Wales were far too strong to be dismissed so easily. But with the Devonian reclassification and the unpacking of the Cambridge collections, the prophetic quality of this sentence soon became apparent. Fossils could justify stratigraphical as well as geographical expansions of the Silurian.

During his Russian tours and slightly later visits to Scandinavia, Murchison added Continental support for the persistence of the Lower Silurian fauna to the base of the geological column. Towards the end of his travels, he found strata containing Lower Silurian fossils resting unconformably on a crystalline basement series without any distinctive fossils of a Cambrian system intervening (Fig. 4.4).

[33] Secord 1982: 426-431, and Stafford 1984. For the quoted passage, [anon.] 1847: 292.

[34] Murchison 1839: 308. The composition of this crucial passage has an interesting history, for the page on which it occurs was reset in proof after objections from Sedgwick on another issue (Murchison's encouragement of Bowman's geologizing in Wales, to be discussed in the next chapter). Afterwards Sedgwick refused to read the rest of the manuscript; had he done so, and learned of the existence of this sentence, he almost certainly would have protested. A similar passage occurs in Sedgwick and Murchison 1839: 259, but this speaks only of drawing the Cambrian-Silurian boundary at a "different" level.

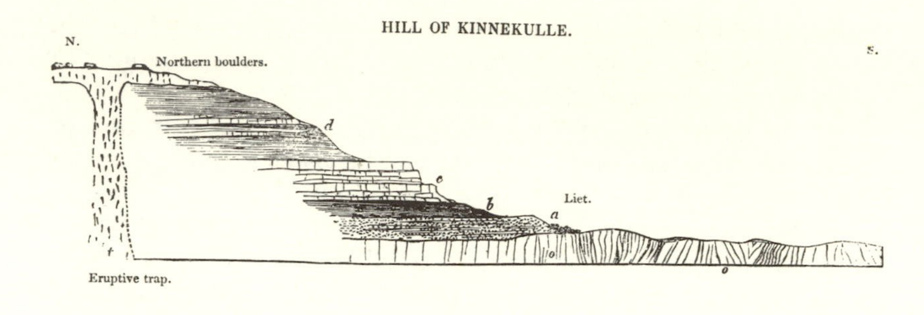

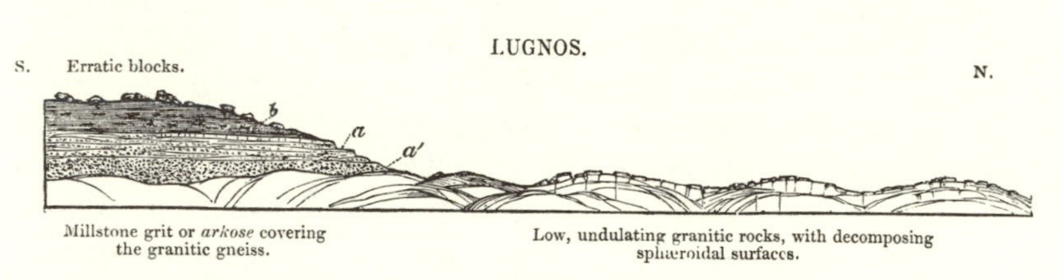

Fig. 4.4. Traverse sections from Sweden showing Murchison's oldest Silurians resting unconformably on granite and gneiss.

In papers of 1844 and 1845 he elaborated the point by showing that these gneissic rocks were stratigraphically below the Silurian. In his view their crystallization had taken place during an "Azoic" period too hot for life, and the Lower Silurian species gradually introduced in the overlying strata thus constituted the earliest created organic forms. Eventually Murchison would be forced to back down from this position after geologists found more ancient and extremely primitive relics of creation in strata that had been definitively excluded from the Silurian. During the 1840s, however, he regarded the "proofs" of the origin of life within his Silurian system with great pride, for the organic remains of his original type area in Wales, like those of Russia, comprised the first unequivocal evidence of a "beginning" ever provided by scientific research.[35]

From Murchison's perspective, the overwhelming mass of evidence accumulated in the two or three years after the publication of the *Silurian System* demanded the complete abandonment of "Cambrian" as an internationally valid term. He expounded this view in his Geological Society addresses, dwelling on Sedgwick's findings

[35] The initial announcement can be found in the *Athenaeum*, 26 Oct. 1844, no. 887, p. 976; see also Murchison 1845. For the origin of life question, see Rudwick 1976c.

as recently announced in the 1841 "Supplement." He showed that the Cambrian had been defined in 1835 and 1838 on sectional rather than fossil evidence, principally as a means of naming the huge thickness of strata underlying the Llandeilo Flags. Geologists had then waited for a peculiar set of organic remains to enliven these ancient rocks:

> In regard, however, to a descending *zoological* order, it still remained to be proved, whether there was any type of fossils in the mass of the Cambrian rocks different from that of the Lower Silurian series. If the appeal to nature should be answered in the negative, then it was clear, that the Lower Silurian type must be considered the true base of what I had named the *Protozoic* rocks [a term used by Murchison in the *Silurian System* to refer to the Cambrian and Silurian systems together]; but if characteristic new forms were discovered, then would the Cambrian rocks . . . have also their own fauna, and the palaeozoic base would necessarily be removed to a lower geological position.[36]

As we have seen, the totality of the Cambrian's eclipse actually surprised everyone. However, in this passage and all his later accounts of the establishment of the joint classification, Murchison reconstructed a situation where the absence of a Cambrian fauna was not only possible, but positively anticipated. By presenting the problem in this light, he could introduce his downward extension of the Lower Silurian as a foregone conclusion. According to the terms of this argument Sedgwick's "Supplement" had answered "the appeal to nature" unequivocally, and Murchison could conclude that the collaborative boundary was henceforth abandoned as a palaeontological marker between the two systems.[37] Murchison still retained a limited use for this boundary in the local geology of Wales, as a convenient indicator of structural and lithological changes; in both the 1842 and 1843 addresses he carefully confined his remarks to the *palaeontological* extension of the Lower Silurian. Because of this limitation, Sedgwick apparently interpreted these speeches as noncontroversial summaries of his "Supplement." Their crucial assumption was unstated: fossils provided the *sole* grounding for classification, and a stratigraphical group without a distinct fauna had no validity at all.

Like almost everyone else at the November 1841 meeting, Murchison ignored Sedgwick's proposal for identifying sub-Silurian

[36] Murchison 1842: 641.

[37] Murchison 1842: 643, also 1843a: 74-75.

groups through the progressive appearance of Silurian fossil species. In his second address, Murchison positively misrepresented (whether on purpose or not) the aim of his friend's search for fossils during the summer of 1842. We have seen that Sedgwick and Salter had entered Wales without hope of finding enough distinct species to give the Cambrian a separate fauna, fully convinced that Lower Silurian forms would persist to the base of the fossiliferous rocks. Instead they had wished (and had momentarily failed) to chart the progressive appearance of Lower Silurian fossils within the Cambrian rocks. From remarks on the subject in letters and in his 1843 address, however, Murchison seems to have thought that his friend had still hoped to give a palaeontological basis to the Cambrian *as a system*. He thus presented the unpublished results from the summer as a repetition of previous failures to find a distinct fauna among the older slaty rocks. Most geologists had no specialized knowledge of these strata and were content to follow his lead in interpreting Sedgwick's findings in this way.[38]

After the success of his Russian expeditions, Murchison found Sedgwick's recourse to the nonpalaeontological characteristics of strata difficult to accept and even to comprehend. In previous years, while conducting the research for the *Silurian System*, he had considered both sections and fossils in constructing his classification, and his scientific territory (like Sedgwick's) had been a limited geographical region defined by the collaborative boundary. But after the book was published and its results widely accepted, the highly variable characteristics of structure and lithology held little further interest outside the Welsh Borders. For the purposes of international correlation these local considerations dropped out of the picture, leaving the fossil contents of the system as a universally applicable stratigraphical key. The collaborative boundary did not hold much interest for Murchison after 1839 except as it marked the type area of the original Silurian fossil species.

This new conception of the boundary is illustrated perfectly by Murchison's brief excursion in North Wales during the summer of 1842, just when Sedgwick and Salter were laboriously comparing the various fossiliferous zones of the region in hopes of finding a progressive pattern in the appearance of the Silurian species. Murchison's task was far simpler. He examined a few localities around Snowdon, found some of the fossils illustrated in the *Silurian Sys-*

[38] E.g. D'Omalius D'Halloy 1843: 531-532; [Owen] 1844: 149-150; Moxon 1842a: vi; [Brewster] 1846: 182.

tem, and on that basis spread the colors of the Lower Silurian to the western coast of Wales. This extended Silurian classification appeared for the first time in print on a small map of England and Wales commissioned by the Society for the Diffusion of Useful Knowledge, published in 1843 and included here in a redrawn version as Figure 4.5.[39] The key on the map shows Cambrian as a local slaty subdivision of the Lower Silurian, without a color of its own. It seems fitting that Keyserling accompanied Murchison on this tour of North Wales, for the palaeontological methods they used to extend the Silurian there were identical with those so recently successful in Russia.

The widespread agreement with Murchison's downward extension of the Silurian is immediately clear from reports of the discussion that followed Sedgwick's "Supplement" in November 1841. The general reaction was one of disappointment and dismay; many geologists had anticipated a full description of the Cambrian as a palaeontologically valid system like the Silurian or Devonian. The young barrister Edward Bunbury expressed the consensus in a letter to Lyell, who was touring America at the time:

> The greatest novelty in the geological world since my return has been Sedgwick's paper, a "Supplement to an Introduction" as he called it, but which supplement seemed to prove that there was nothing for the Introduction to lead to. Everyone was in hopes that we were now at length to have some definite account of the Cambrian system, but as the paper proceeded & one tract of country after another was described & we heard of *Caradoc sandstone* fossils in the very heart of Snowdonia, & first Cumberland & then North Wales seemed to be given up as undistinguishable from the lower Silurian system, every one was disposed to cry out "The Cambrian system where is it?" And Echo answered Where? —for from beginning to end it was not to be found. In the next edition of your Elements your chapter on that subject may be like Bishop Horrebow's on the Snakes of Iceland—"There is *no* Cambrian system."[40]

Lyell had in fact already arrived at this conclusion even before leaving for Boston in July. The first edition of the *Elements* in 1838 had

[39] Murchison 1843b. The library of the British Geological Survey has an early draft version of this map, which unfortunately is without a key. The North Wales excursion is mentioned in Murchison to Sedgwick, 16 Oct. 1842, in Geikie 1875, *1*: 376-382, at p. 379; also pp. 369-370.

[40] Bunbury to Lyell, 13 Dec. 1841, EUL: Lyell 1/462.

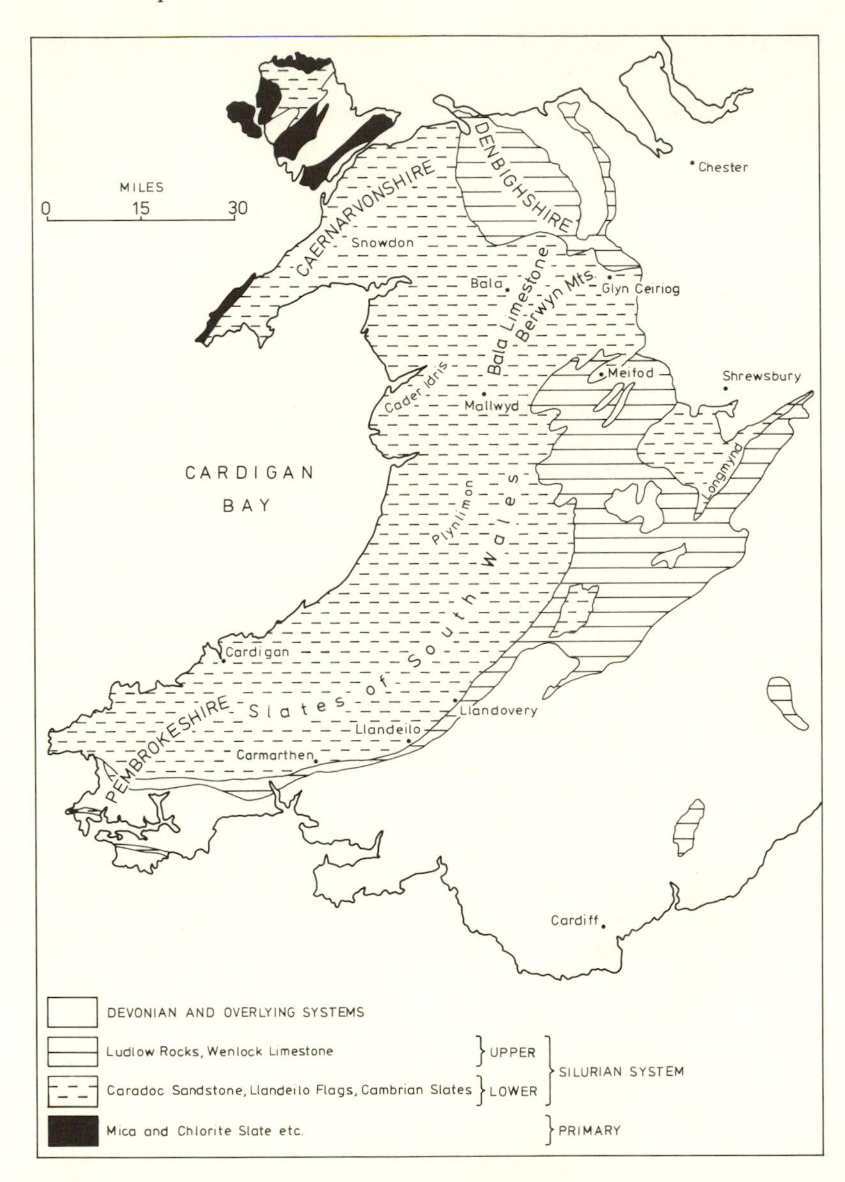

Fig. 4.5. Part of Murchison's map for the Society for the Diffusion of Useful Knowledge, 1843. Note the extension of the Lower Silurian to cover most of Wales.

pictured *Endosiphonites* (Fig. 3.14) as a Cambrian fossil, and while noting the lack of full palaeontological proofs for a sub-Silurian group, he had concluded from the great thickness of fossiliferous rocks below the Llandeilo formation that additional organic remains typifying a lower division would be found. In the second edition of the *Elements*, completed shortly before his departure, Lyell moved *Endosiphonites* to a new Devonian chapter and excised the hopeful passage on the Cambrian.[41] The opinion of Fitton, who gave a long speech at the November meeting, likewise typifies the consensus. Writing a few months earlier in the *Edinburgh Review*, he had recommended that Sedgwick publish the complete palaeontological proofs for the Cambrian as soon as possible. Fossils provided what Fitton called the "*pièces justificatives* of geological history," and until distinct organic remains characterized the Cambrian, its separation from the Silurian anywhere except in Wales was merely a matter of convenience.[42] Given this perspective, Fitton's response to the sad revelations of the "Supplement" is not difficult to imagine.

Although geologists regretted Sedgwick's inability to unveil a Cambrian fauna, they scorned his attempt to use the progressive appearance of Lower Silurian species as a way of defining Upper and Lower Cambrian "groups." Such "speculations," as Bunbury called them, relied upon an unproven and highly theoretical conception of the history of life—scarcely a secure basis for classification even if one shared Sedgwick's progressionist views. Despite the Woodwardian professor's acknowledged position as an authority on such matters, many geologists doubted his claim that the same fossil species extended unchanged through several vertical miles of strata. As Lonsdale reportedly said, "The thing's impossible—so it's no use discussing it."[43] After such a fierce barrage of criticism, it is hardly surprising that Sedgwick spent much of the summer of 1842 checking the accuracy of his old sections in North Wales. For most geologists, however, the main point had already been settled by palaeontology alone. Assuming the accuracy of Murchison's sections in the Borderlands, Silurian fossils automatically defined Silurian rocks. Certainly no one later showed the slightest degree of support for Sedgwick's progressionist scheme

[41] C. Lyell 1838: 456, 464-466; 1841, 2: 176, also p. 160. L. G. Wilson 1972: 516 shows that Lyell had completed the edition before leaving for America.

[42] [Fitton] 1841: 38. For a similar comment from abroad, see Alexander von Humboldt to Murchison, 26 Feb. 1840, GSL: M/H32/1. Of course, there were exceptions to the consensus (e.g. Trimmer 1844), but few within the circle of expert specialists.

[43] Bunbury to Lyell, 13 Dec. 1841, EUL: Lyell 1/462.

for identifying Cambrian groups. Given the response to the "Supplement," one can date the complete abandonment of the Cambrian by the British geological community from Sedgwick's own announcement in 1841 that no distinctive fossil types occurred below the Lower Silurian.

Far from being censured as an unjustifiable trespass, Murchison's extension of the Silurian in 1842 was thus widely accepted as a natural consequence of the disappearance of the Cambrian fauna. In the opinion of most experts, Sedgwick's "Supplement" had already provided, though unintentionally, an obituary for the Cambrian, not only as a system but as an element of British stratigraphical nomenclature. For the rest of the decade Sedgwick would raise one of the few dissenting voices to the extended Silurian, and it is to the alternatives that he and a few others proposed that we now return.

Alternatives to the Expanded Silurian

Although Sedgwick recognized the need for a new classification after the disappearance of the Cambrian fauna, he felt that Murchison's proposed solution represented a giant step backwards. The two geologists first faced the question in private correspondence while working out the consequences of their Continental researches into the Devonian problem. If no distinct fossils were to be found below the Lower Silurian, what was to be done? "There is one way of getting rid of the difficulty of classing the *sub-devonian*," Sedgwick told Murchison in January 1840, "viz. by calling Silurian all the fossiliferous rocks below the devonian. But what then? The term Silurian System then only changes places with fossiliferous greywacké, means neither more nor less, & consequently loses all its value as a term of *distinct classification*."[44] In Sedgwick's view, the hard-won advantages of their joint classification would be thrown away if all was merged within the Silurian. As he wrote in a letter of 1842, an enlarged Silurian would deserve Conybeare's criticism of the old term "Grauwacke": "*Est Jupiter quodcunque vides.*"[45] The Devonian reclassification and the loss of a Cambrian fauna necessitated a thorough rethinking of the classification, not an inflexible insistence on names now shown to be inappropriate.

This emphasis on the need for flexibility was entirely shared by John Phillips, who commended Sedgwick for refusing to take the old classificatory boundaries for granted.

[44] Sedgwick to Murchison, 27 Jan. 1840, GSL: M/S11/167a & b.
[45] Sedgwick to Murchison, 20 Oct. 1842, GSL: M/S11/198a & b.

> I think it is very evident that we must endeavour to unsystematize our minds again; & as I hold geology to be never rightly round when she is forced to bend to conventional plans of temporary importance, and that she is able to produce in all cases sound reason for legitimate conclusions, I wish to put again & to keep for some time, *in doubt*, the exact *lines* of demarcation which best suit the soft *shades* of mother nature. Murchison's Silurian System is a noble assemblage, and he is entitled to every applause that he has got & more; but I hope he will have the magnaminity to permit the whole basis & superstructure of that classification which he has favoured, to undergo a new & searching analysis, for the discovery of perhaps new & important relations.[46]

Unlike most naturalists of the period, Phillips had no independent income; orphaned at eight, he had learned his early science by helping his uncle and guardian William Smith. But he soon went far beyond Smith's simple rule of "strata identified by organic fossils." As an experienced palaeontologist with a strong interest in biogeography, Phillips believed that strata could be grouped only with a thorough understanding of the spacial and temporal distribution of whole populations of extinct organisms. In publications of the 1830s he had developed a highly unusual approach to the classification of the older rocks. Stimulated in part by the contemporary vogue for statistical enquiries, and by Adolphe Brongniart and Augustin de Candolle's palaeobotanical studies, Phillips made elaborate percentage counts of species held in common between adjacent parts of the geological record on a region-by-region basis. "I rather like to see *attempts* to make *classifications* for the districts," he told Sedgwick, "because thus some peculiar local characters come out in bolder relief, & some facility is gained for imagining the ancient local peculiarities of land & sea."[47] These regional studies could then be compared to derive a general picture. Phillips's employment from 1838 as palaeontologist to the embryonic Geological Survey afforded ideal opportunities for the detailed studies so necessary to his approach. Appropriately, Phillips reviewed some of his broadest conclusions in his 1841 Survey memoir on the fossils of Devon and Cornwall. As a first step towards a fully palaeontological clas-

[46] Phillips to Sedgwick, 24 Jan. 1842, CUL: Add. ms 7652ID120a. Biographical information on Phillips, whose work and career would well repay further study, is available in [anon.] 1870 and Morrell and Thackray 1981: 439-444.

[47] Phillips to Sedgwick, 25 Oct. 1845, CUL: Add. ms 7652IE128. For an exposition of the method, see [Phillips] 1840a; for Brongniart and Candolle, J. Browne 1983: 58-85. Further discussion of Phillips and the Survey statistics is in Secord 1986.

sification, he suggested that his method of comparative ratios could be used to subdivide the history of life into three broad groupings: Palaeozoic (ancient life), Mesozoic (middle life), and Kainozoic (recent life). He also indicated that more specific groupings might eventually be justified on this basis as well, noting that the fossiliferous strata below the Old Red Sandstone could be distinguished as the "Lower Palaeozoic."[48] Murchison saw this latter term as a direct threat to the expanded Silurian nomenclature and a personal attack upon himself, and said so in his 1842 address. As he told Sedgwick a few days afterwards, to substitute Lower Palaeozoic for Silurian was "intolerable coxcombry, now that the term of the man who worked the thing is spread over Europe & America." The strength of this reaction led Phillips to have very real fears for his employment prospects in science. But having struggled for independence from Murchison's patronage during the previous decade, he was scarcely willing to see his first foray into theoretical geology crushed in its infancy. As Phillips saw the question, his new arrangement of past life was fully in keeping with his usual circumspect caution in classification.[49] He strenuously repudiated Murchison's subdivision of the worldwide geological record into neat palaeontological systems: although perhaps valid locally in Britain, terms like Silurian and Devonian were at best premature and at worst positively misleading.

In fact Phillips advocated even more caution in classification than did Sedgwick himself. While perfectly willing to start again from scratch, Sedgwick saw no need for any extensive program of empirical research. "I know Phillips contends for more perfect fossil lists before we classify," he told Murchison in 1842. "But by way of reply I state that he (by his more perfect lists) has not shaken a single point in our Devonian classification founded only on a few of the more prominent types."[50] Once the physical groups and sections were in place, Sedgwick could agree fully with Murchison in using the merest handful of fossils as the foundation for a suitable classification. Phillips wanted to wait for years before establishing an alternative to the outmoded Cambrian and Silurian of the previous decade. Sedgwick and Murchison, each in his own way, were ready to begin immediately.

[48] Phillips 1841, esp. pp. 155-182; the initial proposal is made in passing in [Phillips] 1840b.

[49] Murchison to Sedgwick, 26 Feb. 1842, CUL: Add. ms 7652IIID36 (copy); Murchison 1842: 646-649; Phillips to De la Beche, 31 Mar. 1842, NMW. For the relations between Phillips and Murchison, see Morrell and Thackray 1981: 439-444.

[50] Sedgwick to Murchison, "Norwich Friday Morn.," [late 1842], GSL: M/S11/199.

Sedgwick's classification in the November 1841 "Supplement," with its proposed reformulation of the Cambrian through the use of progression, was only the most prominent of several schemes that he put forward in the two years following the Devonian reclassification. He had, of course, already rejected outright the extension of the Lower Silurian to the base of the fossiliferous rocks. In a letter sent to Murchison soon after the reading of the "Supplement," he proposed an alternative and far more radical reclassification of the entire sequence below the Carboniferous. According to this arrangement, which he borrowed from the Belgian geologist André Dumont, the Upper Silurian would be merged with the Devonian and the Lower Silurian would join with the Upper Cambrian:[51]

Carboniferous

{ Devonian
Upper Silurian

{ Lower Silurian
Upper Cambrian

This proposal reflected the palaeontological and physical situation in Germany and Belgium, but it could just as easily have been based on the work being done on the New York State Geological Survey, where James Hall and his coworkers were finding that Devonian fossils and strata graded imperceptibly into those of the Upper Silurian.[52] The existence of such passages led Sedgwick, like Dumont and Hall, to contemplate major changes in the received classification. While Murchison, Lyell, and most other British geologists viewed the extension of the Lower Silurian as a foregone conclusion after the loss of the Cambrian fauna, he was prepared to start afresh.

In insisting on the importance of a philosophically based classification, Sedgwick as yet evinced little concern about keeping his Cambrian nomenclature intact. Unlike Murchison with his strong personal stake in the terminology of the *Silurian System*, Sedgwick at this preliminary stage did not wish to be tied irrevocably to an established set of names. After long harangues on nomenclature from Greenough, Fitton, Murchison, and other Fellows after reading his "Supplement" at the Geological Society, Sedgwick reportedly con-

[51] Sedgwick to Murchison, [Nov.-Dec. 1841], GSL: M/S11/187.

[52] The results of the New York Survey began to be fully published only with J. Hall 1847, but were already becoming known in England in the early 1840s, particularly after Lyell's visit in 1841-1842. Accounts of Hall and his survey are in Clarke 1823 and Merrill 1924.

cluded the discussion by saying that "for what it concerned him, they *might call the rocks what they pleased*."[53] This was not because he thought scientific nomenclature was unimportant—on the contrary, he wished to take such an important step only at the proper time. At this stage of events he was primarily concerned with the accuracy of his structural groups and the possibility of identifying them palaeontologically through the theory of progression. Sedgwick similarly offered to abandon his Cambrian name in privately proposing the creation of "Devonian–Upper Silurian" and "Lower Silurian–Upper Cambrian" groups to Murchison at the end of 1841. As shown in Figure 4.6, he suggested two alternative nomenclatures to match these groups, including one that omitted "Cambrian" entirely. Either alternative A or B could be accepted and the important natural units of structure, lithology, and fossils (as Sedgwick saw them at this time) would be maintained. In his view, the fact that the second alternative "shoves out my Cambrian—altogether as to name" was a consideration of relatively little moment.[54] What he so emphatically denounced in correspondence was Murchison's attempt to join Upper Silurian, Lower Silurian, and Upper Cambrian into a single system (Fig. 4.6, alternative C) merely because fossils from these strata had originally been illustrated together as "Silurian" species.

Original Names 1839	*Sedgwick Alternative A*	*Sedgwick Alternative B*	*Murchison Alternative C*
Carboniferous	Carboniferous	Carboniferous	Carboniferous
Devonian	Silurian	Devonian	Devonian
Upper Silurian			Silurian
Lower Silurian	Cambrian	Silurian	
Upper Cambrian			

Fig. 4.6. Alternative nomenclatures for the older rocks, November 1841.

Murchison's work in Russia soon quieted most of Sedgwick's doubts as to the distinctiveness of the Devonian, and the plan to merge it with the Upper Silurian remained unpublished. But at the same time, Sedgwick became increasingly convinced of the importance of a strong break between the Upper and Lower Silurian. This

[53] Moxon 1842b: 131.

[54] Sedgwick to Murchison, [Nov.-Dec. 1841], GSL: M/S11/187.

break, together with his continuing uncertainty about the precise stratigraphical relations of the rocks in North Wales, provided the principal grounds for his first publicly announced alternative to the expanded Silurian classification.

The first hints of a gap at the base of the Upper Silurian in North Wales came from John Bowman, the provincial naturalist who had corresponded with Sedgwick during the autumn of 1840 and 1841. In letters written just before his death, Bowman had mentioned that the band of Lower Silurian "Caradoc" strata which Sedgwick had placed at the base of the Denbigh Flags in 1834 might be much more closely related to the underlying "Cambrian" rocks just to the south.[55] This interpretation would move the principal unconformity in North Wales from the top of the Upper Cambrian of 1838 to a point within the original Silurian, on a level with Murchison's break between the Lower and Upper Silurian (Fig. 4.7, compare columns a and b).

Sedgwick confirmed the importance of Bowman's gap during his field tour with Salter during the summer of 1842. His acceptance of the new interpretation depended upon an important change in his geological methods. During the early 1830s Sedgwick had spent most of his time constructing sections at right angles to the strike of the strata, a technique well adapted to a rapid reconnaissance. Using this method and guided by the theories of Élie de Beaumont, Sedgwick had constructed a band of Caradoc strata along the Holyhead road at the base of the Denbigh Flags. According to this view, which he adopted immediately after the 1834 joint tour, a major unconformity intervened between the Lower Silurian and the Cambrian. The disjunction was most immediately evidenced by a major change in strike. But in this case Élie de Beaumont's theory had been misleading. During the excursion with Salter, Sedgwick followed (much as Bowman had begun to do) the principal bands of fossiliferous limestone along their outcrops rather than at right angles to them. In other words, what Sedgwick had accomplished in 1832 for the Bala Limestone, he now attempted for some of the less conspicuous calcareous bands of North Wales. By tracing fossiliferous limestones from the supposed belt of Caradoc strata into the so-called Cambrian beds below, he showed that these two groups—although differing in strike—were in fact parts of a single continuous deposit. The principal break in North Wales now moved to the

[55] Bowman to Sedgwick, 25 Aug. 1840, 4 Sept. 1840, 4 Oct. 1841, 15 Oct. 1841, 27 Oct. 1841, CUL: Add. mss 7652ID51, 52, 100, 101, 130b.

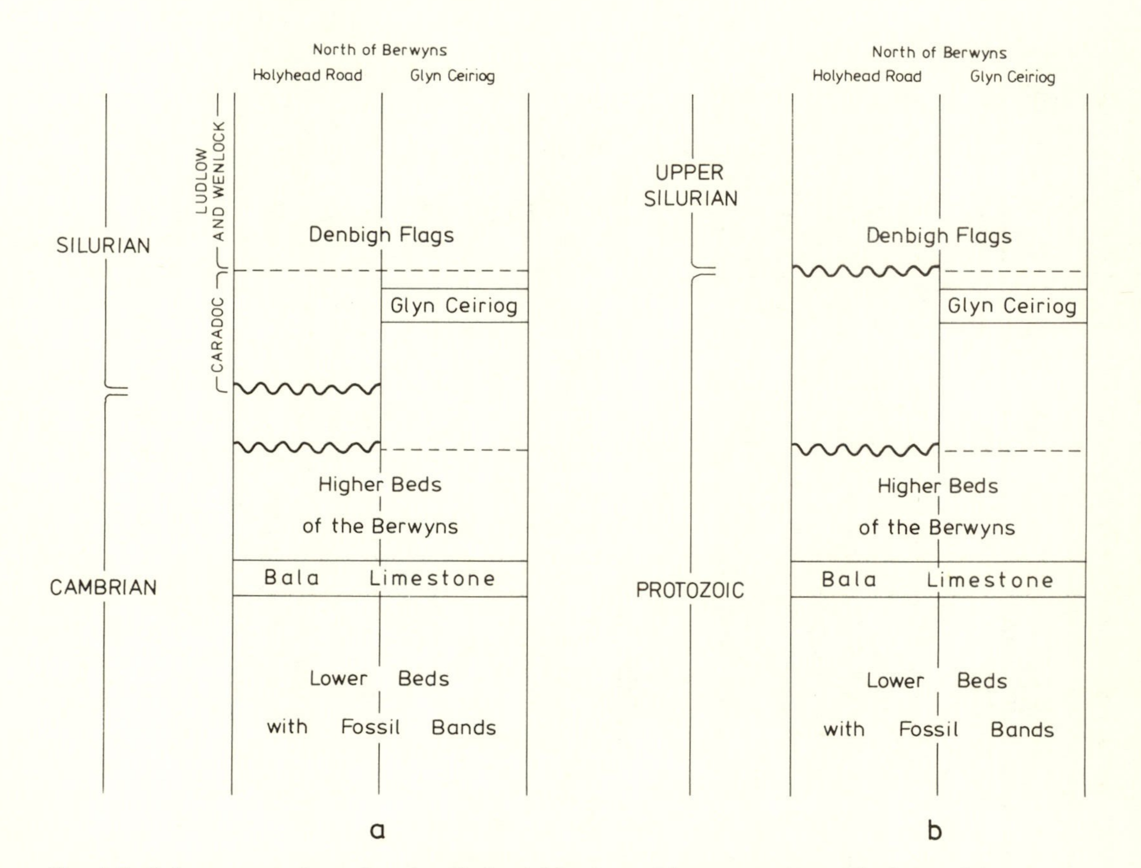

Fig. 4.7. Columnar sections showing Sedgwick's view of the succession at the base of the Denbigh Flags in (*a*) 1834 and (*b*) June 1843.

base of the Upper Silurian Denbigh Flags (Fig. 4.7, column b), and a boundary that had once seemed perfectly natural faded into thin air.[56]

During the remaining months of 1842 Sedgwick puzzled over the structural relationships of the rest of Wales and began to suspect that this gap at the base of the Upper Silurian would be extended elsewhere. On the strength of this evidence, Caradoc, Llandeilo, and the two halves of the Cambrian belonged in one great division below the Upper Silurian as a single group in both a physical and a palaeontological sense. The original Lower Silurian formations of Murchison, in other words, were simply several hundred feet of fossiliferous strata at the top of a descending series several miles thick. "Does it not look in many respects," Salter asked Sedgwick, "as if Murchison had only the thin & refuse edges of your great system and that Nature to make up for giving him so little batter had loaded it with plums—&c—I confess it seems so to me. . . ."[57] From this perspective, the Llandeilo and Caradoc formations had only skimmed the highly fossiliferous surface of a much larger stratigraphical unit.

Sedgwick believed more than ever after his 1842 excursion that a new name was needed to express the intimate union between the former Cambrian and Lower Silurian, and their separation from the overlying Upper Silurian. In a paper on North Wales delivered to the Geological Society in June 1843, he accordingly emphasized the presence of a single fossiliferous unit below the Upper Silurian and offered "Protozoic" as a temporary name until such time as it could be further subdivided.[58] A geological map of North Wales, redrawn here as Figure 4.8, was displayed in the rooms of the Geological Society. Sedgwick had to apologize for his inability to further subdivide his Protozoic on it, as virtually the entire Principality was painted a uniform shade. By comparison the second (1840) edition of Greenough's map, to which he had contributed substantially, must have seemed far more accurate and useful. For the present, however, he wished to keep a single broad grouping, partly to indicate the uncertain structural relations of the Protozoic subdivisions, and partly because he had not yet achieved success with his

[56] Sedgwick to Phillips, 21 Nov. [1842], ff. 29-44, UMO; Sedgwick to Murchison, [dated 1 Nov. 1842 by Murchison], M/S11/203a & b.

[57] Salter to Sedgwick, [1842-1843], CUL: Add. ms 7652ID161e.

[58] Sedgwick 1843b: 221. The earliest use of "Protozoic" in this sense that I have found occurs in Sedgwick to Phillips, 21 Nov. [1842], ff. 29-44, UMO.

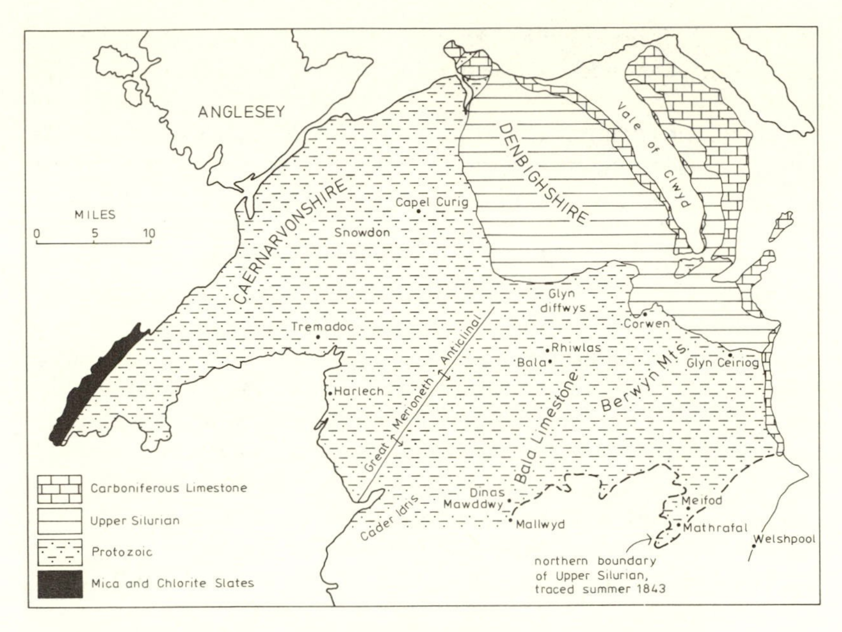

Fig. 4.8. Geological map illustrating Sedgwick's conception of North Wales in 1843. Igneous rocks and most structural information omitted.

progressionist scheme of the 1841 "Supplement."[59] Sedgwick also offered "Protozoic" as a compromise with Murchison; rather than defending some revised form of the Cambrian, he started from scratch with a provisional usage for the entire series. In Phillips's words, he seemed "disposed to make Cambrius and Siluria join in holy matrimony, Siluria giving up *half* her dowry & Cambrius, resigning *all*."[60]

As a term of compromise "Protozoic" held a certain advantage in having been employed previously by both Sedgwick and Murchison, although each had used it in a sense different from that proposed in 1843. Sedgwick had suggested in the original "Synopsis" that the chlorite slates of Anglesey could be called the "Protozoic system" if fossils were ever found in them, while in the *Silurian System* Murchison had proposed that the Cambrian and Silurian sys-

[59] Sedgwick, "On the Physical Structure of the Berwyn Chain . . . ," unpublished ms, GSL; read 29 May 1844.

[60] Phillips to Sedgwick, 29 Nov. 1842, CUL: Add. ms 7652ID152c. For the later statement, see Sedgwick 1855: l.

tems could be considered together as the "Protozoic Rocks." In both cases "Protozoic" suggested "first life," for the two men agreed that these ancient slaty rocks preserved what Murchison referred to as "undeniable proofs of a beginning," obscure evidences of the origin of life on earth. Sedgwick's new use of "Protozoic" thus reemphasized a common ground in this important theoretical point.[61]

On another level, Sedgwick's proposal of "Protozoic" reflects his deep concern about the correct use of the word "system." In particular, he wished to employ this crucial classificatory term in a sense consistent with its meaning in the 1830s, when it had involved not only palaeontological but also structural and stratigraphical considerations. The centrality of the meaning of "system" for Sedgwick becomes evident when the Protozoic subdivision is viewed in the context of his overall classification for the older rocks. In the June 1843 paper, it was one of four major "groups" among these strata, which he arranged according to a scheme modified from the 1841 "Supplement":

Carboniferous	Palaeozoic System
Devonian	
Upper Silurian	
"Great Protozoic Group"	

Throughout his paper Sedgwick stressed the difficulties in distinguishing these four groups from one another. Frequently one division merged conformably into its neighbor, presenting an unbroken transition that made boundary drawing almost impossible. Likewise fossil species often characterized more than one group and added a palaeontological dimension to the problem. As Sedgwick had emphasized in advocating Dumont's classification, in many regions the Upper Silurian and Devonian could be separated only with great difficulty. If Murchison retained these two groups as separate systems, then consistency demanded that Upper and Lower Silurian be so divided as well; for in 1843 the break between them seemed the most conspicuous in the entire series. "If Devonianism makes a system," agreed Phillips, "so does Cambrianism."[62]

[61] Sedgwick 1838: 684, Murchison 1839: 11; for "undeniable proofs," see Murchison et al. 1845, *1*: 1.

[62] Phillips to Sedgwick, 12 Nov. 1844, CUL: Add. ms 7652IE116.

For both Sedgwick and Phillips, however, the Cambrian, Silurian, Devonian, and Carboniferous no longer deserved the status of individual systems. After the Devonian reclassification there were simply too many passages between them to warrant any sharp divisions. Sedgwick, despite his later assertions to the contrary, had been perfectly willing to call these units "systems" before 1839, but after that date he felt that the word could be applied only in a more comprehensive sense. With the ever-increasing evidence for transitions, he believed that the only unequivocal breaks in the entire geological record came at the top of the Carboniferous and the bottom of the Tertiary. These breaks delineated three great "systems" in the history of life: the Palaeozoic, the Secondary, and the Tertiary. As we have seen, Phillips—who from his own perspective shared much of Sedgwick's interest in constructing a philosophically consistent classification—had recently christened three similar divisions from his palaeontological standpoint with the more familiar designations Palaeozoic, Mesozoic, and Kainozoic. According to such a threefold scheme the Palaeozoic concluded with the wholesale extinction of productids, trilobites,and orthoceratites. Since the original "systems" of the collaborative classification had proved to be less distinctive than had originally been supposed, Sedgwick wished to reserve this crucial term for those groups that actually constituted *systema naturae* in a comprehensive Linnaean sense.[63]

By the early 1840s, then, the breakdown of the palaeontological boundary between Cambria and Siluria had led to disagreement on two fundamental issues. First, were systems of British geological nomenclature well enough established to be spread across the world by fossils alone, or should local work in the British type areas continue under temporary designations (e.g. Protozoic) until a new classification could be developed? And second, how could the original meaning of the term "system" best be maintained—by using it for smaller divisions like Silurian and Devonian, or by reserving it for broader groups like the Palaeozoic?

The boundary separating the Cambrian and the Silurian systems was fully breached after the 1842 field season. As Murchison, Sedgwick, Phillips, Lyell, and their contemporaries reacted to the loss of the Cambrian fauna, two independent investigations questioned the very existence of the great thickness of rocks below the Lower

[63] Sedgwick 1843b: 223-224; 1847: 160-161; Phillips 1841: 155-182. The two schemes differed in one important respect: Phillips included the Magnesian Limestone within the Palaeozoic, Sedgwick did not.

Silurian that had played so large a part in all of Sedgwick's successive classifications. Other researchers had begun to tear away the structural foundations of the collaborative classification, suggesting that most of Wales was stratigraphically as well as palaeontologically equivalent to the original Silurian formations.

CHAPTER FIVE

Restructuring Wales

AFTER the resounding success of his Russian campaign, Murchison no longer concerned himself with what he saw as minor squabbles over the structure of Wales or the Lake District. Like most other geologists, he believed that the persistence of the palaeontological types of the Lower Silurian to the base of the fossiliferous rocks had settled the basic issues of classification. For Sedgwick, on the other hand, investigations in these areas remained critically important, for alternatives to the enlarged Silurian could only be elaborated within the type areas of the British Isles. According to his sections, a vast thickness of strata underlay the lowest beds of the original Lower Silurian in Wales. The preliminary dovetailing of 1832, the "Cambrian system" of 1835, the "Cambrian groups" of 1841, and the "Protozoic" of June 1843 had all been proposed in recognition of the immensity of this ancient series. But during the 1840s this supposedly secure basis for classification became a matter for active and often acrimonious debate. In order to defend the structural underpinnings of his views, Sedgwick clashed directly with other students of the older rocks in Britain: in South Wales the official Survey, and in the north a newcomer to Lower Palaeozoic geology, Daniel Sharpe. The incipient debate with Murchison involved the interpretation of palaeontological facts accepted by all; the concurrent dispute over the structure of Wales provides many examples of disagreement about the relevance of observations made by conflicting parties in a controversy.

A SURVEY IN THE SOUTH

After the completion of Sedgwick and Murchison's fieldwork, the southern limits of their collaborative boundary were not examined by any other geologists until 1841. Whatever the situation with regard to fossils might be, the importance of the boundary as an indicator of a particular geological horizon—the lower limit of the Silurian system—remained unquestioned. Although Sedgwick had found anomalous dips during his traverses in 1832, he shared Murchison's view that most of the slates north of the boundary belonged in the Cambrian. This compromise, while corresponding with available information from the field, also rested on the basic unfamiliarity of the sections, for after a few rapid traverses several different boundaries could have been chosen without any undue twisting of the evidence.

In the spring of 1841 the official Geological Survey of Great Britain completed mapping the South Wales coal basin and began advancing section lines into the critical areas for the Cambrian-Silurian boundary. From its origins in the early 1830s as a one-man operation, the Survey had by this time grown into a spirited team of seven or eight geologists under the direction of the founder, Henry De la Beche, who had been Murchison and Sedgwick's principal opponent in the Devonian controversy. Using many of the traditional techniques of the stratigraphical enterprise and by 1841 more detailed and instrumentally precise standards of accuracy, the forty-six-year-old De la Beche and his young assistants were coloring geologically the one-inch-to-the-mile maps of the Ordnance Survey. Besides John Phillips, who provided palaeontological expertise, the Survey corps included W. Talbot Aveline, Trevor E. James, David H. Williams, Josiah Reese, and Andrew C. Ramsay, a recent recruit of particular promise and a figure of great importance for the present controversy.[1] Before joining the Survey at the age of twenty-seven, Ramsay had pursued an unsuccessful career in the Glasgow calico business and had geologized during his holidays as a pleasant avocation. A paper on Arran read at the 1840 meeting of the British Association met with a favorable reception from Murchison, who asked the aspiring beginner to accompany him on his first Russian tour. This plan fell through, however, and he arranged instead for Ramsay's employment on the Survey. "Little did I dream," rhapsodized the grateful Ramsay after his first season in Wales, "that the meeting of the British association at Glasgow would be productive of such happy results to me. Formerly I *existed* now I *live*. I shall not tire you by again expressing how much I am indebted to you for bringing about this agreeable change."[2]

Ramsay soon rewarded his benefactor with an important discovery, the single most consequential finding for the structure of Wales since the unravelling of the passage beds below the Old Red Sand-

[1] The early history of the Survey is recounted in Bailey 1952, Flett 1937, McCartney 1977, North 1932, North 1934, and Secord 1986. Geikie 1895 has further information on this subject and biographical details about Ramsay.

[2] Ramsay to Murchison, 11 Nov. 1841, GSL: M/R1/1; Geikie 1895: 15-33.

Fig. 5.1. Survey geologist in Carmarthenshire with hammer and collecting bag. The government geologists originally wore Ordnance uniforms, but after administrative control passed to the Department of Woods and Forests in 1845, they were allowed to don the more traditional gentlemanly garb of top hat and coat.

stone during the previous decade. In the spring of 1842, at the beginning of his second year with the Survey, he was hammering along the supposed northern limits of the Silurian, where it was thought to occupy a narrow zone just below the Old Red Sandstone. By following his traverses carefully to the north and spending months, not just days, in the district, Ramsay concluded that this lower limit for the Silurian formations was completely chimerical. Murchison's *Silurian System* sections had shown the strata near Llandovery at Noeth Grüg dipping to the south, just as expected on the basis of the joint dovetailing. But Ramsay showed that these strata then arched over to the north and were afterwards repeated in a complex series of undulating folds all the way to Cardigan Bay as illustrated in Figure 5.2. The anomalously dipping black slates glimpsed by Sedgwick in 1832 now served as the foundation for a fundamental reinterpretation of Welsh geology: Murchison's Lower Silurian formations, transformed by slaty cleavage, spread towards the northwest to cover nearly all of South Wales. Ramsay, clearly pleased to contribute to the geographical extent and general importance of the Silurian, summarized his discovery to Murchison in April 1842:

> Now however I have gradually gone over the whole of the ci devant Cambrians between St Davids [Pembrokeshire] & Llandovery, & skirted along your old northern boundary, & I can clearly show . . . that all *your* rocks under a somewhat different form, spread over the surface of the land at least as far as Cardigan. It has been my good fortune to have had a great deal to do with this grand extension of your System & glad am I to have had it in my power to do so by adhering firmly to your advice "continue to work steadily in the same line."[3]

Ramsay may have been following his patron's advice, but as an aggressively ambitious member of an official survey charged with the complete coloring of the Ordnance maps, he proceeded with a thoroughness unmatched even by Murchison. A discovery of this magnitude, while aiding the Silurian, would certainly do Ramsay's own reputation no harm.

Within two months Ramsay and his fellow geological surveyors added palaeontological support to their sectional proof that South Wales was Lower Silurian by finding characteristic Caradoc and Llandeilo fossils deep within the supposed Cambrian, far to the north of the old collaborative boundary. As Murchison learned in a letter from De la Beche at the end of July: "It goes on in great rolls, and *no mistake* a long way beyond the Caermarthen ordnance map sheet.—No want of fossils—in fact organics and sections all going to prove the same thing. . . . It would be a *long story* to go further into the *old story* hereabouts—that your Silurian System must have a jolly extension, at our hands, over the rocks of this land seems certain."[4]

[3] Ramsay to Murchison, 7 Apr. 1842, GSL: M/R1/2.

[4] De la Beche to Murchison, 31 July 1842, EUL: Gen. 1425/389; also De la Beche to Ramsay, 10 June 1842, ICL(R).

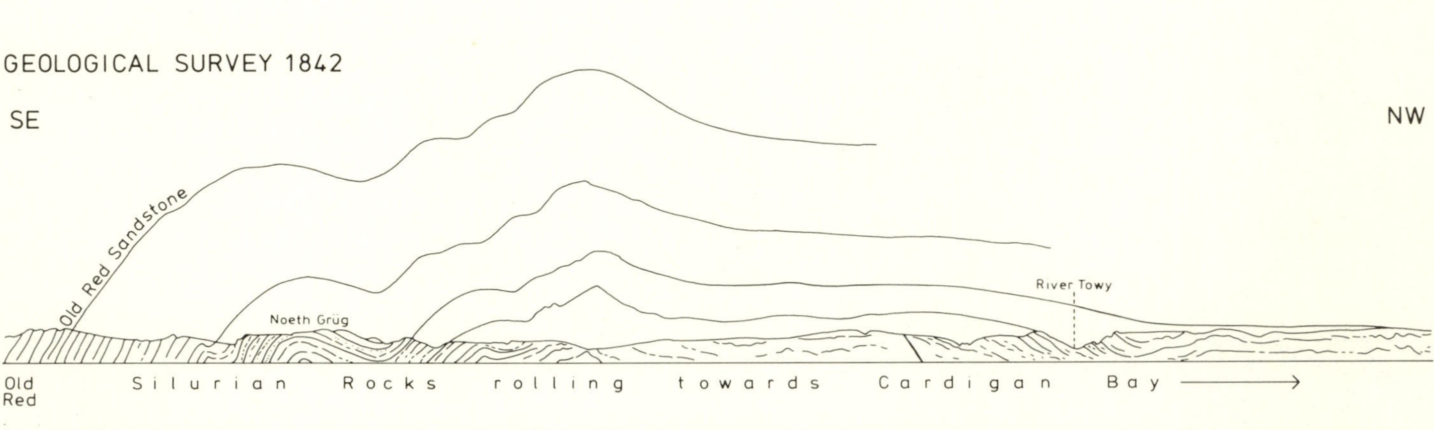

Fig. 5.2. Murchison's traverse section from the Old Red Sandstone into older Cambrian strata, contrasted with the Survey section (here drawn on the same scale) indicating a great "roly-poly" across the same region.

One would expect Ramsay to welcome such "jolly extensions" to Murchison's system; that De la Beche also did so may seem rather surprising. After all, the term "Silurian" originated during the Devonian controversy, a dispute that had nearly deprived him of his livelihood as a government scientist. In his official *Report on the Geology of Cornwall, Devon, and West Somerset* of 1839, De la Beche had understandably preferred the noncommital name "Grauwacke group" for the strata that Murchison and Sedgwick had called "Silurian" and "Cambrian." These latter terms, he argued, were perfectly acceptable as provisional names for a series of formations in a limited area, a single country for example. But to imply that they were potentially applicable everywhere else raised the old spectre of universal formations, making one small district the pattern for the whole world.[5] Like Phillips, who studied the Survey's fossils from a similar perspective, De la Beche stressed the importance of environmental factors in determining the sediments deposited in specific localities. He used minutely detailed studies of local successions to reconstruct ancient environments, hoping to demonstrate that modern causes failed as explanations for all past geological phenomena, and thus attempting (in part) to beat Lyell at his own game. From such a particularistic point of view, Murchison's imperialist stratigraphy and Lyell's strict uniformity seemed equally objectionable and overgeneralized. It was, as Phillips wrote in 1842, "a poor & low philosophy, which . . . makes a Saxon Mountain the world, an English oolitic hill the measure of a cycle of time, a modern earthquake the standard of ancient revolutions."[6] For all his agreement with these sentiments, however, De la Beche had little to gain and everything to lose from continued opposition to the Silurian. When the opposing parties in the Devonian controversy moved towards reconciliation in the early 1840s, his oft-stated willingness to accept the local validity of the Murchisonian classification conveniently paid off. Since the Survey was just entering the very region in which the system had been defined, De la Beche could consistently maintain his own principles of classification without continuing to alienate his former enemy, whose support in government circles was absolutely essential for the Survey's continued expansion. No one, not even Phillips or De la Beche, could object to "Silurian" rocks linked so directly by sections to Siluria itself.

[5] De la Beche 1839: 37-41.

[6] Phillips to De la Beche, 23 May 1842, NMW. Also De la Beche 1846, esp. pp. 1-20; De la Beche to Ramsay, 28 May 1843, and other letters in ICL(R). For the anti-Lyellianism, see McCartney 1977, Rudwick 1975, and Secord 1986.

Perhaps more than any other single factor, their decision to introduce Silurian nomenclature into all official maps sealed its widespread adoption among British geologists, because it thereby obtained sanction from public documents.

Ramsay's discovery of the great "roly-poly," as De la Beche came to call it, received the immediate approval of the geological community, indicating that the Survey methods were already beginning to achieve a thoroughness difficult for individuals to challenge outright. Even Sedgwick, who doubted that the extensions of the Lower Silurian would proceed so far north as the Survey believed, fully agreed that the original Cambrian-Silurian boundary in South Wales must be abandoned. Murchison, who first publicly announced the Survey discovery in his 1843 presidential address, welcomed Ramsay's finding as further proof of the importance of his system.[7] The confidence with which he extended the Silurian, both in that address and in his little map for the Society for the Diffusion of Useful Knowledge, owed a great deal to the Survey findings—even if his own expansions required only evidence drawn from fossils. After 1842 the penetration of Silurian strata past the collaborative boundary was taken as a starting point for any picture of the geology of South Wales, however much one might dispute the precise extent of the phenomenon.

A Newcomer in the North

While the Geological Survey threatened the southern boundary of Sedgwick's territory in Wales, a geological interloper named Daniel Sharpe was hammering in the very heart of Cambria. Son of a brewer, nephew of the poet Samuel Rogers, Sharpe was orphaned in 1806 before his first birthday and raised by a relative. Eventually he became a shipping merchant in partnership with his elder brother, which led to his spending the late 1830s in Portugal on business. In his spare time he pursued studies of Iberian geology, and after his return to London read several papers on the subject to the Geological Society.[8] Casting about for a fresh topic, he soon learned that the unsettled relations between the Cambrian and Silurian systems were ripe for independent investigation. At the same time that the Survey was demonstrating that the South Wales

[7] Murchison 1843a: 75-77. For the "roly-poly," see De la Beche to Ramsay, 23 Aug. 1842, ICL(R).

[8] For Sharpe's life, see Portlock 1857: xlv-lxiv, and *DNB*.

slates were parallel with the Llandeilo and Caradoc, Sharpe hoped to show that all the fossiliferous strata of North Wales were merely expanded equivalents of various portions of the Silurian system. Unlike the Survey's detailed mapping, however, his work could only point the way towards such a radically new interpretation. But it came at a critical juncture and held important consequences for the emerging controversy.

As the first step in "Silurianizing" North Wales, Sharpe entered the field intent on reassigning the Bala Limestone to the stratigraphical level of Murchison's Lower Silurian. Acting on a suspicion arising out of his earlier studies of the Lake District, he focused his Welsh tour of 1842 on two critical sections—one to the west of the Berwyns in the Bala region and the other to the south of the chain. Both (if accepted) would indicate that the Bala Limestone belonged at the top of the Lower Silurian. According to these sections the Berwyn strata were below the Bala Limestone rather than above where Sedgwick had always placed them. The Berwyns thus formed the type of a "Cambrian" utterly without fossils of any kind.[9] Sharpe quickly completed this initial stage in his plan to pack North Wales into the Silurian, and he prepared to announce his conclusions at a Geological Society meeting in November 1842.

For many years everyone familiar with the older rocks had known that the Bala fossils were virtually indistinguishable from those of Murchison's Lower Silurian, but the sectional equivalence claimed by Sharpe was completely unexpected and extremely controversial. As Phillips told Ramsay the day before Sharpe's announcement, the new paper showed "*not by fossils,* (*that we have all known for 10 years at least*) but by measures, sections, dips, &c. that the Bala Limestone is stratigraphically identical with the Caradoc."[10] The evidence for a grand sequence of fossiliferous strata below the Lower Silurian appeared increasingly questionable, for Sharpe's equation of the Bala beds and the Caradoc pointed in precisely the same direction as the Survey's more authoritative placement of the slates of South Wales in the Lower Silurian. It seemed entirely possible that the disappearance of the Cambrian fauna might be followed by a corresponding disappearance of any sub-Silurian fossiliferous rocks. The gauntlet was down. How would Sedgwick respond?

[9] Sharpe 1843. The earlier study of the Lake District is Sharpe 1842; see also Marshall 1840.

[10] Phillips to Ramsay, 29 Nov. 1842, ICL(R).

Defending the Protozoic Sections

Sedgwick first learned of the two-pronged threat to his sections from Murchison in October 1842. The Survey, Murchison reported, had found Lower Silurian strata rolling over to the north in great waves; similarly Sharpe's summer fieldwork purported to "show that the Bala limestone is nothing more than a calcareous course in the middle of the Caradoc sandstone." Although for his own purposes Murchison's only concern was whether Sedgwick had found a distinctive set of fossils in North Wales, his upcoming Geological Society presidential address also required comments on the sectional findings of Sharpe and the Survey. In his letter, Murchison parenthetically asked if Sedgwick still argued for a thick sequence of strata below the Llandeilo formation, or if he now agreed that both Silurian fossils *and* Silurian rocks persisted to the base of the fossiliferous column.[11]

By this date Sedgwick had no further hope of finding anything but Lower Silurian species in North Wales. He had even failed to define Cambrian "groups" through the theory of progression. However, he believed that his evidence for fossiliferous strata below the Silurian remained substantially intact: in his view North Wales held a thick succession of rocks, with the Snowdon beds only the lowest of several major fossil zones. Many thousands of feet above were successively higher limestones, particularly at Bala and (just below the Upper Silurian) at Glyn Ceiriog (Fig. 4.7*b*). The only important change in Sedgwick's picture of the structure of North Wales during his 1842 tour had been his elimination of the Caradoc strata under the Denbigh Flags. But even after this adjustment, the Glyn Ceiriog and Meifod beds continued to hold their high place in the succession, and the overall thickness of the rocks remained unchanged.[12]

After his receipt of Murchison's letter, Sedgwick naturally began to wonder if his upper beds would require repositioning. The Survey had created his greatest dilemma, for it was questioning one of the keys to the dovetailing of the previous decade—the relation of the Bala Limestone to the South Wales slates. Ever since 1832 he had paralleled these slates with the higher beds of the Berwyns and had

[11] Murchison to Sedgwick, 16 Oct. 1842, in Geikie 1875, *1*: 376-382; a copy is at CUL: Add. ms 7652IIID37, although the relevant portions are printed fully by Geikie.

[12] Immediately after returning from the field, he gave a clear exposition of his views in Sedgwick to Phillips, 28 Sept. [1842], ff. 22-28, UMO. It should be noted that before seeing the details of Sharpe's work, Sedgwick mistakenly thought their views were almost identical: see Sedgwick to Murchison, [?22 Nov. 1842], GSL: M/S11/205.

shown Bala plunging underneath them both; Murchison's formations were placed far above, on a different horizon altogether. Given these views, Sedgwick understandably found the Survey's shortening of the sequence extremely improbable. He wrote to Phillips for confirmation.

> All I presume you contend for is that as fossil systems they cannot be separated and that the S. Welsh slate mountains are expansions (in *descending order*) of those beds wh. are described by Murchison. Or do you hold, that Murchison's *very beds* are by undulations repeated over again and again towards the west, so as to be stranded on the Eastern flanks of the Cader Idris chain?—If you say yes to the latter hypothesis then I do not know that the Bala limestone is a day older than the lower beds of Horderly or May Hill [both considered typical representatives of the Caradoc at this time].

Phillips's answer was unequivocal: the South Wales slates were nothing other than palaeontological and stratigraphical repetitions of Murchison's strata. They might well roll on all the way to the base of Cader Idris, although the Survey sections did not yet extend that far. As he remarked, "it remains to be seen what is the series of older strata in Wales which you can rescue from the downrushing Caradoc."[13]

Only a fresh examination of the sections could fortify Sedgwick's defenses against these challenges, and so after proposing the Protozoic in June 1843 he left with Salter for another field excursion in Wales. As an attempt to counter the Survey, however, the "rescue" mission remained incomplete, for time ran out before they could travel very far south of the Berwyns. As a result Sedgwick refrained from discussing the South Wales slates in print until January 1846. Although he entertained opinions on the subject in private correspondence, all his publications in the intervening years drew a line due west from Welshpool to Mallwyd and left the region to the south as an unknown quantity.[14] After the summer of 1843 Sedgwick felt that his views on the relation between North and South Wales were simply too speculative for public display. Any debate with the Survey was thus delayed while he gathered ammunition from the field.

[13] Phillips to Sedgwick, 29 Nov. 1842, CUL: Add. ms 7652ID152c; for the previous quotation, Sedgwick to Phillips, 21 Nov. [1842], ff. 29-44, UMO.

[14] E.g. Sedgwick 1845a, with a hint of his views on South Wales at p. 8. An explicit statement can be found in Sedgwick to Phillips, 10 Oct. [1843], ff. 46-53, UMO.

In contrast with his failure to penetrate the mysteries of the South Wales slates, Sedgwick believed that his work in the north had been a complete success. From his return from the field in 1843 to the commencement of the third phase of the Cambrian-Silurian controversy in 1852, he fought against Sharpe's encroachments on his sections with skill and tenacity, opposing attempts to pack North Wales into the limited vertical range of the Silurian. His first response came at a Geological Society meeting in November 1843. The most comprehensive view of northeastern Wales he ever assembled, this paper presented not only new information gathered in the summer, but also details that had languished in his notebooks for over a decade.[15] The published versions featured sections and a colored map, significantly the first illustrations included with any of his essays on North Wales. The prospect of being forestalled by an outsider (not to mention the inevitable onward march of the Survey) was enough to goad even Sedgwick into presenting results in written form.

Read in isolation, the paper has always been seen as a purely descriptive account of the geology of North Wales. In reality the map and sections form a running dialogue with Sharpe, and the text itself argues implicitly (and occasionally explicitly) against that geologist's views in a muted tone appropriate to the publications of the Geological Society. Any contemporary expert would have recognized the paper for what it was, a skillful effort to quash an opponent by a striking display of superior knowledge.

Sedgwick's dispute with Sharpe hinged on what might seem a basic fact—the direction of dip of the Bala Limestone. This was not just a matter of a single measurement, however; it was a complicated issue. Despite its straightforward appearance on geological maps (and in my own account up to this point), the Bala Limestone could not be observed in the field as a continuous bed with a single obvious orientation. Rather, like most inland strata in Wales, it was visible only as a set of temporary quarries and discrete outcrops, assembled by the geologist according to a set of tacit rules and grouped under a single name. For a full understanding, one had to be taken over the ground by someone already familiar with the exposures. The notion of an identifiable "Bala Limestone" thus depended upon a host of prior assumptions and theoretical precepts. Many of these—the very existence of strata, for example—appear as obvious to us as they did to both Sharpe and Sedgwick; but as Roy

[15] Sedgwick 1844, 1845a.

Porter has shown, in earlier periods of earth science even this basic notion would not have seemed commonsensical.[16] On other issues, especially which outcrops to emphasize or include in determining dip, there still remained ample room for disagreement.

Sharpe argued that the Bala Limestone dipped *west*. The most important section in his paper of November 1842 showed it dipping in that direction and then reappearing near the northwestern side of Bala Lake at the tiny village of Rhiwlas (Fig. 5.3*a*). With this pair of limestone outcrops thus connected by a trough-like syncline, the strata of the Berwyns suddenly ceased to intervene between the Upper Silurian and the Bala Limestone. In consequence the latter automatically moved onto the horizon of the Caradoc formation, perfectly in accordance with Sharpe's overall scheme.[17] Sedgwick, on the other hand, was equally adamant that the Bala Limestone dipped *east*. His interpretation of the region (Fig. 5.3*b*) was already familiar from the 1841 "Supplement," which spoke of a series of strata dipping eastward, with younger beds in the east (notably the Bala Limestone) followed by successively older ones to the west (such as the limestone at Rhiwlas). In his reply Sedgwick suggested that Sharpe's section crossed a minor repetition of the Bala Limestone with an anomalous westward dip; his own traverses showed the main body of the stratum plunging in the opposite direction and under the Berwyns. As for the Rhiwlas Limestone, he believed that Sharpe had confused it with yet another calcareous zone at Glyn Diffwys. The Glyn Diffwys beds were true equivalents of Bala, but in Sedgwick's view Rhiwlas most definitely was not, and he presented further sections to substantiate this important stratigraphical separation.[18] In this way he tried to dispose of the claim that all the limestone bands on the western side of Bala Lake were of the same age, nipping in the bud Sharpe's audacious attempt to pack the rest of Wales into the Silurian. At the same time, it is important to note that his own previously simple picture of the Bala succession was becoming more complex.

Sedgwick's reaction to Sharpe's revisions elsewhere in Wales further illustrates the ambiguities involved in ascertaining apparently straightforward facts about the strata. At the southern end of the Berwyns, for example, Sharpe had shown the Bala Limestone passing directly into rocks of the Upper Silurian as indicated in Figure

[16] Porter 1977, esp. pp. 118-122, 176-183; also Greenough 1819: 1-90.

[17] Sharpe 1843: 12-13; for a report of the meeting, Salter to Sedgwick, 2 Dec. 1842, CUL: Add. ms 7652ID161a.

[18] Sedgwick 1845a: 5-11; for his earlier views, Sedgwick 1841: 551.

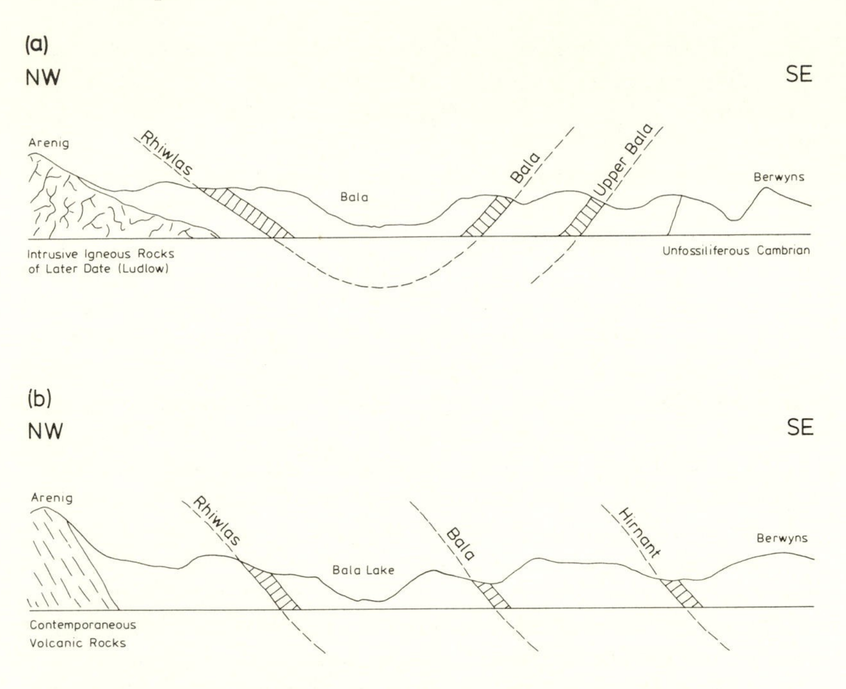

Fig. 5.3. Traverse sections across the Bala region comparing the views of (*a*) Sharpe and (*b*) Sedgwick.

5.4*a*. Although Sedgwick and Salter adopted Sharpe's novel extension of the Upper Silurian into that region, many miles west of the original boundary, they interpreted this junction with the underlying strata as an unconformity (Fig. 5.4b).[19] Separated by such a gap, the beds of Bala age remained safely out of the Caradoc's hungry grasp. Moreover, this new unconformity perfectly matched the one they had traced in 1842 at the base of the Denbigh Flags. If similar breaks below the Upper Silurian proved widespread, Sedgwick would have evidence not only against Sharpe, but also against the unified Silurian of Murchison himself.

One suspects that this search for stratigraphical breaks was of prime importance in leading Sedgwick to another of the principal recorrelations of the summer tour in 1843, a linkage between the Bala Limestone and its counterpart near Meifod. A Bala-Meifod cor-

[19] Sharpe 1843; Sedgwick 1845a: 11-12, which explicitly mentions this particular difference in interpretation.

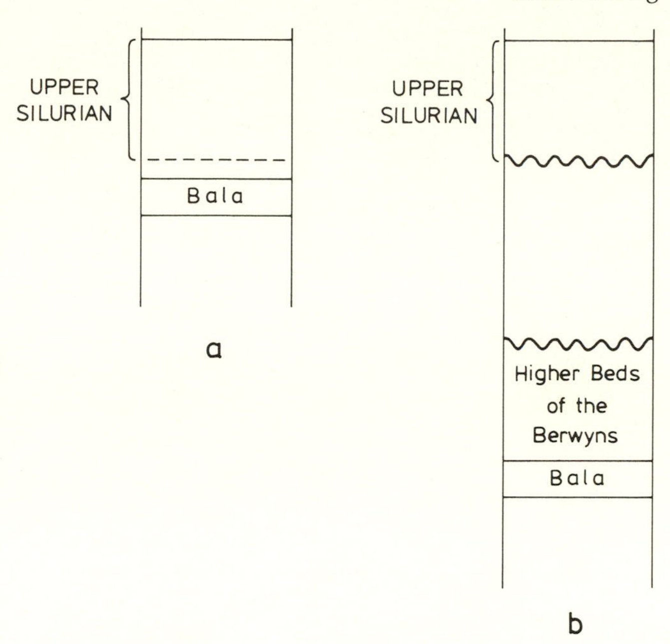

Fig. 5.4. Schematic columnar sections comparing (*a*) Sharpe's and (*b*) Sedgwick's views of the relation of the Bala Limestone to the Upper Silurian rocks near Mallwyd.

relation had potential advantages even on the 1834 dovetailing tour, but it had also seemed to demand the radical conflation of Sedgwick and Murchison's groups shown in Figure 5.5*a*. Consequently Meifod had been placed in the Caradoc formation, with Bala far below (Fig. 5.5*b*). But during the autumn of 1842 Sedgwick had developed a new solution to this problem. As he had told Murchison by letter, there was "one way of getting out of this cleft stick": rather than raising Bala to the level of Meifod, he would lower Meifod to the level of Bala and out of the Caradoc entirely.[20] The two limestones could thus be correlated without lessening the overall thickness of the sequence (Fig. 5.5*c*). Not mentioned in his letter, but apparent from Sedgwick's concerns at this time, is the fact that this hypothetical linkage would greatly extend the unconformity between

[20] Sedgwick to Murchison, 20 Oct. 1842, GSL: M/S11/198a & b.

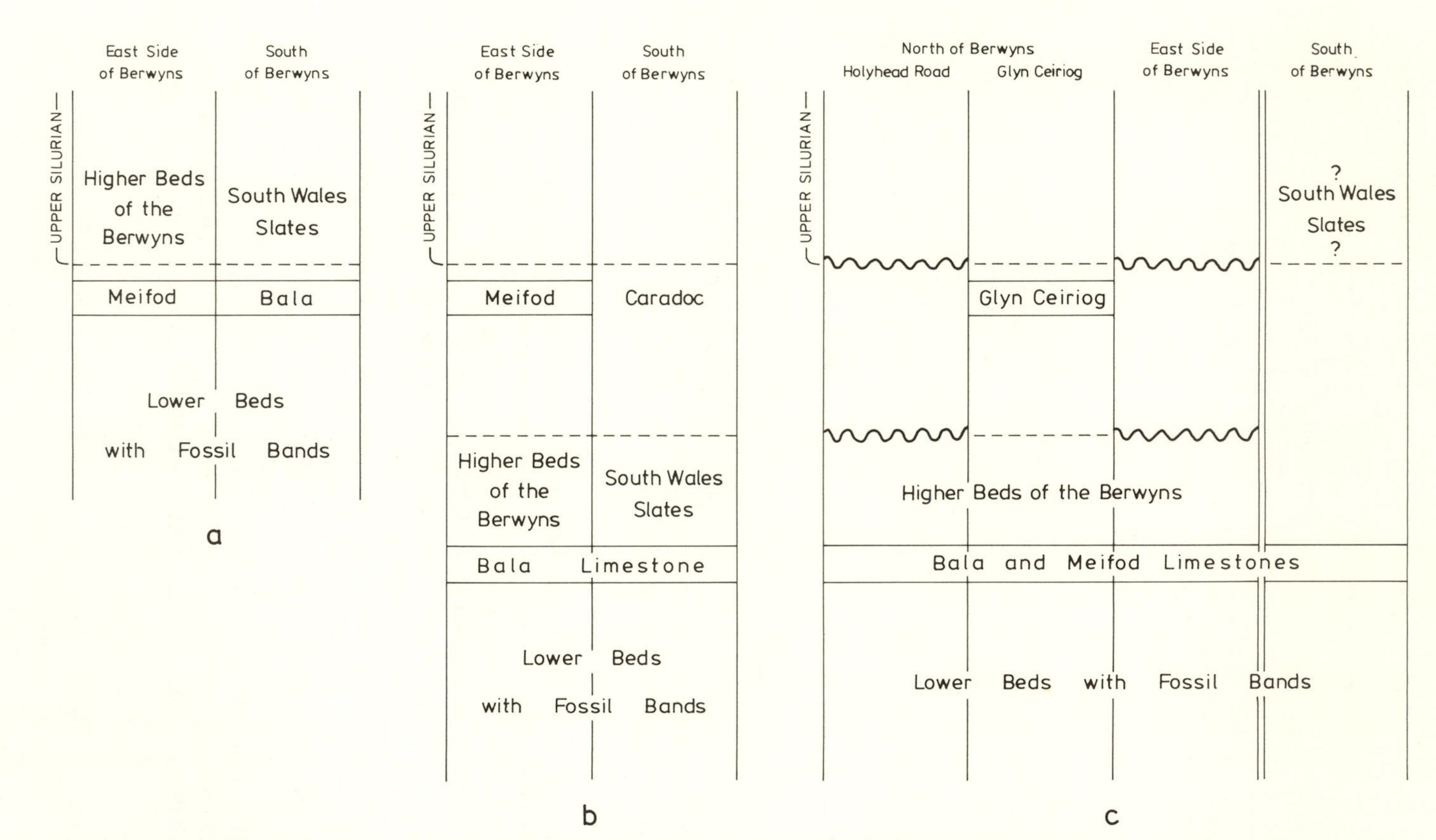

Fig. 5.5. Columnar sections illustrating three potential solutions to the problem of correlating the Bala and Meifod limestones.

Murchison's Upper and Lower Silurian. Under the circumstances Sedgwick entered Wales in 1843 only too willing to see stratigraphical gaps where Sharpe would have expected smooth conformities.

These details suggest how the data on which Sedgwick (or any other geologist) based a classification remained open to a great deal of reinterpretation and controversy. The field could be viewed as an almost inexhaustible armory of conflicting information, from which parties to a dispute could find ammunition suited to their particular needs. But the qualifying "almost" is essential. Even in a region as little explored geologically as North Wales then was, the rocks were not always amenable to the immediate expectations of scientific combatants. Although Sedgwick in the summer of 1843 confirmed the unconformity in the Meifod region necessary to his new scheme, he also found at least one instance near Mathrafal (close to Meifod) where a fossiliferous limestone did appear to pass directly into the Upper Silurian. This was just what Sharpe would have predicted. Fortunately Sedgwick could base his interpretation of this on the sequence at the northeastern end of the Berwyns, where the limestones at Glyn Ceiriog also passed directly into the Upper Silurian. According to Salter, the fossils in both of these calcareous passage beds conveniently matched their high position and differed from those at Bala.[21] Thus even if traverses in the field occasionally threw up anomalies, these could usually be harmonized with an immense quantity of information gathered over many years. Indeed, Sedgwick believed that Sharpe erred above all in relying too much on one or two sections considered in isolation. Most contemporaries would have agreed.

After his fieldwork in 1843 Sedgwick thought that the Protozoic was more secure than ever, and he continued using the term, notably in his November paper. Although particular correlations of the 1834 dovetailing were now abandoned, the aggregate thickness of the sequence in North Wales remained precisely what it had been. From Sedgwick's point of view his "Protozoic" represented by far the best solution to the naming of this huge series. The increasingly widespread unconformity at the base of the Upper Silurian also indicated the need for a name other than Murchison's expanded Silurian. In the longer term, this classification promised to facilitate comparisons between the chief physical groups in the Lake District and Wales, and his brief trip to Westmorland at the end of the 1843 tour was only one of several efforts to correlate the rocks of the two regions.[22]

[21] Sedgwick 1845a, esp. Tables I-III.

[22] Sedgwick 1845d, 1846.

In the context of Sedgwick's entire career, the attempted packing of North Wales into the Silurian obviously required immediate refutation, for it cut directly at the heart of his whole conception of the older rocks. After the initial failure of his progressionist scheme in the summer of 1842, Sedgwick had let the fossil barriers between Cambria and Siluria fall without complaint. But on the fundamental evidence of sections he felt sure of his ground and tenaciously defended his descending series of strata in North Wales from the onslaughts of Sharpe. He had willingly abandoned his nomenclature, his hopes for a peculiar Cambrian fauna, and even his correlations between individual strata; but establishing the sequence of the older rocks was the greatest achievement of his career in research and could not be abandoned without a fight.

The Sedgwick-Sharpe dispute illustrates the interpretive process involved even in the neutral descriptive statements of a field science. But the difficulty in determining the dips of the Bala Limestone should not obscure another and ultimately more consequential aspect of this episode: in terms of the wider stratigraphical enterprise, the arguments had brought new information about North Wales into print and into the public arena of the Geological Society. Sedgwick, sparked by a competitor and the encroaching Survey, had published his first substantial papers on the Principality, thereby making available findings that would otherwise have been confined to his notebooks. More important, both he and Sharpe returned again and again to the field during their dispute, each time bringing new evidence to light that not only supported their own interpretations but also added to the generally accepted picture of Welsh geology. Certain points remained controversial, but others did not: Sharpe's extension of the Upper Silurian to the southern end of the Berwyns, for example, and Sedgwick's conformable passage at Mathrafal. Even in the Bala region several new limestone zones were described and located, even if agreement on their stratigraphical position seemed impossible to achieve. In spite of its lack of resolution, the dispute between Sedgwick and Sharpe thus proved extremely fruitful. In many respects it was the chief impetus to discovery in the older rocks of Britain during the early 1840s.

Trespass and Territory

In defending his Protozoic sections Sedgwick was by implication also reasserting his authority as the sole qualified interpreter of the geology of North Wales. Just as Murchison had redefined his scientific property in palaeontological terms after the collapse of the collaborative classification, Sedgwick had redefined his own territory on a largely structural and sectional basis. But while Murchison's fossils made the world his potential domain, Sedgwick's concern with geological "physics" focused his attention ever more closely on the crucial geographical regions that he had studied for so long. In consequence his defense of the sections and his desire to maintain his research areas inviolate against intruders continued to be as closely linked as they had been from the beginning. Without the independent testimony of a substantial published work comparable to Murchison's *Silurian System*, Sedgwick could do only two things to maintain his priority: he could overwhelm opponents with new discoveries and his superior command of the subject, as shown in his papers against Sharpe, or he could attempt more directly to discourage scientific trespassers.

Sharpe was not the first outsider to hammer in North Wales. The potential for conflict over territory had already become evident several years earlier during an exchange over John Bowman's holiday excursions. Sedgwick, while reading the introductory chapter of the *Silurian System* in manuscript in the spring of 1838, had found an inserted page announcing that Bowman had found Upper Silurian rocks in Denbighshire. He was furious, and understandably so. Although unfamiliar with the sections, Sedgwick could accept the finding; but for Murchison to have acquired information on the sly from other investigators seemed a gross violation of their confidence. Murchison's express intention of visiting North Wales himself only made matters worse. "On putting these things together," Murchison later told Whewell, "Sedgwick took fire—and I grant that he had reason to do so, *if* as he supposed I was endeavouring to supplant him or to rob him of one jot of his well earned, though (*as I have* always said) ill requited labours." In response to a letter couched even by Sedgwick's own standards "in very violent language," Murchison sent a fulsome apology, cancelled the offending note, and explained that he had nothing to do with the unauthorized entry.[23]

The difficulty arose in this instance through inevitable discrepancies between the collaborative boundary defined stratigraphically

[23] Murchison to Whewell, 16 Apr. 1838, TC: Add. ms a.209[103]. For "very violent language," see Sedgwick to Murchison, [pmk. 27 Mar. 1838], CUL: Add. ms 7652IC13. Also valuable are Murchison to Sedgwick, 7 Mar. 1838, 19 Apr. 1838, CUL: Add. mss 7652IIID15, 16 (copies).

on the one hand, and as a limit to two separate field territories on the other (Fig. 3.9). Murchison and Sedgwick had labored to make these definitions mutually supportive, but this was not always possible. Thus when Bowman met with Upper Silurian strata along the Denbigh coast in 1837, he wrote to Murchison as the acknowledged authority on rocks of this age, rather than to Sedgwick, who actually held responsibility for the area in question. Sedgwick subsequently accepted Murchison's apologies for failing to inform him of the letter from Bowman and made his own for the pungency of his original reply. But at the same time, he had been put on his guard, and he asked his friend to alter portions of already printed pages of the *Silurian System* dealing with the all-important dovetailing tour. The new version would show that Murchison had not followed Silurian rocks through their full course in Wales because certain districts were reserved for Sedgwick; it would serve equally as a kind of private contract, to "remove all possibility of mistake between ourselves." Sedgwick also refused to review any further proof sheets, for he wanted to forget the entire affair.[24] But just two weeks later, a second flare-up made his continued sensitivity to Murchison's lack of respect for established boundaries only too clear. "Having skimmed to your heart's content the whole region S of the Severn," he bitterly declaimed, "is but a poor reason for beating up for recruits in order to make an onslaught on me in Denbighshire." With matters threatening to take an ugly turn, Whewell intervened as a mutual friend to effect a reconciliation, although he admitted that "Sedgwick is certainly still a good deal troubled about what he conceives [to be] interference with his ground."[25] Whewell's old fear that Sedgwick and Murchison might turn their hammers from the rocks to one another over the boundary had come perilously close to fulfilment.

Throughout this early episode, of course, Sedgwick worried about Bowman only as an agent of Murchison; from the perspective of the elite, a provincial of such lowly scientific status could never have presented a direct threat. Like most zealously deferential "cultivators of science," Bowman placed his results entirely at the disposal of the famous geologists and avoided becoming embroiled in their disputes. As Murchison wrote in explaining away yet another incident involving his encouragement of Bowman's hammering in

[24] Sedgwick to Murchison, [pmk. 29 Mar. 1838], CUL: Add. ms 7652IC15.

[25] Whewell to Murchison, 17 Apr. 1838, GSL: M/W4/21. Sedgwick's outburst (here with italics removed) is quoted in Murchison to Whewell, 16 Apr. 1838, TC: Add. ms a.209[103]. The original letter was presumably destroyed.

North Wales, "Mr B is a man of detail & a minute philosopher with whose services you will in the end be satisfied, whenever you gather in your disjecta membra."[26] Although Sedgwick must have found such smooth disavowals somewhat unsatisfactory, he took the hint, and (as we have already seen) plans for a joint tour in 1842 were ended only by Bowman's untimely death.

Even if Bowman, for all the significance of his researches, could be condescendingly dismissed as a provincial fact-collector, Sharpe most definitely could not. Although still outside the London scientific elite in the early 1840s, he had published important papers during his residence abroad and had been a Fellow of the Geological Society since 1827. In his view Sedgwick's claim over the older rocks had waned with the passage of time. Brash, sarcastic, and self-assured, Sharpe chafed under the unwritten ban on entering another man's field of research without permission. The vehemence of Sedgwick's response to his incursions into Wales and the Lake District should already be evident, and contemporaries fully acknowledged the legitimacy of this concern for scientific territory. Francis B. Danby, fossil collector and informant in Kendal, was especially angry because Sharpe's publications on Westmorland had reduced the potential importance of his own contribution to geology, which he had long reserved for Sedgwick's book. "It annoys *me* much," he told Sedgwick, "that every year somebody comes, lights his rushlight at yr lantern, & works out a patch of our Silurian district."[27] As this comment suggests, Sharpe's unseemly behavior threatened a carefully balanced system of paternalistic exchange between provincial amateurs and Geological Society experts. Puns on the intruder's name ("the 'Sharp' fellow," "a sharp move," "too sharp for us") abound in the correspondence of the period. North Wales in particular, as Murchison pointedly if somewhat hypocritically said in his 1843 address, was "the very tract in which Professor Sedgwick has so long worked."[28] Although most geologists followed

[26] Murchison to Sedgwick, 26 Sept. 1840, CUL: Add. ms 7652IIID48 (copy), passage omitted in Geikie 1875, *1*: 306-308. See Sedgwick to Murchison, 13 Sept. 1840, GSL: M/S11/174 for further details of this later incident. Also Bowman to Sedgwick, 4 Sept. 1840, CUL: Add. ms 7652ID52. For cultivators of science, see esp. Reingold 1976, Morrell and Thackray 1981; and for the relations between provincial and central science, Inkster and Morrell 1983.

[27] Danby to Sedgwick, 17 Sept. 1842, CUL: Add. ms 7652ID158a. Sedgwick's relations with the Kendal men are discussed in Speakman 1982: 94-97.

[28] Murchison 1843a: 71. For the puns, Murchison to Sedgwick, 16 Oct. 1842, in Geikie 1875, *1*: 376-382, at p. 379; Phillips to Ramsay, 29 Nov. 1842, ICL(R); Phillips to De la Beche, 29 Nov. [1842], NMW.

Murchison in extending the Lower Silurian to the base of the fossiliferous rocks, they believed that Sedgwick should be allowed to publish his matured views on the structure of North Wales in peace, and condemned Sharpe's interference on the eve of the Survey's entry as an opportunistic attempt to reduce the reputation of a pioneer. Once the Cambrian no longer held a place in the standard nomenclature, they lauded Sedgwick's work as an exercise in the determination of local structure, precisely the element of his accomplishment that Sharpe now challenged. Salter expressed a generally held view when he told Sedgwick in 1843: "Perhaps it is no business of mine, but I feel a little indignant at a person who has so lately & so little examined Cambria, attempting to subvert upon so little evidence the results of so many years labour."[29] These widespread sympathies undoubtedly gave Sedgwick a certain edge in the debate. For example, Murchison secretly used his position as Geological Society president to send Sharpe's Bala manuscript to his friend before its official refereeing; and although Salter identified fossils for Sharpe, he admitted to Sedgwick that he felt "obliged to be reserved in a way I don't like only it would not do to help him fight you."[30] Sharpe lacked comparable resources and later felt that his work had been unjustifiably ignored. He did, however, have the advantage of Bowman's colored field maps, obtained from his son William after Sedgwick rather carelessly failed to exhibit any interest in them.[31]

The men of the Survey shared the general dislike for Sharpe's ill-mannered trespass, and not for Sedgwick's sake alone. Throughout the early 1840s, Phillips pressed De la Beche to run a quick section from Bala to the Menai straits, so that they might bask in the glory of "the only remaining Crown which English Geology can offer," the correct determination of the sequence in North Wales. As delay followed delay in their entrance into the region, Phillips feared that "the Corps" and its "Chef de Bataille" would lose ground to the individual researchers busily hammering in Cambria. After considering the matter, however, De la Beche decided with characteristic thoroughness to wind up work in South Wales and the Borders be-

[29] Salter to Sedgwick, 24 Jan. [1843], CUL: Add. ms 7652IE24.

[30] Salter to Sedgwick, 10 Jan. [1844], CUL: Add. ms 7652IE83a; for the preview of Sharpe's paper, Murchison to Sedgwick, 26 Feb. 1842, CUL: Add. ms 7652IIID36 (copy).

[31] For the maps, Sharpe 1846; for his sense of injustice, Sharpe to Sedgwick, 9 Nov. 1854, CUL: Add. ms 7652IIL6.

fore moving north.[32] He undoubtedly felt that the Survey's reputation would gain from care and method what it lost in immediate priority; a hurried section into Snowdonia might lead to prematurely announced conclusions (another Devon Coal plants fiasco) and at the very least would be open to easy attack. This strategy also had the advantage of retaining Sedgwick's goodwill by respecting his boundaries for as long as was practicable. Perhaps De la Beche, still sensitive about the social status of government science, did not want the Survey to play the game by the opportunistic rules of a mercantile man.

Responsibility for Error

Whether one accepted Sharpe's claims or not, it was already clear as the summer of 1842 drew to a close that the dovetailings of the previous decade had involved serious errors of correlation. Although the Survey results were still incomplete, Ramsay's "roly-poly" invalidated the old Cambrian-Silurian boundary almost entirely. Similarly, in North Wales the boundary had been superseded by Sedgwick's fieldwork along the base of the Denbigh Flags, even if one discounted his opponent's more speculative conclusions. With widespread acknowledgement of these errors, the issue of responsibility for them assumed an important place in the emerging debate. If, as Charles Gillispie has written, scientists compete for priority by "scratching their initials upon the Pantheon of nature with eponymous passion,"[33] they appear even more eager to erase them once a result is called into question.

Dispute over the respective roles of Sedgwick and Murchison in creating the collaborative boundary first arose out of an incidental comment in Sharpe's paper on the Bala Limestone. Although Sedgwick had been familiar with the scientific contents of this paper before it was read in November 1842, he missed the meeting in question and did not see the printed version in the Geological Society *Proceedings* until the following autumn. He could hardly believe his eyes. In an introductory paragraph, Sharpe stated that Sedgwick had put the Upper Cambrian below the Lower Silurian and that Murchison had subsequently adopted this view on his authority.[34]

[32] Phillips to De la Beche, 14 May 1843, NMW; De la Beche to Ramsay, 11 Sept. 1842, 23 May 1843, ICL(R).

[33] Gillispie 1981: 78.

[34] Sharpe 1842: 10. Sharpe thought he was merely reproducing earlier statements

All blame for the boundary was thus laid at Sedgwick's feet; it was bad enough that Sharpe should invade North Wales presuming to correct his work with a few hasty sections, but to receive sole responsibility for a joint error was more than Sedgwick could bear. Soon afterwards he told Murchison of his angry surprise in reading Sharpe's abstract. "It contains a mass of most consequential swaggering & one queer statement, if I remember right, that *I persuaded you* to split off *South Wales* from your Silurian!"[35] At this point Sedgwick was perfectly willing to accept full liability for the mistaken boundary in the north, but he had only seen South Wales during his brief traverses in 1832 and could scarcely be blamed for errors there.

From Sedgwick's perspective Sharpe's distorted history required immediate correction in the *Proceedings*. From its establishment in 1826, this journal had been intended as a document of the history of British geology, and the misleading account thus possessed something of an official character until corrected. The annual anniversary meetings provided the usual forum for setting such matters straight, and a timely comment in Murchison's 1843 address would have undone most of the damage. This at least was Sedgwick's view of the matter; but he learned with dismay that the speech contained no such disavowal and in later years became virtually obsessed with Murchison's failure to speak out from the public prominence of the president's chair. Judging from incidental comments in the address, however, his friend tended to agree with Sharpe's apportionment of blame. While Sedgwick disowned any responsibility whatsoever for South Wales, Murchison believed that the mistakes there had to be shared jointly.[36]

Since Murchison missed this supposed opportunity to rectify Sharpe's misstatements, Sedgwick took it upon himself to do so, and in several manuscript abstracts of his own papers for the Geological Society disclaimed sole responsibility for errors in the original boundary. He initially attempted to reject Sharpe's account in his November 1843 paper on North Wales, only a few weeks after finding the offending passage. As Sedgwick presented the issue, the basic reason for the error in South Wales was simple. Murchison had stopped his survey short and left it only partially linked to the

by Sedgwick and Murchison, and in a certain sense he was; see Sharpe to Sedgwick, 9 Nov. 1854.

[35] Sedgwick to Murchison, 25 Oct. [1843], GSL: M/S11/219.

[36] Murchison 1843a: 76, also p. 73. An early expression of Sedgwick's complaint is in Sedgwick to Phillips, 25 Oct. 1844, ff. 71-98, UMO; for later ones, see Sedgwick 1854c: 505-506, and 1855: lxxi-lxxiv.

older rocks. To have finished the job properly "would have required at least two years of hard labour," a task that Murchison had evidently been unwilling to undertake.[37] These brief comments, presumably delivered to the Geological Society along with the rest of the paper, were the germ of literally dozens of similar disavowals in letters, speeches, and prefaces. Questions of responsibility for errors in boundary drawing, of only minor importance in 1843, would eventually assume a critical place in the controversy.

The early date of Sedgwick's response is well worth noting. Previous students of the Cambrian-Silurian dispute, even those unabashedly biased in Sedgwick's favor, have always been puzzled by a two-year delay in his repudiation of responsibility for the error in South Wales. But the manuscript for the November paper shows that Sedgwick did reply at the first available opportunity: his apparent silence is due not to any reticence on his part, but rather to the Geological Society's refusal to print his rebuttal. Events surrounding the passage of the paper through the press make this censorship abundantly clear. Following established custom, Sedgwick had sent a written script of his lecture to the Society. Editing was then undertaken by the current president, Henry Warburton, who had performed similiar chores many times before.[38] Indications of his changes are evident in the manuscript of the paper, which still exists in the cut and pasted form sent to the printers. In general, Warburton's redaction accurately reflects its scientific contents, although the original is more detailed. But as Sedgwick would later lament, common names were misspelled, traverse lines were shown on the map but not described, and the woodcut sections were printed in an almost microscopic size. The map itself was greatly reduced and was not from Sedgwick's field copies, but rather from a crude sketch made for temporary exhibition at Somerset House. Thus the accumulated geological work of a decade appeared on a reduction of a topographical map intended for tourists.[39]

These errors (together with one of far greater magnitude which

[37] Sedgwick, "Continuation of a paper (read June 21 1843) entitled an outline of the Physical Structure of N. Wales. This continuation was read on Wed. Nov. 15 & Wed. Nov. 29 1843," GSL. This is the ms of Sedgwick 1845a, with much cutting and pasting; hereafter "North Wales ms."

[38] Warburton's editorial activities are mentioned in Sedgwick 1855: li, Murchison 1843a: 151.

[39] A list of these complaints is included in Sedgwick 1855: xlix-li; see also the papers themselves, Sedgwick 1844 and 1845a.

will be discussed in a later chapter) resulted almost entirely from an unusual circumstance surrounding the paper's publication. Although they corresponded about discrepancies in Sedgwick's use of words like "group" and "subdivision," and about details of some of his geological sections, Warburton resolutely refused to let Sedgwick correct proofs even after he specifically asked to see them. In doing this Warburton probably acted within his rights, for the manuscript became the Society's property from the moment it was received, and substantive alterations were usually prohibited to preclude the postdating of priority claims. His motives for denying this request are obscure, however; he had allowed Sedgwick to correct proofs of earlier papers. There is no evidence suggesting interference from Murchison himself. Rather, Warburton was apparently afraid that Sedgwick would reinsert his reply to Sharpe. In any event, when his paper finally appeared in print Sedgwick was particularly disturbed by the suppression of his account of the drawing of the boundary.[40]

Early in 1845 Sedgwick received an opportunity to publish an uncensored version of his remarks. Ansted, formerly his assistant in Cambridge, was anxious for his old teacher to receive full justice. In his new post as Geological Society curator Ansted presumed that he was well positioned to set matters right, for he superintended the editorial work for the recently instituted *Quarterly Journal*. The first number, which contained a second publication of the November 1843 paper, had just returned from the printers, and Ansted knew that Sedgwick would find Warburton's excisions highly unsatisfactory. Another paper by Sedgwick was scheduled for early publication, and Ansted offered to make amends for the earlier article's mutilated state. "If you can give a tolerably short abstract yourself it shall not be altered," he promised in February 1845. "This you may be sure of & therefore you will have an opportunity of correcting the one just published. This is *entre nous* & because I am afraid you will find it necessary."[41] Sedgwick welcomed this chance for an airing of his grievances. His new paper (delivered orally to the Society in May 1844) had already been intended as a scientific reply to Sharpe, for it repeated his criticisms of the latter's structure for the

[40] Sedgwick, North Wales ms; an early complaint about the matter is in Sedgwick to Phillips, 10 Oct. 1846, ff. 120-129, UMO, although he had not yet looked at the published version closely. Correspondence about the ms includes Sedgwick to Warburton, [1844-1845], 20 Jan. [1845], GSL.

[41] Ansted to Sedgwick, 8 Feb. 1845, CUL: Add. ms 7652IE113.

Berwyn chain; in these circumstances the inclusion of the censored historical passages would be highly appropriate. He also added a fuller statement of the justification for "Protozoic" than was ever to appear in print, and directed his disagreement over the word "system" explicitly against Murchison—this time in a strongly polemical tone.[42]

Despite Ansted's assurances, however, this paper was never published, and its existence has previously been known only through a title listed in the *Quarterly Journal*. The circumstances leading to its total suppression are obscure, although comparison of the surviving manuscript with surrounding papers suggests that Sedgwick's polemical denunciations would have ill-fitted the staid empiricist tone of the new periodical. The Geological Society wished to present a united front to the world in its publications, however acrimonious the meetings might be on occasion. From Sedgwick's perspective, the effects of the rigid application of this policy carried out at this time were highly unfortunate, for Sharpe's history remained uncorrected until the end of 1846. Whether justified or not, Sedgwick was unable to deny responsibility for an error that he believed belonged mainly to Murchison.

Throughout the early 1840s Sedgwick remained near the peak of his intellectual powers. As he began his sixtieth year in 1845, he was a leader of British geology, a great teacher, and one of the most dynamic figures on the scientific stage. In that year alone he geologized for months in the Lake District, opened the geological section at the meeting of the British Association in Cambridge, and launched an eighty-five page critique of the *Vestiges of Creation* in the *Edinburgh Review*—all this in addition to the considerable duties of his prebendal stall at Norwich and his professorial chair at Cambridge.[43] At the same time, however, difficulties and debate beset his research on all sides. By the middle of 1843 Sedgwick had become embroiled in three related disputes over the geology of the older rocks, each an outgrowth of the failure of the collaborative boundary. First, he profoundly disagreed with Murchison's palaeontologically based extensions of the Silurian across the world and to the base of the fossiliferous rocks. The temporary designa-

[42] Sedgwick, "On the Physical Structure of the Berwyn Chain . . . ," GSL. Ms of paper delivered to the Geological Society 29 May 1844; the title alone was published as Sedgwick 1845b.

[43] Clark and Hughes 1890, 2: 80-101.

tion "Protozoic," however unacceptable to his contemporaries, was for Sedgwick a far better indicator than an enlarged Silurian of the provisional state of Lower Palaeozoic geology. A second dispute pitted him against Sharpe's views of the structure of Wales. This disagreement, which was of fundamental importance in clarifying the sequence of rocks included within the Protozoic, required renewed proofs of a vast fossiliferous succession below the lowest of the original Silurian formations. Third, Sedgwick disagreed explicitly with Sharpe and implicitly with Murchison himself as to the relative responsibility for the mistakes uncovered in the now-abandoned collaborative boundary. In each of these concurrent debates (and in spite of widely expressed sympathy) Sedgwick suffered from a serious handicap: almost all his contemporaries believed that the failure to identify a separate Cambrian fauna had immediately doomed any alternative to the Murchisonian classification. This failure turned Protozoic into a synonym for Lower Silurian and made the dispute with Sharpe seem a minor skirmish of importance only for the local geology of Wales. Moreover, Sedgwick's grievances over the unjust assignment of responsibility for error remained within the walls of Somerset House, as unknown as the Protozoic classification was misunderstood or ignored. On the principal geological issues to which he had devoted his life, Sedgwick was becoming an isolated dissenter beleaguered by competing investigators, poor health, and the widespread desire for an unchanging classification useful in international correlation.

Murchison, in contrast, seemed to pass from strength to strength in his advocacy of a taxonomy for the strata based entirely on fossils. In the summer of 1845 the Russian tours with Keyserling and Verneuil bore fruit with the delivery to the Czar of preliminary copies of two bulky quarto volumes, entitled *The Geology of Russia in Europe and the Ural Mountains*. The first volume, by Murchison himself, discussed the gently dipping strata of western Russia. On the basis of fossil comparisons with England and Wales, he elaborated a sequence that conformably ascended from the earliest strata immediately above the "Azoic" crystalline schists, through the series of "systems"—including the new addition above the Carboniferous called the "Permian"—and so on up to the most recent unconsolidated deposits. (The second volume, written by Verneuil, figured and described the organic remains.) In an introductory chapter Murchison reiterated the principles that had led him to abandon the system below the Silurian, recounting the results of Sedgwick, the Survey, and his own research which pointed to the persistence of Lower Silurian types to the base of the fossiliferous rocks.[44]

If reviewers of the *Geology of Russia* occasionally disagreed with particular theoretical interpretations in the book, they had nothing but praise for its achievement as a scientific description. Whewell, Horner, Forbes, Conybeare, and other contemporary critics all spoke in glowing terms of Murchison's "monster publication" and the wide-ranging program of research that had produced it. David Brewster, writing in the *North British Review*, even went to the extreme of comparing Murchison with Copernicus and claimed that the work on the older rocks represented "an achievement of the same order as that which placed the sun in the centre of our system." Such critical effusions, combined with foreign military orders, a snuffbox from the Czar, the Copley Medal of the Royal Society, and a knighthood, naturally gave Murchison an exaggerated idea of his own importance. He became increasingly impatient of criticism and reluctant to consider any changes to a classification that had brought order to the rocks and honor to himself and his country.[45]

Despite the deepening contrasts on important issues, active debate between Murchison and Sedgwick was still confined to private correspondence and verbal discussions at scientific meetings. On only one occasion, in brief reports of the 1844 York meeting of the British Association, did any explicit disagreement find its way into print.[46] With the exception of these few lines in the *Athenaeum*, the positions of Sedgwick and Murchison were never juxtaposed; their published papers rarely mention the author of the view being opposed, and in fact the very existence of alternatives is often only implied. This oblique method of friendly disputation, typical of most scientific literature in the Victorian era, maintained debates as gentlemanly affairs governed by an elaborate code of conduct. Without a knowledge of unpublished correspondence and an awareness of the more public phases of the controversy, one would be hard pressed to prove any active confrontation during this period. Even the existence of the Sedgwick-Sharpe debate is not im-

[44] Murchison et al. 1845, esp. pp. 0-6*. An excellent account of the publication of the book is provided by J. C. Thackray 1979.

[45] Whewell 1857, *3*: 442-443; Horner 1846; [Forbes] 1846; [Conybeare] 1846; [anon.] 1847; for the quoted phrase, [Brewster] 1846: 181. See also Geikie 1875, *1*: 356-357, *2*: 50-54.

[46] *Athenaeum*, 12 Oct. 1844, p. 930; 26 Oct. 1844, p. 977.

mediately evident from a straightforward reading of the relevant papers. Not surprisingly, readers familiar only with the published writings analyzed in the last two chapters have assumed that Sedgwick acquiesced in the expanded Silurian until 1846. As the Scottish geologist James Nicol wrote in 1854, he "clearly let the case go against him by default of appearance in court." During the more openly controversial stages of the dispute, Murchison took advantage of the early lack of explicit disagreement and claimed that Sedgwick opposed his classification only after a decade of quiet acceptance.[47] But once the underlying framework of debate is recalled from the obscurity sanctioned by contemporary practice, the publications of both men appear in a new light, and apparently straightforward scientific facts emerge as parts of a continuing argument.

[47] Murchison 1847b: 170. The quotation is from Nicol to Murchison, 6 Nov. 1854, EUL: Gen. 1999/1; the best example of such a reading is Dana 1890, esp. pp. 173-175.

CHAPTER SIX

Revivals of the Cambrian

THE *Geology of Russia* was a difficult book to ignore. Tipping the scales at thirteen pounds, filled with fine lithographs of fossils and scenery, it was just the sort of comprehensive monograph that most geologists dreamed of producing. These qualities combined with Murchison's persuasive advocacy to give his classifications an air of permanence and finality. With publication of the book in England looming on the horizon for most of 1845, Sedgwick must have realized that his options would soon become severely limited. For all its flexibility and breadth of reference, "Protozoic" was confessedly a provisional term for the rocks below the Upper Silurian and required replacement as soon as possible. Most geologists believed that the need for a secure nomenclature far outweighed any potential advantages of continued caution. Clearly Sedgwick had to do something soon or be left behind. His response finally came in 1846, with not one but two successive revivals of the Cambrian.

A Preliminary Skirmish

In January 1846, just two weeks after the Russian volumes went on sale in London, Sedgwick read a paper to the Geological Society that finally lifted the Protozoic veil to reveal a new classification for the rocks below the Devonian. Almost three years after abandoning the Cambrian, Sedgwick called it back into existence. But this version of the Cambrian stood in stark contrast to the original "system" of the 1830s: it was a Cambrian without a peculiar fauna of its own, relying on the progressionist scheme of the 1841 "Supplement," and tactfully including only strata below the formations described in the *Silurian System*. Although unquestionably motivated by a desire to challenge the Murchisonian stranglehold on Lower Palaeozoic geology, this new proposal also had a firm basis in the same program of research that had produced the Protozoic in June 1843. In this sense it can best be seen as Sedgwick's last attempt to produce a compromise acceptable not only to himself, but also to Murchison and the rest of the geological community. In the end, however, this limited revival never got off the ground, to such an extent that it affords an excellent example of a failure of strategy in scientific debate. Supporting evidence existed, and often in abundance, but the January paper failed to marshall it into a form even potentially convincing to contemporaries.

One problem is evident at a glance. The bulk of the paper argues against Sharpe on the basis of a reexamination of the critical sections in the north of England, with discussions of several hitherto undescribed districts—but all without any indication of the surprise ending. At the last moment, Sedgwick suddenly turns from the Lake District to Wales and just as abruptly offers a new classification to replace the Protozoic.[1] This scheme divided the sub-Devonian into three separate "groups": an "Upper Silurian," a very narrow zone of "Lower Silurian" passage beds, and a "Cambrian" for all other strata to the base of the fossiliferous rocks. The reintroduction of the Cambrian nomenclature is obviously the crucial step. But it also is unheralded, unexpected, and only briefly explained, so much so that the revival appears almost arbitrary. Sedgwick's new arrangement actually depended on a chain of arguments based on specific details of the geological succession. The proposal also relied on broad methodological principles, although consideration of these is best reserved for a discussion of his paper of the following December. For the moment I will reconstruct the technical rationale behind the rebirth of the Cambrian.

Sedgwick began by basing his "Upper group" or "Upper Silurian" on sequences in North Wales and Westmorland. He argued that these were more typical of this part of the succession, particularly in lithological terms, than were Murchison's generally accepted Welsh Border sections. Siluria, he wrote, "is not the true mineral type either for England, Wales, or Ireland," for in his view the anomalous superabundance of limestones precluded using the Welsh Border as a type, however common fossils might be in that region.[2] Having established his heterodox standard for rocks on the Upper Silurian horizon, Sedgwick then made a startling conjecture about the age of the South Wales slates, a subject on which he had maintained public silence for over two years. He told his incredulous audience at Somerset House that nine-tenths of these slates would prove to be neither Upper Cambrian (as determined in the early 1830s) nor Lower Silurian (as the Geological Survey believed),

[1] Sedgwick 1846, esp. pp. 125-131.
[2] Sedgwick 1846: 126.

but *Upper* Silurian (Fig. 6.1). Sedgwick had arrived at this remarkable conclusion during his 1843 field trip, and subsequent investigations undertaken by Salter on his behalf had only strengthened it.[3]

But what of the supposedly authoritative results of the Survey, which showed the South Wales slates as undulating repetitions of the *Lower* Silurian, most probably of the Llandeilo formation? "The answer to this question," Sedgwick wrote, "involves another—what is the age of the Llandeilo Flag?" Here he announced another result of Salter's fieldwork in 1845. In South Wales and on the east side of the Berwyns, the young palaeontologist had found that beds originally included in the Llandeilo actually appeared to be *above* rather than below the Caradoc (Fig. 6.2). Given such a high position for the Llandeilo, Sedgwick thought it unsurprising that the Survey should find traces of Llandeilo fossils and strata among what he now saw as the "the Upper Silurian rocks of South Wales."[4]

As the next step in Sedgwick's argument, the repositioned Llandeilo Flags joined other beds to form the basis for his "middle group" or "Lower Silurian." This narrow zone comprised all the rocks occupying the few hundred feet immediately below the Upper Silurian—strata like those at Glyn Ceiriog, Mathrafal, the Caradoc Sandstone of South Wales and the Malvern Hills, and the Coniston Limestone of the Lake District. These, of course, had previously been established as passage beds, and Sedgwick now linked them to parts of Murchison's Wenlock formation on the basis of lithology and fossils.

The upward shift of the Llandeilo to the top of this intermediate series had a consequence particularly crucial for the resurrection of the Cambrian. Hitherto the characteristic Llandeilo fossil, the trilobite *Asaphus Buchii* (Fig. 3.3), had been given a very low position in the Welsh sequence. But according to Salter, the specimens of this trilobite from the 1834 Berwyn traverse, from the Rhiwlas beds, and from the areas farther to the west explored by Bowman were all either unidentifiable or belonged to other species.[5] That such an important change could be considered after so many years of careful work speaks volumes about the sparse evidence available to Victorian palaeontologists. In this instance, reclassifying these obscure

[3] Sedgwick 1846: 119, 127. Also Sedgwick to Phillips, 10 Oct. [1843], ff. 46-53, UMO. For Salter's work in Wales, see his correspondence with Sedgwick at CUL: Add. ms 7652, esp. IE171f, "Notes on S. Wales J.W.S.," and his ms field maps (SM). Sedgwick (1873: xxv) later thought that these materials had been destroyed.

[4] Sedgwick 1846: 127, 128.

[5] Sedgwick 1846: 128.

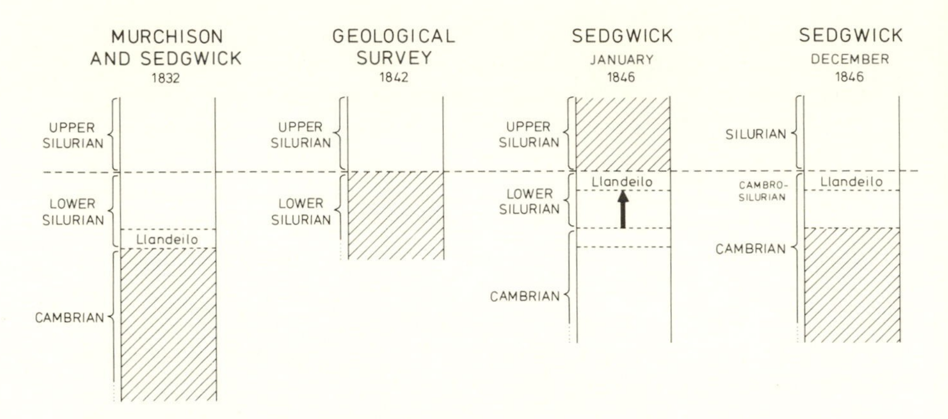

Fig. 6.1. Interpretations of the position of the South Wales slates (shown by shaded parts of each section), 1832–1846.

trilobite fragments deprived the rocks below the Caradoc of an extremely important species. Suddenly all North Wales was without the most typical Llandeilo fossil. Here at last was just the kind of negative datum required for the success of the progressionist scheme proposed in the 1841 "Supplement." With fossil evidence finally matching his physical groups in North Wales (Fig. 6.2), Sedgwick could reassert a Cambrian "group" comprehending the bulk of the Protozoic.

Sedgwick thus reclaimed the Cambrian name as the successful outcome of a search for fossils that had proceeded throughout the early 1840s. Although his debate with Sharpe had necessitated a focus on structural questions during these years, he had always expected his physical groups to obtain a fossil basis through the progressionist scheme. The initial failure of this idea had led him to create the Protozoic in 1843; however, in returning to Wales and the Lake District in the summer of that year, Sedgwick had gradually achieved some success in his search for progression. The November 1843 paper on North Wales showed him (with expert help from Salter and Sowerby) already tracing groups of fossils to match the various horizons below the Upper Silurian (Fig. 6.3). During the next two years the tracing of progression had continued as Sedgwick compared his Welsh sequence with its counterpart in the north of England.[6]

Another issue, and one closely related to progression, concerned the suitability of Sedgwick's sections for proving the origin of life. Presented with Murchison's claim that Sweden and Russia held

[6] Sedgwick 1845a: 7, 15-17, and Table I; Sedgwick 1845d: 583-584.

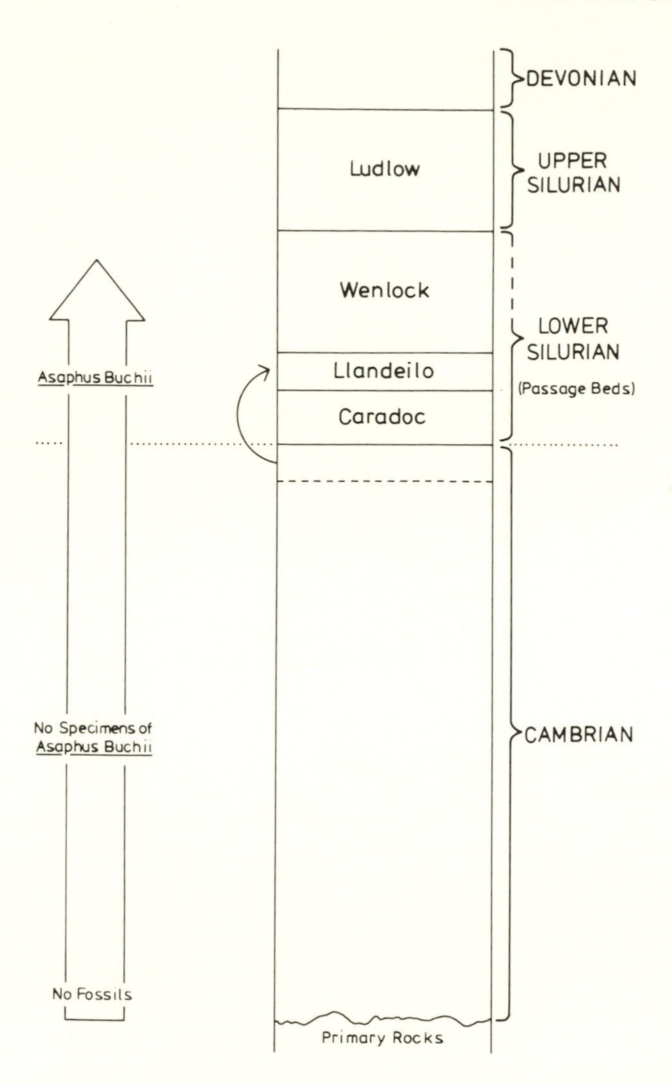

Fig. 6.2. Sedgwick's classification of the older rocks of England and Wales in January 1846; an interpretative diagram showing the progressive appearance of Lower Silurian fossils within the major physical groups.

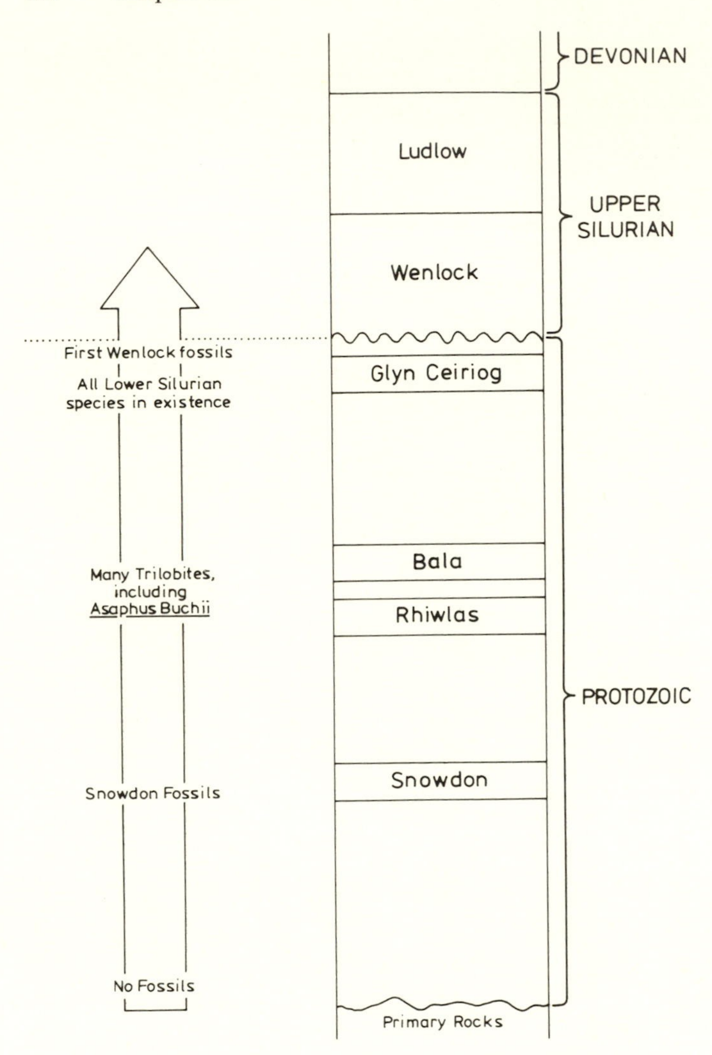

Fig. 6.3. Sedgwick's classification of the older rocks of England and Wales in November 1843, showing the progressive appearance of Lower Silurian fossils within the major physical groups.

traces of a beginning of organic creation, Sedgwick had argued at the York British Association meeting in 1844 that the successions there were incomplete and hence inherently unsuitable for such purposes. "Wherever we have an *unconformable junction*," he pointed out in a letter of June 1846, "something *must be wanting*."[7] His own sections in the Lake District seemed much more promising candidates for the complete sequence that was needed. In any case he was now not so sure that the problem of the early history of life would find an easy solution. A similarly sceptical view was held by Lyell and Leonard Horner, who drew on Edward Forbes's dredging work in the Aegean and concept of a deep-sea "azoic zone" to develop an environmental explanation for the absence of fossils at the base of the geological column.[8]

The terms in which Sedgwick first reestablished the Cambrian clearly reveal the dependence of his own methods on negative evidence, the absence of certain fossils rather than the presence of a distinct fauna.

> Now that I have no evidence of the existence of *Asaphus Buchii*, and other Llandeilo characteristic fossils [which were not specified] in this vast group, I am no longer embarrassed for its name. I cannot speak of a *Cambrian system*, with peculiar fossils found in no other; but I may speak of the lower or great *Cambrian group*. . . .

He then articulated, for the first time in any of his published works, the picture of the history of life behind his search for a progressive and orderly appearance of Lower Silurian fossils in the ancient strata of the British Isles.

> In this great Cambrian group began the lowest fossil species we know in the British Isles. Many of the lowest species lasted throughout the whole Lower Silurian period; but new species were added, as conditions gradually changed, during the epochs marked in the ascending sections; so that the lower fauna reached its maximum of development in the Caradoc sandstone and Llandeilo flagstone. Afterwards [during the Upper Silurian] the fauna underwent a much more rapid change, certain tribes of *Brachio-*

[7] Sedgwick to Murchison, 16 June [1846], GSL: M/S11/231a-c. This letter also mentions the comments at York, which were not printed in the reports of the discussion in the periodical press; see *Athenaeum*, 26 Oct. 1844, p. 977. The criticisms of the Russian sections were made in print in Sedgwick 1850: lviii-lix.

[8] C. Lyell 1851: 360; Horner 1846: 160-161. Forbes's work is described in Rehbock 1979, 1983, and Mills 1978.

> *poda* diminishing in numbers, and being replaced by other forms, while, as far as our evidence goes (at least in the north of England), the *Lamellibranchiata*, though beginning low in the Cambrian group, also formed a more important part of the fauna of the Upper Silurian rocks.[9]

It is essential to remember that during the first half of 1846 Sedgwick never spoke of a distinct Cambrian fauna. His progressionist scheme involved the gradual appearance of *Lower Silurian* fossils from the first stirrings of life on earth, in a succession he thought best illustrated by British sections.[10] But to name all the ancient fossil-bearing rocks after the Silurian beds at their summit was to let the tail wag the dog. In contrast, Sedgwick believed that his new threefold classification—Upper Silurian, the passage beds of the reconstituted Lower Silurian, and Cambrian—effectively meshed the limited evidence provided by palaeontology with the physical groups of his descending sections.

Busy with foreign geological work and dinners honoring his recent knighthood, Murchison apparently learned of the revival of the Cambrian only in May through letters from Élie de Beaumont and Verneuil, several months after Sedgwick's memoir had been read. "I at once saw," he told Sedgwick, "that either I must defend the positions taken up in the opening chapter of the work on Russia . . . or allow Geologists to think that the shot you had fired in the eleventh hour was effective in breaking up all my entrenchments."[11] He accordingly replied at the first available opportunity, at the end of a paper on the Silurian rocks of the Baltic region read in early June. This paper provided an appropriate forum for rebuttal, because it discussed a succession with particularly good transitions from the Lower to the Upper Silurian and thereby supported the unitary character of the system.

Murchison's memoir attacked three of Sedgwick's most vulnerable points: first, the placement of the South Wales slates in the Upper Silurian; second, the proposal to create a series of passage beds linking the Wenlock with the Llandeilo and Caradoc formations; and third, the construction of a Cambrian "group" containing many Lower Silurian fossils but lacking *Asaphus Buchii*. The first of these points needed little comment, for Sedgwick had abandoned his hypothetical placement of the South Wales slates almost immediately

[9] Sedgwick 1846: 129.

[10] Sedgwick to Murchison, 16 June [1846], GSL: M/S11/231a-c.

[11] Murchison to Sedgwick, 19 June 1846, CUL: Add. ms 7652IIID52, G19 (copies).

after the January meeting. The Survey men had responded angrily to this claim; as Ramsay told Aveline soon afterwards, Sedgwick "is not content with the Cambrian, and so, gulping it down, he wheels about ten times, and turns it all in Upper Silurian."[12] For his own part Sedgwick belatedly regretted the printing of these passages and stressed their tentative character. "Some things at the end of my paper I threw out as *tubs to be tossed*," he told Murchison in June, "& in hopes of getting up De la Beches views of S. Wales."[13] The confident assertions effective in debating an individual geologist like Sharpe simply did not work against the corporate might of the Survey. In order to challenge its interpretations Sedgwick needed to speak from a much greater familiarity with the sections in question.

Sedgwick's other revisions received a sharp rebuttal from Murchison. In part, he welcomed the evidence for passage beds between the Wenlock and the Lower Silurian. "I rejoice in such discoveries," Murchison told the Geological Society, "because they still better *link the two groups together in one indissoluble natural system*."[14] Sedgwick had proposed a reclassification of the Wenlock as part of his effort to create the narrow zone of passage beds between distinctive Cambrian and Silurian groups; Murchison turned the same evidence against him, to argue that both belonged in a vastly enlarged Silurian system. Here Sedgwick's attempt to compromise by using "Lower Silurian" for these passages played directly into Murchison's hands, for as long as Sedgwick called *any* strata by that name, Murchison could use their palaeontological contents to justify the extension of Siluria to the base of the fossiliferous rocks. From Murchison's point of view, the findings exhibited by Sedgwick only strengthened the case for joining Upper and Lower Silurian into a single unit. At the same time, he objected to any attempt (however tentative) to shift the dividing line between these two halves of his system by grouping the Wenlock with the Llandeilo and Caradoc formations. This was a point well taken, and Sedgwick backed down on it even before leaving for his summer field tour.[15]

The resurrection of a Cambrian, even in a limited form, clearly posed the greatest threat to the integrity of Siluria. Because Murchison dismissed the physical evidence on which Sedgwick's new scheme primarily depended, he naturally found the fossil basis for

[12] Ramsay to Aveline, 31 Jan. 1846, in Geikie 1895: 77.

[13] Sedgwick to Murchison, 16 June [1846].

[14] Murchison 1847a: 44.

[15] Sedgwick to Murchison, 16 June [1846]; for his proposal, Sedgwick 1846: 125-126; for the objections, Murchison 1847a: 41-44.

reviving the Cambrian pitifully inadequate, as indeed it was when considered in isolation. Even if one accepted the need for repositioning the Llandeilo, the absence of a single fossil species scarcely argued for a wholesale separation of strata from the main body of the Silurian. With scarcely veiled sarcasm Murchison showed that any number of separate "systems" could be established on this basis. The lowest Russian strata held an unusual number of cystoids: should they become the basis for a "Petropolitan system"? The black schists and limestones of Scandinavia held a few species unknown in England: should an "Odinian system" be established? Lyell had found strata on the same low horizon in North America: were these the types of a "Canadian system"? As Murchison pointed out, each of these mock alternatives to the Silurian—including the Cambrian itself—relied on some local peculiarity in the fossil population and failed to account for the situation in rocks of similar age throughout the world. For example, he showed that *Asaphus Buchii* often occupied a low position in areas outside Britain, even if it should prove to be absent from the oldest fossiliferous rocks of Wales. Moreover, certain equally characteristic Llandeilo fossils (such as the trilobite *Asaphus tyrannus*) were found deep within the so-called Cambrian strata in Wales itself. By using such arguments to highlight the severely limited palaeontological support for the Cambrian, Murchison turned Sedgwick's own chosen evidence against him, effectively blocking any name for the older rocks save "Silurian."[16]

The first round in open debate ended without doubt as to the victor. Sedgwick was recuperating from gout at Harrogate spa in Yorkshire when Murchison read his rebuttal to the Geological Society in June. During the discussion Buckland spoke up in favor of the Cambrian, but by 1846 his support was worth little. As Forbes told Ramsay in his account of the exchange, Murchison "replied rather fiercely that he had lately observed symptoms of the Dean having forgotten all his geology." This reference to Buckland's recent preferment to the deanery of Westminster, he added, "caused some amusement to all but Bucky."[17] Without exception those geologists really familiar with current work on the older rocks—including De la Beche, Sharpe, and Forbes—supported Murchison's enlarged Silurian. Lyell, just returned from a second tour in America, confirmed that the oldest strata there held Lower Silurian types and

[16] Murchison 1847a: 37-38, 45.
[17] Forbes to Ramsay, 20 June 1846, ICL(R).

failed to buttress Sedgwick's case for a Cambrian "group."[18] Even Salter, who had supplied so much of the fossil and field data for Sedgwick, abandoned the latter's principal conclusion; although he maintained that Murchison had erred in classing certain strata with the Llandeilo, he publicly disassociated himself from the arguments for a Cambrian separate from the Lower Silurian. "I added too for justice sake," he told Sedgwick afterwards, "what I often told you I thought that there were hardly fossil grounds for this division." Salter was just transferring from Sedgwick's service to a better paid job with the Survey, a position that would permit him to marry and settle down.[19] Murchison's convincing attack was the first paper he had ever heard at the Geological Society, and his inaugural speech was so blatantly in its favor that his immediate superiors condemned him for too sudden a switch in loyalties. Given his circumstances, however, it is not surprising that Salter hesitated to continue in opposition. "Of course I did not volunteer to gild gold," he nervously told his old patron, "or defend the Cambrian system."[20] Needless to say, neither did any other active geologist or palaeontologist. By all accounts Murchison's reply went unchallenged. "The speakers at the Society were so decidedly in favour of no interference with the terminology & classification now so generally adopted," Murchison told the recuperating Sedgwick, "that I had nothing to do but express my great regret at your absence & still more for the cause of it."[21]

For all this, it would be a mistake to conclude that the sheer weight of factual evidence simply compelled adoption of the expanded Silurian. The problem was not a dearth of facts favoring the Cambrian—the field was much too rich a resource for that—but rather the particular ones that Sedgwick had chosen to stress.[22] As the preceding discussion demonstrates, the case for the Cambrian suffered severe defects of presentation. In particular, by latching on to the insecure and poorly validated "fact" of the absence of *Asaphus Buchii*, Sedgwick offered his opponents an easy target. In this respect his attempted compromise had badly backfired. Similarly, the

[18] Forbes to Ramsay, 20 June 1846; also C. Lyell 1849, 2: 263.

[19] Biographical information in Secord 1985b; for the quotation, Salter to Sedgwick, 18 June 1846, CUL: Add. ms 7652IE171fi.

[20] Salter to Sedgwick, 18 June 1846, CUL: Add. ms 7652IE171fi. As Forbes told Ramsay (20 June 1846): "Salter is a most valuable man but wants ballast & training."

[21] Murchison to Sedgwick, 19 June 1846.

[22] A helpful perspective based on a close study of a modern example is provided by Law and Williams 1982.

concluding paragraphs of his paper, with their hypothetical placement of the South Wales slates in the Upper Silurian and reliance on negative evidence in defining a Cambrian "group," had been easily dismissed. For any hope of success, a much more forceful approach was needed, one emphasizing the strength of the structural arguments for a fossiliferous group below the Silurian. The entire issue merited above all a more coherent treatment than could be presented in a few closing remarks. With his lack of concern for literary form, Sedgwick was often his own worst enemy.

Cambria Confronts Siluria

Sedgwick returned to North Wales in the summer of 1846 determined to gather proofs for the existence of the Cambrian. The resulting paper, read to the Geological Society in December, was one of his best in many years. He continued to stress the evidence of sections and physical structure, but for the first time couched the palaeontological aspect of his argument in terms fully comprehensible to his contemporaries. No longer did the definition of the Cambrian rely on the absence of a single species; he now used his vast physical groups in Wales to argue that *all* the fossils found in the narrow zone of strata originally in the Lower Silurian should be considered as the final manifestations of a *Cambrian* fauna. The time for compromise was over. According to Sedgwick, if the word "system" was to refer to lesser divisions than the Palaeozoic itself, then there were two systems in the older rocks, Silurian *and* Cambrian. "I mean to stand up for my own system and my own name," he told Jukes in October 1846, "to call the vast undulating system (both of S. & N. Wales) *under* the Llandeilo & Caradoc group . . . by the name of Silurian is rank nonsense."[23] Sedgwick's vigorous new paper cut right at the palaeontological heart of Siluria, and Murchison responded with a full statement of his views at the next meeting of the Geological Society early in January. Murchison found his old friend's attack extremely agitating, but offered in response what he considered to be a "brief and calm exposé." "In short, the sudden *check* is quite Hebrew to me," he told De la Beche. "The only rationale is that he is *in love* & has lost his head like many a great man before him. So says rumour!"[24]

With this exchange in December 1846 and January 1947 the con-

[23] Sedgwick to Jukes, 12 Oct. 1846, CUL: Add. ms 7652IIIE2 (copy).

[24] Murchison to De la Beche, 7 Jan. 1847, NMW; Murchison 1847b.

sequences of the breakdown of the collaborative boundary became painfully obvious and the debate assumed many of its characteristic features. Although Sedgwick and Murchison later altered their classifications, their basic positions remained in many respects unchanged after this first episode of open conflict. Years of implicit disagreement were brought to a head and the pattern was set for countless similar exchanges in the following decades. How does one best construct a "natural classification" of the geological record? Should sections and structure be of primary importance, or should palaeontology hold undisputed sway? Although only one of many possible perspectives on the dispute, this was the fundamental question of method at stake.

Throughout the debate, Sedgwick had argued that any classification must take into account the immense thickness of the fossiliferous strata underneath those described in the *Silurian System*. As a preliminary to taking over the "Lower Silurian" fauna as the fossil basis for a Cambrian system, he needed to show once again that the Llandeilo and Caradoc formations were only the uppermost beds of a huge underlying succession (Fig. 6.4). The first part of his paper in December 1846 thus continued the earlier debate with Sharpe and the Survey. The need for such preliminaries was more acute than ever, for Sharpe had just published a paper extending his controversial results throughout North Wales. After a tour of a few weeks in 1843, he had come before the Geological Society with a paper modestly entitled "Notes on the Geology of North Wales." This actually put forward an audacious reinterpretation of the structure of the entire Principality, moving all the fossiliferous rocks below the Old Red Sandstone up to the level of the original Silurian formations. The Cambrian became a barren group of slaty rocks typified by the core of the Berwyn range and the Longmynd in Shropshire.[25] Sharpe displayed his truncated succession in a map (Fig. 6.5) published in the Geological Society's *Quarterly Journal* for 1846.

Sedgwick was absolutely furious. "By the way," he told Jukes after returning from his own field tour in 1846, "what an insufferable hash Mr. D. Sharpe has made of N. Wales ! Have you seen his map? It is very discreditable to him & to our journal. *He* has also looked through an inverted telescope—& instead of raising his mind to the grand conception of the Cambrian development, has hacked & mangled the whole system in the hopes of packing it

[25] Sharpe 1846.

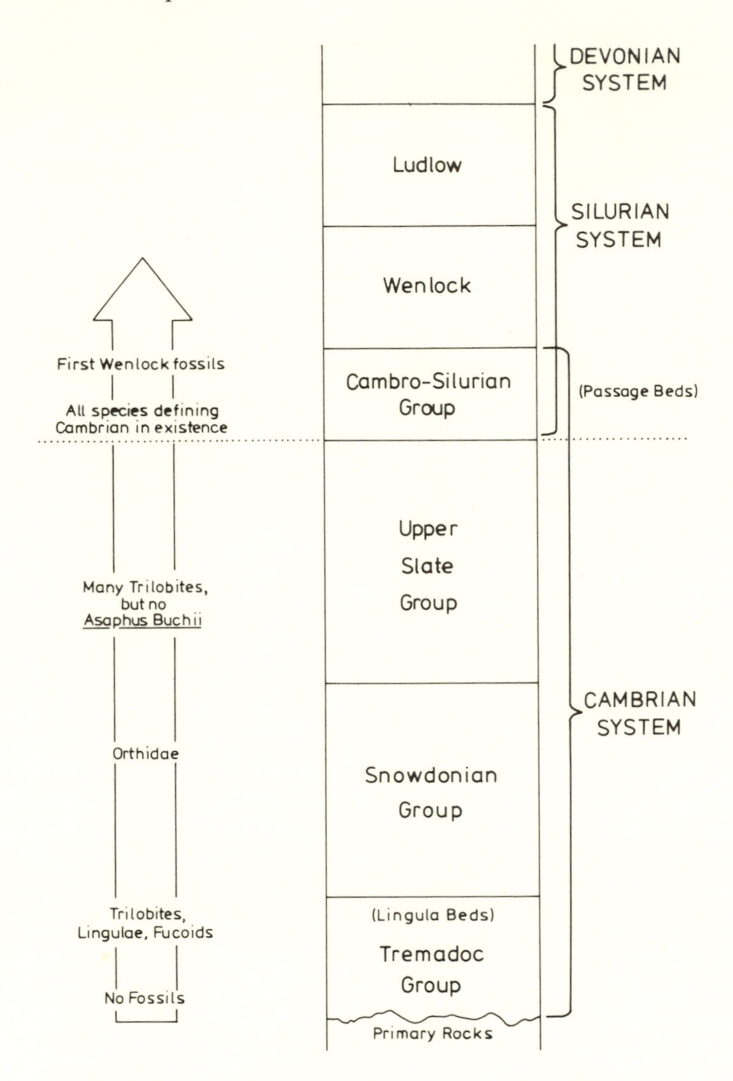

Fig. 6.4. Sedgwick's classification of the older rocks of Wales in December 1846, showing the major physical groups and the progressive appearance of fossils—now renamed "Cambrian"—within them.

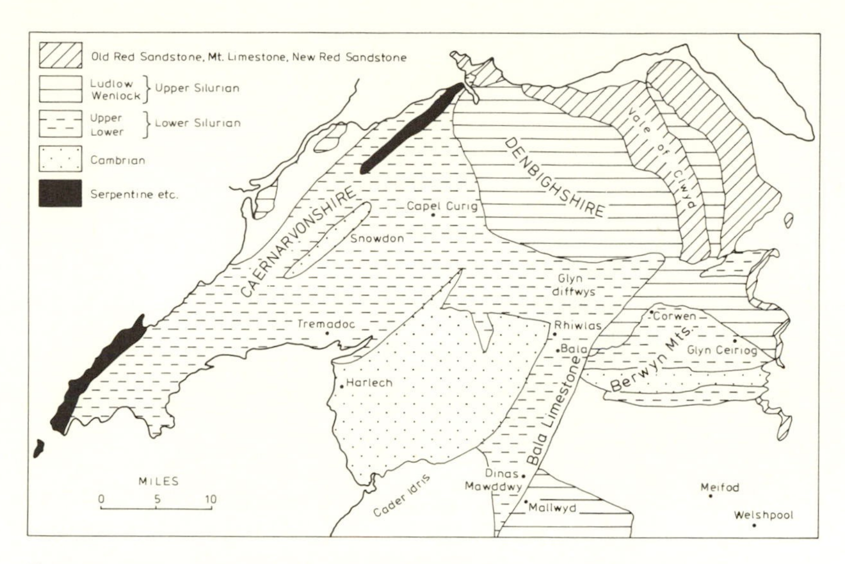

Fig. 6.5. Geological map illustrating Sharpe's conception of North Wales in 1844. Igneous rocks omitted.

within his own preconceived compass."[26] Almost all of the sections in Sedgwick's December paper argue against this heresy. The most powerful new arguments came from northwest Wales. In previous years Sedgwick had always viewed the Snowdon beds as the lowest fossil zone in the Principality, although he knew of even older (but apparently barren) strata near Tremadoc. In 1845 a local fossil collector, John Edward Davis, had found fucoids in these rocks, and even more significantly a brachiopod species that Sedgwick's newly appointed palaeontologist Frederick McCoy christened *Lingula Davisii* (Fig. 6.6). Sedgwick hastened to visit Davis and his brachiopods during the summer of 1846, for the discovery of specimens so far below the level of the Snowdon beds greatly extended the thickness of the fossiliferous succession. The December paper established the position of these "Lingula beds" with great skill, detailing their occurrence at various points near the western coast of Wales. From these strata, through Snowdon to the Arenig chain, and finally to the summit of the Berwyns, Sedgwick could march his listeners through a section more extended than ever before.[27]

[26] Sedgwick to Jukes, 12 Oct. 1846. For his summer tour, see Sedgwick to Murchison, 7 Aug. [1846], GSL: M/S11/235, partly printed in Clark and Hughes 1890, 2: 104-105; also Sedgwick to Phillips, 10 Oct. 1846, ff. 120-129, UMO.

[27] Sedgwick 1847: 136-150; Davis 1846.

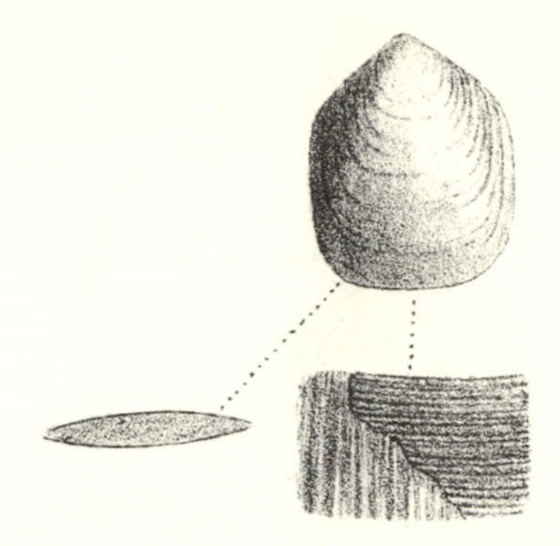

Fig. 6.6. *Lingula Davisii,* named by McCoy in honor of the provincial naturalist John Edward Davis, its discoverer.

Sedgwick next turned his audience's attention to South Wales, which he had revisited for the first time since his ill-fated traverses at the end of his 1832 tour. In the brief time at his disposal, he had only repeated this earlier work with the supposed goal of understanding the thorough descriptions of De la Beche and his assistants. But despite fulsome praise for the government geologists, Sedgwick once again arrived at a conclusion markedly different from theirs. Rather than viewing the South Wales slates as repetitions of strata on the horizon of Murchison's Llandeilo and Caradoc formations, he argued that here too the rocks extended downwards in a continuous section thousands of feet thick (Fig. 6.1). He was willing to admit that the passage beds of the original Lower Silurian did extend for some distance past the limits of the collaborative boundary, but contended that almost all the older rocks of the region were far below this level. For a man who only eleven months before had maintained that South Wales was *Upper* Silurian, this was a remarkable about-face. Sedgwick prudently emphasized the preliminary nature of these findings, although he was eager to use them in conjunction with his work in North Wales as part of a single argument.[28] In opposition to Sharpe, the Survey, and a growing body of doubters, he still believed that Wales held a succession of strata unparalleled anywhere else in Britain.

As might be expected, Murchison's reply to these arguments minimized the importance of physical groups as a basis for any change in classification. He rehearsed once more the familiar tale that ended with the Silurian fauna extending to the base of the fos-

[28] Sedgwick 1847: 150-155.

siliferous rocks. "I firmly thought," he wrote, "that however thick, however diversified in mineral characters, all Welsh and British as well as all foreign strata, in which certain typical forms prevailed, would be included in the Lower Silurian group."[29] None of this was new; Murchison had used similar arguments to justify the extended Silurian classification ever since his 1842 presidential address and had argued the point at length in the *Geology of Russia*. At the same time, however, his criteria for accepting a sub-Silurian system were becoming noticeably more stringent. Most geologists continued to believe that the discovery of a distinctive set of fossil species in rocks below the Lower Silurian would justify a revival of the Cambrian in some form. For example, in a letter of December 1846 to Murchison, De la Beche left open the possibility that enough species other than those of the Silurian might be found in the Lingula beds for them to deserve a separate name on the Survey maps.[30] Murchison's June rejoinder to Sedgwick had made a superficially similar announcement that he would accept the Cambrian if an underlying fauna could be found distinct from the one illustrated in the *Silurian System*. But his comments seven months later indicate that his requirements for such a fauna had become formidably rigorous. Unlike De la Beche, Lyell, Phillips, and many other geologists, who would willingly accept an underlying group if sufficient numbers of distinctive species were found, Murchison had tightened his criteria for a sub-Silurian fauna almost to the extreme of impossibility. In his view no specific or even generic differences in the fossil groups would provide grounds for breaking up the Silurian. As he told the Geological Society, "it is not on the duration or passage of *species* from the one group to the other, that I alone depend for the conservation of the zoological unity of my system. The qualifications and character of what I term a system are chiefly based on the assemblage of its *classes* of animals."[31] Such a comprehensive definition rendered the defenses of the expanded Silurian virtually impregnable, for to supplant any part of it, an alternative would have to contain types of animals other than those such as trilobites, orthid brachiopods, graptolites, and cystoids. This was a tall order indeed. By encompassing all the major classes of fossils useful for correlation within the Silurian, Murchison had taken a step beyond what was required for the immediate defeat of the Cambrian, thereby

[29] Murchison 1847b: 170.

[30] De la Beche to Murchison, 29 Dec. 1846, in Geikie 1895: 93-94.

[31] Murchison 1847b: 176; my italics. For his earlier view, see Murchison 1847a: 45-46.

preempting attempts to use species counts or comparative percentages to undermine the unity and the territorial extent of his system. This tendency to push his position to extremes became typical of Murchison's later defense against the Cambrian and ultimately led to an outpouring of sympathy for his opponent. But this lay in the future. Throughout the 1840s Silurian fossils and Silurian names extended to the base of the fossiliferous rocks for almost everyone save Sedgwick.

Even considering the case on physical evidence, Murchison harbored grave doubts as to the need for a Cambrian system. Despite laudatory comments for Sedgwick's efforts in unravelling the structure of North Wales, he had evidently begun to believe that Sharpe's conclusions came closer to the truth. Even so early as the 1843 presidential address, he had accepted the intruder's westward dip for the Bala Limestone, a dip that opened up the possibility of the rest of Wales being stratigraphically equivalent to the Silurian. Moreover, as Murchison had recognized in a footnote, this dip directly contradicted his friend's view of the matter. This willingness to entertain opinions opposed to those of Sedgwick continued, as evidenced in his June 1846 reply, which had implied not only that the rocks of North Wales contained the same fossils as those in the Silurian region, but also that they occupied the same geological horizon.[32] Of course, this was Sharpe's opinion, and Sedgwick accordingly accused both men of viewing the older rocks "through an inverted telescope." Although Murchison denied in correspondence that he had accepted Sharpe's conclusions, his published papers strongly hint precisely the reverse. Thus his January reply does not even mention Sedgwick's revised—and highly idiosyncratic—view of the South Wales slates, but categorically states that the Survey had already paralleled them with the Lower Silurian. As to North Wales, Murchison was willing to wait until the Survey finished its mapping, although with permission from De la Beche he reported indications that the Bala Limestone belonged somewhere in the Lower Silurian on sectional as well as palaeontological grounds.[33]

Considerations involving geological structure, however, were

[32] Murchison 1847a: 44; also 1847b: 170. For the anniversary address, see Murchison 1843a: 72, and the reaction in Sedgwick to Murchison, [18 Feb. 1843], GSL: M/S11/212.

[33] Murchison 1847b: 170-175, esp. p. 176. Limited permission to use the Survey results was obtained after an exchange of letters at the end of 1846 (printed in Geikie 1895: 92-94, except for Murchison to De la Beche, 23 Dec. 1846, NMW). The "telescope" is mentioned in Sedgwick to Jukes, 12 Oct. 1846.

clearly peripheral to Murchison's concern with the fossil basis for the expanded Silurian. Consequently, as long as Sedgwick allowed any Lower Silurian *strata* in Wales, as he had in first reviving the Cambrian, Murchison had been able to use their organic remains to extend his system to the base of the fossiliferous rocks. In December Sedgwick attempted to cut off this route of escape once and for all. By renaming the narrow zone of "Lower Silurian" passage beds "Cambro-Silurian," he transformed their fossil types into the basis for a Cambrian fauna of North Wales. As the vigor of Murchison's response suggests, this change had profound consequences for a palaeontologically based nomenclature. For the first time Siluria stood challenged on its own ground. "New indeed it is & with a vengeance!" he complained to De la Beche of Sedgwick's revised classification, "for his memoir of last year only treated of Upper & Lower Silurian throughout N. Wales!"[34]

From Sedgwick's point of view, though, the change merely carried his charting of progression within the older rocks to its logical conclusion. As illustrated in Figure 6.4, the December paper showed how the fossils that Murchison had called "Lower Silurian" gradually appeared in ascending through the Cambrian sections. Sedgwick had emerged from the field with a new confidence in his progressionist scheme, in which the major classes of invertebrate fossils appeared in regular order, with *Lingula Davisii* coming near the beginning. In the passage beds at the very top of the sequence one began to find *Asaphus Buchii* (which he continued to position high above the Caradoc formation) and traces of some characteristic species of the Upper Silurian. The fossils of these intermediate "Cambro-Silurian" passage beds, he argued, had their principal affinities with the underlying Cambrian strata. (This reversed his view in the January paper, where they had been bracketed with the Wenlock formation.) In essence Sedgwick now claimed that the entire series from the base of the fossiliferous rocks to the midst of the passage beds constituted a single "system" in both a physical *and* a palaeontological sense, a system whose fauna had been mistakenly described by Murchison along with that of the Silurian proper. Among the older rocks of Britain, Sedgwick wrote, there were "two great physical systems and two nearly coordinate palaeontological systems. . . . One is the system of Cambria, the other is the system of Siluria."[35]

[34] Murchison to De la Beche, 7 Jan. 1847, NMW.
[35] Sedgwick 1847: 160.

After his objections to the Murchisonian use of the term "system" in earlier papers, it may seem surprising to find Sedgwick embracing Cambrian and Silurian as "systems" with such apparent enthusiasm. However, his adoption of this controversial term in the December paper appears to be motivated more by a wish to be understood (and accepted) than by any basic change of principles. A footnote reiterated his earlier scheme which reserved "system" for the largest divisions in the geological record, such as the Palaeozoic, and referred to lesser units such as Cambrian, Silurian, and Devonian as "groups." He explained that he had been forbidden to follow out "a discussion which goes beyond the immediate objects of this communication" and in consequence refrained from quarrelling with received usage in the main body of his paper. In any event, when Sedgwick resumed his anti-Silurian campaign during the 1850s, he once again referred to Cambrian and Silurian "series" or "groups."[36]

Sedgwick's December paper certainly indicated that the Cambrian and Silurian were not demarcated "systems" in the specific, theoretical sense he had adopted in the "Synopsis" of the preceding decade. Indeed, his latest research in Wales had strengthened his conviction that the two groups graduated into one another by means of transitional beds similar to those characteristically found between all the lesser subdivisions of the Palaeozoic. Although Sedgwick still believed in an unconformity at the base of the Upper Silurian in a few parts of Wales, the evident preponderance of the "Cambro-Silurian" passage beds made it difficult to pinpoint a precise boundary. No longer could Sedgwick say—as he had in 1843 when naming the "Protozoic"—that the two major divisions among the older rocks "interchange hardly any fossil species," for Glyn Ceiriog, Mathrafal, and Coniston all exhibited a thorough intermixing of Cambrian and Silurian types. As a result he could not rely on any abrupt change to separate the two systems, such as might be revealed by percentages of species held in common. Instead, he now argued that the physical groups of Cambria held fossils essentially different from those found in Siluria, whatever intermixing might be present at their juncture.[37] Given these sudden shifts in Sedgwick's position on unconformities in Wales, it is not surprising that many geologists found his views difficult to follow. In France, for example, Élie de Beaumont had welcomed Sedgwick's tracing of

[36] Sedgwick 1847: 160-161. He returns to this argument in Sedgwick 1852d: 144.

[37] Sedgwick 1847: 151, 154, 158. The original gap is described in Sedgwick 1843b: 224.

a widespread gap in 1843, considering it good evidence both for a separate Cambrian and for his own theories of mountain building. Now, just three years later, he found Sedgwick pressing for passage beds instead![38] Needless to say, Murchison made effective use of these new-found transitions. Relying on his own researches in northern Europe, as well as Sedgwick's inopportune bracketing of the Wenlock with the Llandeilo and Caradoc in January, he showed that a large proportion of species passed between the two halves of his Silurian.[39]

According to Sedgwick the great error of the *Silurian System* had been the inclusion of these transitional beds and their fossils within the Silurian. He was perfectly willing to admit the existence of passages, but complained that Murchison had originally drawn his Silurian boundary too low, cutting off the very top of the physical and palaeontological system that occupied North Wales. Sedgwick thus argued that the old collaborative boundary had been almost entirely mistaken, for it ceded to Siluria rocks that actually belonged to Cambria.[40] His dismissal of the boundary, however, did not depend on any supposed overlap or physical equivalence between the two systems as originally defined. We have seen that Sedgwick maintained the essential accuracy of the stratigraphical separations established during the dovetailings of the early 1830s, in opposition to Sharpe, the Survey, and Murchison himself. The changes to his interpretation had involved internal readjustments within a sequence that had maintained its overall thickness almost unaltered since 1832. Sedgwick felt that the mistaken boundary, rather than raising structural difficulties of overlap or equivalence, presented a more straightforward problem in determining the limits of the two great systems in the older rocks.

This concern brought forward one of the most fundamental aspects of Sedgwick's argument. He believed that by using the fossils of the passage beds to incorporate all the underlying Cambrian rocks into the Silurian, Murchison had not only compounded his initial mistake in boundary drawing, but had also violated the basic principles of geological nomenclature. On one level this expansive use of "Silurian" necessitated an anomalous interpretation of geological history, with newer deposits giving names to their predecessors:

[38] Élie de Beaumont to Sedgwick, 15 Mar. 1846, CUL: Add. ms 7652IF102.

[39] Murchison 1847b: 175-177. His own researches are reported in Murchison 1847a.

[40] Sedgwick 1847, esp. pp. 158-159.

> To describe these groups, in the technical language of the Silurian system, would therefore be nothing more or less than this—viz. to describe a series of old rocks of enormous thickness, and with well-defined characters, by words which only acquire their meaning by reference to groups of strata of a later period . . . and assuredly to name an old group of rocks by technical terms drawn from a newer group is a process of nomenclature as incongruous as ever was attempted.

But the chronological inappropriateness of the expanded Silurian was overshadowed by a wider question of "geographical propriety" in stratigraphical naming. The term "Silurian" had been proposed in 1835 according to the established canons of geographical nomenclature in geology, with groups of strata named after the "type area" in which they were best exposed. Since that date, Sedgwick argued, Murchison had abandoned these principles. If Sedgwick's own views of the structure of Wales were correct, Siluria—the region that had served Murchison as a type—contained only a minute proportion of the oldest fossiliferous rocks. By extending his system downwards, Murchison had encompassed a vast group not found in the Welsh Borders, thereby calling "Silurian" a sequence fully developed only in Cambria:

> On this scheme the great Cambrian system . . . is to have neither name nor colour on our geological maps; and all the groups above described (down to the *Lingula* beds inclusive) are to come under the colour of the lower group of the Silurian system; though that group (in the typical country of Siluria) represents only the fourth and highest group of the great Cambrian series of deposits. Why are geographical terms to be retained when we deprive them of their geographical meaning? A good geographical term in geology must refer us to a country which contains a good type of the series of rocks designated by such term. This I consider a perfect axiom in nomenclature. The country described in the Silurian system does not answer this essential condition. The term Lower Silurian, as applied to the older groups of the Cambrian series, cannot therefore be retained with any propriety of language.[41]

When the outlines of the fossil record were understood and the groups of organic remains coordinated with the sections, then—and only then—could Sedgwick accept a classification based en-

[41] Both quoted passages are from Sedgwick 1847: 162.

tirely on fossils. In such a stratigraphical utopia, geographical names like Cambrian and Silurian would disappear from all but local descriptions, and a nomenclature referring to palaeontology alone could stand as the sole determinant of international classification. Opposing Murchison and almost all of his contemporaries, Sedgwick argued that such a situation remained only a distant prospect. "Geology," he said at the end of his speech, "is not however yet ripe for a mere palaeontological nomenclature."[42]

Murchison, of course, had already adopted "a mere palaeontological nomenclature" and saw his type area in the Welsh Borders as nothing more than a storehouse of typical fossils to be used in spreading his classification. Because of this, he could not see any reason to limit the downward extension of the Silurian simply because of Siluria's original geographical boundaries. For Sedgwick, the Silurian type area comprised a series of rocks and fossils on a particular horizon, while the Cambrian type area included a separate series of rocks and fossils on a lower horizon. In his view, the use of a geographical nomenclature automatically defined the lower limits of the Silurian sections by a geographical boundary—namely the boundary between Siluria and Cambria. From Murchison's perspective, this reference to the old collaborative boundary seemed outdated and irrelevant. Ever since the early 1840s he had insisted that this boundary had been adopted merely as an arbitrary line between two areas of investigation, a Silurian region whose fossils had been published in the *Silurian System* and a Cambrian region that had remained for the most part a palaeontological terra incognita. In a letter of December 1846 to George W. Featherstonhaugh, Murchison used an analogy that nicely sums up his view of the artificiality of the boundary. According to the analogy, in the 1830s Sedgwick had taken possession of a "Cambrian manor," while the "Silurian manor" was Murchison's own. The fossils, or "game," of Cambria had been neglected in comparison with those of Siluria, which were rapidly identified and widely publicized. Moreover, Murchison had felt "bound in *honour*" to refrain from "poaching" in the hunting country marked out for exploration by his friend. "Further researches have shewn," he told Featherstonhaugh, "that the Cambrian manor contains after all no animals but what I have got in my preserves & which have therefore very naturally got my names."[43] The unstated if somewhat illogical conclusion of the anal-

[42] Sedgwick 1847: 164.

[43] Murchison to Featherstonhaugh, 15 Jan. 1847, CUL: Add. ms 7652IIKK68.

ogy, at least in geological terms, was that a separate "Cambrian manor" did not exist and that Sedgwick's strata would henceforth have to be colored Silurian on geological maps. Through this characteristic comparison with polite landed society, Murchison brought forward those elements of the original Cambrian-Silurian line that had involved territoriality and social convention, while at the same time casting parallel geological meanings into the shade.

Murchison and Sedgwick's contrasting interpretations of the collaborative boundary lead to a final issue. Although one now claimed that the limit had been wholly territorial and the other argued that it had been wholly stratigraphical, both men agreed that the boundary had been misplaced. Since the case for a Cambrian system depended so heavily on Murchison's overstepping of the real base of the Silurian into a lower system, Sedgwick took pains to disassociate himself from any unjust imputation of error. And this time his disavowals were printed; at long last, three years after his attempted reply to Sharpe, he finally broke through the censorship in the *Quarterly Journal*. In the context of later developments, it is important to note that although Sedgwick assigned the boundary-drawing in South Wales entirely to Murchison, he still accepted responsibility for placing the line too low in the north. Morever, he did not believe the boundary to have been far off the mark, rejecting not only Sharpe's results but also the full extent of the Survey roly-poly. Given these views, the physical masses of the two systems continued to define regions roughly equivalent to the original Cambria and Siluria. Sedgwick even emphasized the degree to which his argument from geographical propriety depended on the broad accuracy of the collaborative boundary.[44]

Murchison continued to assign blame much as he had done in 1843. He gave Sedgwick all responsibility for North Wales and left the situation in the south more ambiguous. In one passage he even went so far as to imply that the boundary had everywhere been Sedgwick's doing, thus undercutting the argument that the geographical limits of Siluria should define the sectional base of the Silurian system.[45] But by this time he was more concerned with boundaries of an entirely different scale. If Sedgwick stoutly defended his sections and the associated geographical boundaries of Cambria, Murchison maintained with equal tenacity the broad limits defining his extensions of the Silurian classification across the world. He concluded his reply by outlining the subtractions from

[44] Sedgwick 1847: 161.

[45] Murchison 1847b: 173.

Siluria's acreage that the adoption of the Cambrian would require. In England, Wales, Ireland, Russia, Scandinavia, and America his system would be drastically reduced from its currently accepted dimensions. Murchison drove the point home in a letter to the American geologist James Hall, written within a week of the reading of Sedgwick's paper:

> It would surely, therefore, be a singular recompense for all my labours if the Silurian System should be so attenuated, so deprived of its larger part *as scarcely to be recognizable on general maps*.
>
> Referring to general maps, it does not even now occupy one-half of the area of the Devonian System in Russia or Scandinavia and not a twentieth part of the area of that system in Germany. But if the Lower Silurian be abstracted, the *very name* would be driven from the mainlands of Russia and Sweden and confined to the Baltic Isles; whilst in England it would be a mere band.[46]

Murchison panicked at the mere thought of such dire possibilities, and to prevent them he penned similar letters to Agassiz, Whewell, De la Beche, and presumably a host of other men of science. To dismember his accomplishment by giving up the fossiliferous rocks below the Upper Silurian was totally objectionable. As he told Featherstonhaugh, "if I cut off my bottom I know that I should soon expire."[47]

Constructing a Natural Classification

The first major confrontation between Sedgwick and Murchison provides a convenient vantage point for surveying the methodological arguments raised by the controversy. Insofar as it concerned the proper method for constructing a natural classification, the dispute bears a close affinity to other conflicts throughout the whole range of the life and earth sciences, both in the nineteenth century and up to the present day. Should organisms be classed primarily by mature or embryonic forms? Should minerals be ordered according to chemical composition or external configuration? Should plants be categorized by their reproductive parts or by some other combination of characters? The answers to such questions were far from trivial, for they demanded fundamental choices about the ordering of

[46] Murchison to Hall, 23 Dec. 1846, in Clarke 1923: 158-162.

[47] Murchison to Featherstonhaugh, 15 Jan. 1847; also Agassiz to Murchison, 28 Apr. 1847, NMW; Murchison to Whewell, 20 Dec. 1846, TC: O.15.48[54]; and Murchison to De la Beche, 23 Dec. 1846, NMW.

nature. In this sense, classifications functioned as primary tools of scientific perception, opening up certain ways of seeing the natural world while closing off others.[48] Sedgwick's use of the perceptual metaphor of an inverted telescope to describe his opponents' way of seeing was thus in many respects highly appropriate. Differing methods of classification led Murchison to see hundreds of feet of rock in Wales where Sedgwick saw several miles, one system of life where Sedgwick saw two intermingled groups, a superseded territorial limit where Sedgwick placed an important geological boundary.

Edward Forbes, the most wide-ranging "philosophical naturalist" of the younger generation, pointed out some of the similarities between the present controversy and contemporary disputes in zoology and botany. He recognized that the analogy between "natural classifications" in the life and earth sciences remained imperfect. In his view, natural groups of species in zoology shared the same plan or ideal type, while natural groups in palaeontological stratigraphy were based on "assemblages of a certain number of organized species within an arbitrarily assumed portion of geological time." This did not mean, however, that classifications in geology could be dismissed as arbitrary human inventions. Drawing upon the work of Leopold von Buch, he emphasized that the entire fossil fauna of each geological period possessed its own uniquely characteristic "facies." Obviously, this was not a modern palaeoecological definition of the term; rather, Forbes was referring to an overall similarity or "aspect" shared by all the organisms that had lived and died within a particular geological epoch. It was "as if the mighty Maker of all things. . . had stamped each age of life with a seal that would distinguish it whilst the fragment of one of its organisms remained." As an extreme idealist and thoroughgoing advocate of a purely palaeontological classification, Forbes held views on classification that not all his contemporaries would have been willing to second. But almost all shared his belief that geologists would eventually uncover a truly natural classification for the strata.[49]

For all the later vehemence of the Cambrian-Silurian debate, most of the classificatory issues it raised were already present in the win-

[48] Important historical case studies include Winsor 1969, 1976; J. Dean 1979. Among many other works on the subject, Mayr 1982: 147-397 gives a convenient discussion of biological classification, while Farber 1976 discusses the significance of the type concept in the life sciences.

[49] [Forbes] 1854b: 390-391. For illuminating comments on Forbes, see esp. Mills 1984, and Browne 1983 and Rehbock 1983.

ter of 1846-1847: the number of separate faunas in the older rocks, the geological structure of Wales, the choice of a sequence to illustrate the origin of life, and the accuracy and significance of the old collaborative boundary. At a deeper level, though, each of these disagreements originated in distinctive ways of seeing the geological record, differences implicit even in the drawing of the collaborative boundary in the 1830s and fully revealed only in controversy. After Sedgwick encompassed the "Lower Silurian" fauna in his December 1846 paper, the tensions present in the original definitions of Cambria and Siluria widened into open conflict. From the first, Sedgwick had emphasized the structural basis of the Cambrian and Murchison had stressed the palaeontological underpinnings of the Silurian, but both men always expected that in the end these criteria would agree. In the event, this had not proved to be the case, at least not in any way that met with universal approval. As a result, after the breakdown of the collaborative boundary their classifications began to present the Victorian geological community with alternative methods of reading the record of the rocks. "Geology," Sedgwick had written in January 1846,

> tells us of the successive revolutions and changes in the crust of the earth. Organic changes are our surest guides in making out this history; but they form only a part of our evidence, and the great physical groups of deposits, however rude and mechanical, are historical monuments of perhaps equal importance in obtaining any true and intelligible history of the past ages of the earth; and after we have descended through a certain number of stages, they become indeed our only monuments and indexes of past events.[50]

All the elements of Sedgwick's argument, from his emphasis on the original boundaries of the type areas to his encompassing of the "Lower Silurian" fauna with the underlying rocks of Cambria, depended ultimately on the primacy of physical geology. From his perspective, any viable classification needed to take geological structures and sections into account as the principal data. In reply, Murchison compared the stratigraphical record to a set of books, and asked readers to choose between his volume and that proffered by the Cambridge professor. Sedgwick's slaty unfossiliferous tome was vast, but almost unreadable, he claimed, while his own had been written according to "nature's clear and normal types" in the

[50] Sedgwick 1846: 129.

straightforward language of palaeontology. "The geologist," he wrote, "therefore ought I think to prefer this simple and unbroken legend of primaeval life, to that which, however voluminous it may be, is interleaved with numerous blank, torn and ruffled pages, and whose earlier leaves are so nearly illegible, that they have not yet been deciphered."[51] As suggested earlier, the methodological differences separating Sedgwick and Murchison reflect the particular social and intellectual circumstances in which they practiced geology. The "geometrical" and structural approach appropriate to Cambridge contrasts with the palaeontological methods more commonly pursued in London and Oxford. Murchison, with his military background and interest in imperial expansion, had become an outspoken advocate of the use of fossils.

In a variety of guises and widely different circumstances, a parallel contrast in methods has resurfaced throughout the history of the earth sciences. Martin Rudwick has pointed to the importance of this issue in the Devonian controversy, in which De la Beche emphasized physical evidence and Murchison (as in the present instance) the palaeontological.[52] When Murchison eventually crossed swords with Jukes over these same rocks during the 1860s, the contrast emerged again. In a famous American controversy over the older strata that in many ways perfectly parallels the Cambrian-Silurian dispute, Ebenezer Emmons fought for a "Taconic system" from a perspective closely akin to that of Sedgwick. Later in the century the cephalopod expert Edmund Mojsisovics von Mojsvar argued with his fellow Viennese Alexander Bittner over the classification of the Triassic rocks of the eastern Alps. Like similar disputes in Sweden, Bohemia, and Belgium, these controversies often involved tectonically disturbed areas that contained a scanty fossil fauna. Such areas, typical of the older rocks, presented the geological community with what were seen as serious ambiguities. The standard exemplars for classification had been developed in highly fossiliferous successions found in the Tertiary and Secondary rocks. In similarly clear sequences among the older strata, a consensus could often be achieved without great difficulty. But sequences perceived to be structurally complex, especially when poor in fossils, almost always brought conflicting methodological priorities into prominence. In constructing their classifications, geologists perforce stressed one or another of these elements.[53] It should be em-

[51] Murchison 1847a: 45.

[52] Rudwick 1985.

[53] For Jukes, see C. A. Browne 1871 and Herries Davies 1983; for Emmons and the

phasized that in any of these disputes a variety of viewpoints could be cogently defended, just as Sedgwick argued for the Cambrian in 1846. But even if there was no universal triumph of one approach over another during the century, there can be little doubt that the palm of victory generally went to those who took a broadly palaeontological point of view.

In mid-century Britain the choice of methods was usually obvious. Most researchers reached unhesitatingly for Murchison's slim volume of fossil types and relied entirely on palaeontological evidence in determining the boundaries between the major divisions of the geological column, even if they used more practical criteria in the field. In the absence of possible candidates for a fauna lower than that described in the *Silurian System*, virtually every active geologist rejected the Cambrian. When Murchison confessed fears for his system to Ramsay over coffee early in 1847, the latter could only confide to his diary the obvious: "He is in a state of needless alarm." Geologists around the world added their unequivocal support in response to Murchison's anguished letters. "As for your Silurian system," wrote Agassiz from America, "you need not have the slightest apprehension; not only your friends but every one who has the least feeling for justice will sustain your claims to the full establishment of the geological and palaeontological character of the oldest fossiliferous strata as *one* great system, (un et indivisible) as the Swiss republic has for some time been called. The same opinion is prevailing among geologists here." No one questioned the brilliance of Sedgwick's paper, which lacked the loopholes that had allowed Murchison's easy escape the previous summer. Its arguments, however, were ultimately based on structure, thickness, and lithology and fell on deaf ears. "If we had come to a division among those geologists present who understand the subject," Horner wrote after the December meeting, "I suspect that the learned Professor would have gone forth alone, notwithstanding his eloquence and his long established authority."[54] After 1846 Sedgwick did not resume full-scale combat for another five years, but his lonely battle for the Cambrian had just begun.

Taconic, Schneer 1969b, 1978; for the Triassic, Tozer 1984; for other debates, see Zittel 1901, esp. pp. 446-450, 482-492.

[54] Horner to C. J. F. Bunbury, 20 Dec. 1846, in K. M. Lyell 1890, 2: 109-110. The other two quotations are from Agassiz to Murchison, 28 Apr. 1847, NMW; and Ramsay, Diary, entry for 6 Jan. 1847, ICL(R): KGA Ramsay 1/8, f. 10r.

CHAPTER SEVEN

Professional Geology and the Quest for Priority

After the first public outburst of the controversy, any paper on the older rocks could lead to a discussion of the opposing classifications. To settle the matter once and for all, the Geological Society devoted an entire meeting early in 1852 to the question. But rather than resolving the dispute, this exchange opened four years of angry debate, which rapidly moved outside Somerset House and led to a bitter estrangement between Murchison and Sedgwick. As before, the arguments centered on the correct use of nomenclature in geology, the validity of purely palaeontological classifications, and the unity of the Silurian fauna. But this time there was a difference. The official Geological Survey, working together as a cohesive team of professional geologists, was now pursuing the stratigraphical enterprise with an unexampled thoroughness. Universal acceptance of the basic sequence shown on the Survey maps of Wales ended substantive debate over its structure, and for a short while also quieted doubts about the persistence of the "Lower Silurian" fauna to the base of the fossil-bearing rocks. In 1852 Murchison and Sedgwick relied upon contrasting interpretations of these authoritative results and couched their debate in terms of historical justice and priority of discovery. Insofar as it concerns the two principals, the dispute from this point onwards provides a revealing and characteristic instance of priority conflict in science.

The Rise of Professional Geology

The publication of the Survey one-inch-to-the-mile maps of North Wales in the late 1840s and early 1850s (Fig. 7.1) profoundly altered the course of the controversy. After the Survey completed its work, individual gentlemen geologists like Sedgwick, Sharpe, or Murchison could no longer present interpretations of the structure of the entire Principality after tours of a few weeks or months. It is already evident just how thoroughly the discovery of the South Wales roly-poly undermined Sedgwick's attempts to prove a vast thickness be-

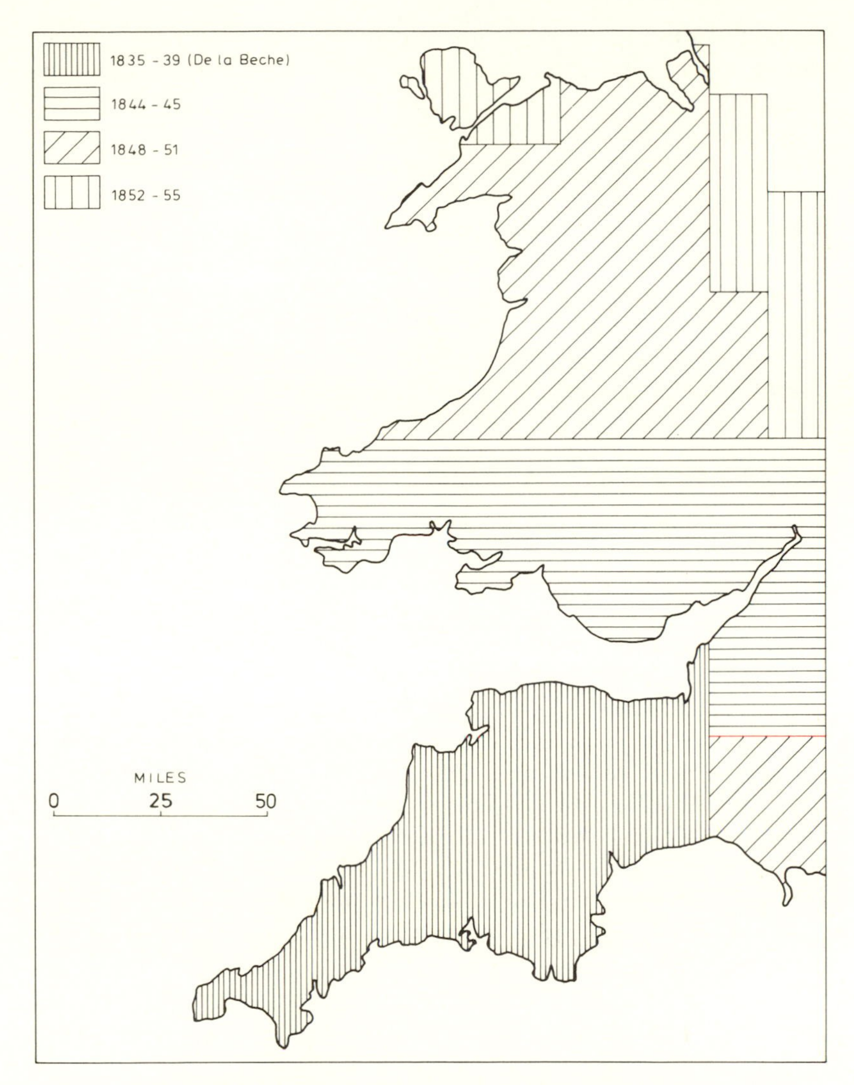

Fig. 7.1. Progress of the Geological Survey in Cornwall, Devon, and Wales, 1835–1855.

low the Silurian. With the completion of the one-inch maps, all parties in the earlier disputes accepted the official picture of the succession of rocks in Wales, despite continuing disagreement on matters of classification, nomenclature, and detail. The reasons for this deserve detailed examination, especially as these changes typify a widespread shift in the methods and standards of accuracy in geological work throughout Europe during the second half of the nineteenth century. In Britain, as in several other countries, the emergence of geological surveying as a paid profession was instrumental in producing these changes.[1]

In an important article, Roy Porter has argued a contrary view: that the Survey had almost no effect at all on the daily practice of British geology. The members of the Survey, he says, continued the old "romance of the field" and followed the traditional goals of the stratigraphical enterprise with an earnest zeal already pioneered by the previous generation of specialist geologists. They mimicked the "big wigs" at the Geological Society with stratigraphical songs and poems, rebelled against bureaucratic forms, and dined with one another in a cosy male confraternity. Their intellectual aims also exhibited important continuities, with strata mapping and classification as the *raison d'être* of the organization. Broadly speaking, the picture drawn in Porter's essay portrays undoubted elements of truth, particularly in its emphasis on the research-orientated ethos of pre-Survey geology. To be a gentleman geologist of the calibre of Murchison, Fitton, Sedgwick, Buckland, or Lyell had demanded the devotion of a career even if it did not provide the emoluments of a paid profession.[2] But in emphasizing this point, it is essential not to lose sight of the marked differences in quality, scale, and scope that separate the Survey work from its predecessors. The official maps and memoirs produced after the mid-1840s possessed a thoroughness and a finish virtually impossible for practitioners of the older individualistic tradition to rival. After the Survey published its findings, geological study in an area usually entered what Murchison and his generation saw as "pottering times, the chief work being over."[3] Professionalization, even in the form of paid career structures, made a major difference.

The rise in standards, however, did not occur overnight. In effect the Survey originated as a means of allowing De la Beche, one of the

[1] The situation in France, as Gillispie 1981 makes evident, was substantially different. Important studies of professionalization in nineteenth-century science include S. F. Cannon 1978: 137-200, Allen 1985, O'Connor and Meadows 1976, Porter 1978, Turner 1978 and 1980; and a particularly useful study by Reingold (1976), with references to the earlier literature.

[2] Porter 1978, esp. pp. 825-829. Morrell 1976: 139-142 gives an interpretation closer to that adopted here, although with reference to an earlier period; see also North 1932 and 1934.

[3] Murchison, quoted in Geikie 1875, 2: 151. The same trend towards increasing detail is evidenced in a different way in D. Bassett 1963 and 1967, which give thorough bibliographies for nineteenth-century Welsh geology.

elite circle of gentlemen specialists, to continue a program of mapping already begun with private funds and on his own accord. As one might expect, the quality of the documents he produced during his solo survey of Devon and Cornwall was very much of a kind with unofficial work done elsewhere. De la Beche's status as a government employee even led to occasional suspicions that his work was less than first-rate, as illustrated by accusations of jobbery leveled during the Devonian controversy.[4] Certainly his statements commanded no immediate assent from the geological community.

But ever the enterprising bureaucrat, De la Beche had wider plans for his Survey, which he began to implement in the late 1830s and early 1840s as the mapping moved across South Wales. Most important, the organization grew. By the time Ramsay joined in 1841, four other assistants were in the field, and Richard Phillips as chemist and curator and Trenham Reeks as his assistant made up the London staff. Only a handful of paid posts in Victorian England existed for those who needed money for their science, and so De la Beche could obtain the very best of the new generation even on a limited budget. Perhaps the most promising recruit of all was the brilliant and sociable young naturalist Edward Forbes, who shared rooms with Ramsay in London. Born on the Isle of Man in 1815 and educated at the University of Edinburgh, Forbes was especially known for his pioneering expeditions in offshore dredging and his interest in the geographical distribution of animals. As curator of the Geological Society after Lonsdale's retirement in 1843, he had gained some experience in invertebrate palaeontology, and less than a year later De la Beche put him to work identifying the Survey's Welsh fossils. Forbes found the task tedious, but admitted that "the gov'ner" also gave time for theoretical pursuits, and he rapidly directed his thoughts to the distribution and history of Lower Palaeozoic life. From 1846 he had the expert assistance of Salter, who at the age of twenty-six already possessed a practical familiarity with the ancient fossils unrivalled in Britain. Another appointee in the same year was Sedgwick's old pupil Beete Jukes, the son of a Midlands manufacturer whose early years had been spent geologizing in Newfoundland and Australia. The fiery-tempered Jukes had become a particularly fine field surveyor and a close friend of Forbes and Ramsay. Together these three constituted what one biographer called the "triple brotherhood" of the Survey. Although Jukes's Survey salary supplemented a small private income,

[4] Morrell 1976: 141-142 and Rudwick 1985.

he needed the money nonetheless, and his acceptance of dreary exile as Irish director in 1851 was motivated almost entirely by financial necessity. Thomas Oldham, Jukes's predecessor in Ireland and later head of the Indian Survey, and Alfred Selwyn, director of colonial surveys in Australia and Canada, both spent their early years under De la Beche. Joseph Hooker, Lyon Playfair, and Thomas Huxley were among many appointees of later note (although of less relevance to the present account) who likewise used the Survey as an essential step along the long road to professional security in science.[5] In the middle of the century this string of impressive appointments made "Sir Henry & Co." (or "that ill-paid enthusiastic band of peripatetic *savans*" as Forbes called it) into the most exciting center for theoretical and descriptive natural history research in the British Isles.[6]

The Survey's standards in its primary task of documenting British geology underwent an equally dramatic change from the days when De la Beche hammered almost entirely on his own. Indeed, the increasing specialization and technicality rightly noted by Porter as characteristic of late Victorian geology resulted in large part from the very Survey whose professionalism he so deemphasizes. The most significant improvement took place on its published geological maps. Admittedly, the one-inch maps of the Ordnance Survey which provided their topographical base had in many cases been available for years, and gentlemen geologists such as Sedgwick, Bowman, and Murchison had used them wherever possible. On many occasions they had obtained advance copies or pencil tracings directly from the Ordnance authorities at the Tower of London. The utility of the one-inch maps is evidenced by the early work in Wales.[7] Much of Murchison's territory had already been covered by the early 1830s, and the topography of the map in the *Silurian System* had an enviable permanence. On the other hand, North Wales had not been finished, and the unfortunate Sedgwick had colored in most of its geology on less accurate unofficial maps. Writing to Phillips in 1843, he lamented that much of his early work had been superseded merely by the publication of the Ordnance Sur-

[5] For Forbes: G. Wilson and A. Geikie 1861; for Salter: Secord 1985b; for Jukes: C. A. Browne 1871 (which speaks of the "triple brotherhood") and Herries Davies 1983; the others are all in the *DNB* and *DSB*.

[6] [Forbes] 1854b; "Sir Henry & Co." is taken from Phillips to Ramsay, 10 July 1842, ICL(R). For the Survey's theoretical work, see Secord 1986.

[7] The progress of the Ordnance Survey is described in Close 1926 and Seymour 1980.

vey.[8] Bad base maps had rendered many of his findings useless for the new generation: the visual language of his work could not be translated from one set of documents to another.

Although Sedgwick and Murchison had obviously been aware of the superiority of the Ordnance maps, like most researchers prior to the Geological Survey they had used them as a means to producing an end product much reduced in scale. Even when the maps appeared at full size (as had those of De la Beche in Devon and Cornwall) no attempt was made to provide the detailed information that the larger scale permitted. After the early 1840s, in contrast, the men of the Survey colored their maps with much greater care. In particular they stressed the tracing of contacts between individual formations, a method that determined the position of the strata much more precisely than did the rapid traverses undertaken by their predecessors. Ramsay, Jukes, Forbes, and other young members of the Survey corps particularly favored the new emphasis on accuracy and frequently wished to include more detail than time, money, or De la Beche would allow. Nothing better suggests the rise in standards than their eventual scorn for De la Beche's earliest work, which had once seemed perfectly adequate. They looked with particular envy across the Irish channel, where the Survey's mapping was conducted at the exacting scale of six inches to the mile. The difference in scales between the two islands was something of a historical accident—Ireland having been topographically surveyed only at the six-inch scale, England and Wales only at the one inch—but the consequences were significant. One-inch maps demanded fairly high levels of interpretation and extrapolation; the six-inch scale could be used to record the characteristics of individual outcrops as seen in the field.[9]

Even with this limitation, the Survey registered major improvements in Wales. In part, horizontal sections constructed on the six-inch scale compensated for the lack of similarly detailed base maps (Fig. 7.2). These sections, which illustrated the succession along particularly revealing traverses, greatly enhanced the authority of the Survey work after the early 1840s. De la Beche adopted the idea (albeit in a modified form) from William Logan, a Canadian-born manager of a Swansea mining enterprise who aided extensively in

[8] Sedgwick to Phillips, 10 Oct. [1843], ff. 46-53, UMO. For Murchison, see Geikie 1875, *1*: 205, 207.

[9] Complaints by the younger generation are evident in Forbes to Ramsay, [1854], ICL(R); Jukes to Ramsay, 5 Nov. 1857, ICL(R); Ramsay to Salter, [n.d.], in Geikie 1895: 206. The Irish mapping is well described in Herries Davies 1983.

the South Wales coalfield mapping.[10] Notably, the six-inch sections were measured instrumentally with theodolite and chain—the first time such methods had been systematically applied in English geology. Murchison extolled their accuracy in his 1843 presidential address to the Geological Society and remarked that the Survey sections possessed an excellence unobtainable by an individual naturalist. "I am convinced," he wrote, "that it will not only act directly as a great national benefit, in making more correctly known the structure of the subsoil, in a manner beyond the reach of private enterprise, but that it will materially tend to elevate Geology, by connecting it in a permanent manner with Physical Science."[11] The detailed maps and measured sections gave the Survey publications an air of exactitude like that in other contemporary government reports, as De la Beche and his assistants provided the British public with a "census" of the British strata.

So effective had these methods become by the end of the 1840s that the completion of the official survey of North Wales immediately stopped all further debate between Sharpe and Sedgwick on the overall thickness of the Welsh succession. As Sedgwick had said in December 1846, "a little army of good observers is now marching over the country, who will settle points of difference, and give a finish to details, utterly beyond the physical powers of one who entered the country single-handed, and while it was unknown in all its peculiarities of physical structure."[12] But the Survey did not so much vindicate his work (or that of Sharpe) as supersede them both. The field geologists of the Survey found the earlier papers useful in guiding attention to particular points of interest—fossil localities, small-scale faults, and important limestone quarries. Nonetheless they usually started almost from scratch in interpreting the phenomena on a larger scale and sometimes seem to have read the relevant papers by Sedgwick and Sharpe at a relatively late stage in the mapping.[13]

The Survey resolution of the long-standing debate over the dip of the Bala Limestone and the position of its alleged counterpart at Rhiwlas provides an excellent example of the relation of its work to that of the earlier investigators. Sedgwick had said that Rhiwlas was

[10] Bailey 1952: 32-34; Geikie 1895: 44.

[11] Murchison 1843a: 79; see also Horner 1847: xxvi.

[12] Sedgwick 1847: 135-136.

[13] For the Survey's use of earlier work, see Ramsay to Aveline, 23 July 1846, f. 46, BGS: GSM 1/420(A); Jukes to Ramsay, 23 May 1847, ICL(R), printed in C. A. Browne 1871: 300-302.

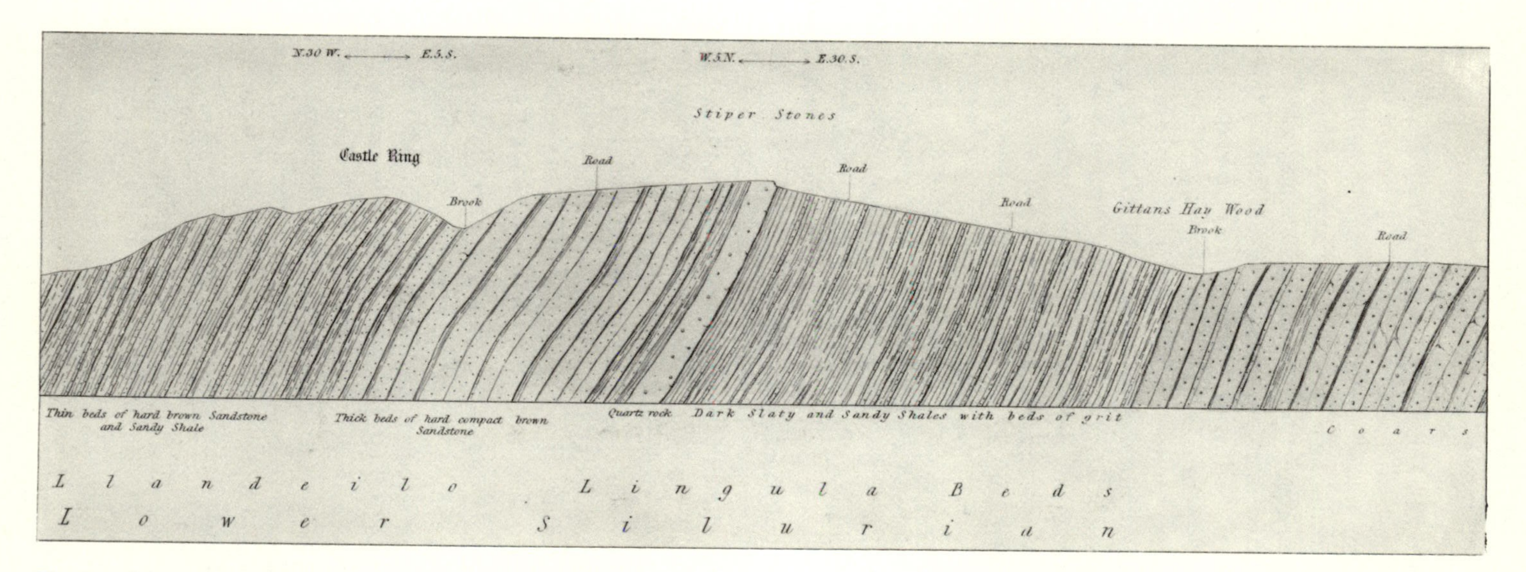

Fig. 7.2. Six-inch-to-the-mile horizontal section across the Stiperstones as surveyed by Aveline; note their stated equivalence to the Lingula Flags of North Wales.

a lower bed than Bala (Fig 5.3*b*), while Sharpe had placed them on the same horizon and connected them by means of a synclinal trough (Fig. 5.3*a*). Jukes, who was responsible for the official one-inch mapping of the Bala Lake region, took yet a third position: Sharpe had been correct in correlating Bala and Rhiwlas, but the repetition occurred through faulting—a category of interpretation often advocated by Sedgwick—rather than by means of a hypothetical trough. Jukes pointed to the wider importance of his finding in April 1847. "How capitally, however, such an example as this shows the necessity of our survey!" he told Ramsay. "Sharpe, or even sharp-eyed old Sedgwick himself . . . might have gone over here with their hasty traverses a dozen times, and fired scores of papers at the Geological Society, and only succeeded in puzzling themselves, perplexing their hearers, and utterly flabbergasting their palaeontological brethern. . . ."[14]

The overall interpretation of Wales offered on the early printings of the Survey maps is shown in Figure 7.3. Superficially the results appear to corroborate Sharpe's views of the succession, with Bala, Rhiwlas, Glyn Diffwys, Glyn Ceiriog, and even Snowdon all on approximately the same stratigraphical parallel rather than on widely separated horizons as advocated by Sedgwick. Particularly as a result of the startling rise of the fossiliferous zone at Snowdon, the Survey had cut the thickness of strata below the Denbigh Flags down to four or five miles, far less than Sedgwick's estimate of ten or fifteen. But this was still a much greater sequence than Sharpe had envisaged, and most geologists felt that the Survey work broadly vindicated Sedgwick's views. In characteristically pungent language Jukes summarized his own opinion of the differences between the work of Sedgwick, Sharpe, and the Survey:

> I have just read over Father Adam's paper on North Wales, and I see he has looked on the physical structure of this country with his accustomed accuracy and sagacity. His "enormous dislocations" are positive facts, only, being more enormous than he imagined, or at least being more frequent, he has taken for different beds what are in reality repetitions of the same. Of course he could not discover his mistake without going into the details as we have done. I have looked also at Sharpe's damned nonsense & can't make top or tail of it.[15]

[14] Jukes to Ramsay, 28 April 1847, ICL(R), partly printed in C. A. Browne 1871: 293.
[15] Jukes to Ramsay, 23 May 1847; see also Jukes to Ramsay, 3 Sept. 1848, ICL(R).

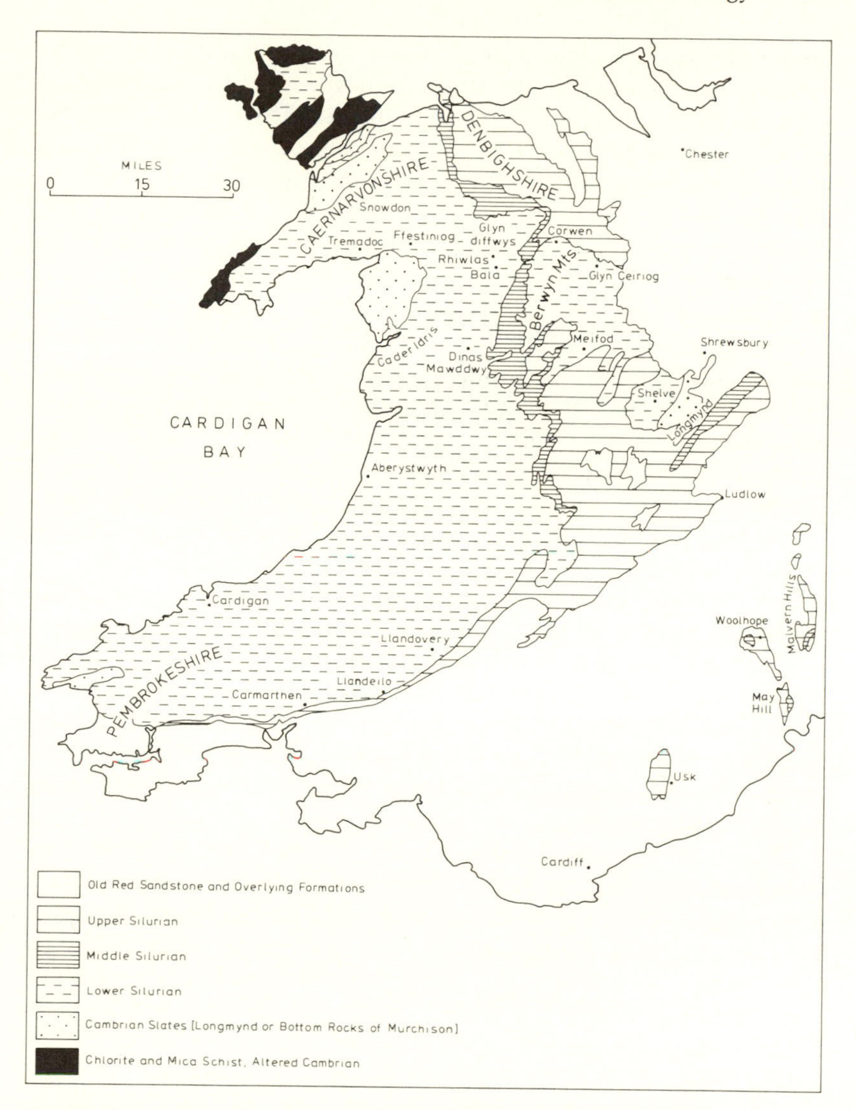

Fig. 7.3. Geological Survey map of North Wales circa 1852, before revisions.

De la Beche and Phillips shared Jukes's opinion of the relative merits of the two pioneers, and an announcement of the Survey's results at a Geological Society meeting in 1848 also singled out Sedgwick for special praise.[16] In this case as in others, however, one senses a residual antipathy towards Sharpe as an upstart trespasser.

In any event the Survey men had strong incentives to distinguish their work from that of all their predecessors. It was in their interest to present new interpretations, to settle long-standing disputes, to harvest unknown facts from the most heavily worked ground. Thus Jukes's untangling of the situation at Bala and Ramsay's work on the South Wales roly-poly are both depicted as independent discoveries rather than minutely detailed elaborations of earlier work. Partly to obtain government funding, but more importantly to retain the respect of the specialized community of geologists, the Survey needed to produce innovations. While mapping Secondary sections along the southern English coast, an area even then in danger of being hammered away to sea level by geologists, Forbes exulted in his discovery of a major stratigraphical error. "The 'geology of England' may be 'done' by the old fellows," he told Ramsay, "but it is not overdone yet."[17] By comparison with such areas, North Wales remained a veritable geological wilderness, bursting with potential discoveries that could enhance a developing reputation. The men of the Survey thus saw the strata not only with more elaborate techniques, but also with an eye for novelty.

The classification and nomenclature adopted by the Survey can be outlined briefly, for it was one area in which innovation played only a small part. In 1846 De la Beche had assured Murchison that the Survey would be guided in such matters by palaeontology alone. When its inquiries concluded in Wales five years later, the organic remains illustrated in Murchison's works seemed to extend to the base of the fossiliferous rocks. Even the Lingula Flags and Tremadoc beds, which the Survey (like Sedgwick) had located on a horizon lower than any in the *Silurian System*, were thought to contain an overwhelming preponderance of Lower Silurian species. The official maps therefore naturally adopted the expanded Silurian. In many respects this acceptance of fossils as the sole basis for the taxonomy of the older rocks marked a further shift from De la Beche's position in the 1830s, when he had explicitly refused to use Mur-

[16] De la Beche to Ramsay, 4 Oct. 1846, ICL(R); Phillips to Ramsay, 14 Oct. 1846, 10 Feb. 1847, ICL(R); Jukes and Selwyn 1848.

[17] Forbes to Ramsay, 28 Oct. 1849, ICL(R).

chisonian terminology anywhere outside the original type area. Now Silurian was employed wherever the appropriate fossils ("the same *kind* of life") could be found.[18] Private conversations with Forbes, whose theoretical views De la Beche deeply respected, seem to have been particularly important in bringing about this change; in addition the proven heuristic value of the palaeontological method was obviously impressive, and as a good administrator De la Beche did not wish to shipwreck his Survey on an issue that no longer held significant dangers for his environmentalist approach to stratigraphy. He did, however, adjust Murchison's classification in two ways. First, the Survey maps placed the Caradoc formation in an intermediate "Middle Silurian"; and second, they used "Cambrian" as a term for the very oldest stratified rocks, then thought unfossiliferous save for a few primitive forms. Although this so-called Cambrian bore strong similarities to that proposed by Sharpe, the Survey men viewed it as a concession to Sedgwick: "Father Adam to have his pristine share of Mother Earth," as Salter told Ramsay in 1848.[19] Of course Sedgwick remained implacable on this point, while Murchison had dropped the troublesome word entirely and spoke instead of "Longmynd strata" or "bottom rocks." With these exceptions De la Beche and his entire staff finished mapping North Wales fully convinced of the validity of Murchison's classification.

Before turning to the effect of these results on the Cambrian-Silurian controversy, two points must be emphasized. First, although the men of the Survey immediately established a consensus about the order of the Welsh rocks, they had by no means said the last word on the subject for all time. Even the specific success trumpeted forth by Jukes—the link between Bala and Rhiwlas—would eventually be challenged by early twentieth-century investigators.[20] Professional geology provided the basis for all future mapping in Wales; it had not ended the possibility of practicing the science there. Second, it is important to recognize that the official classification did not receive the universal approval given to the sequence illustrated in the new maps and sections. One could accept the structure of Wales outlined by the government geologists, and even their views on the persistence of the so-called Lower Silurian fauna, without adopting the accompanying scheme of classification and

[18] De la Beche to Murchison, 29 Dec. 1846, in Geikie 1895: 93-94; and De la Beche 1846, which uses the Silurian classification throughout.

[19] Salter to Ramsay, [1848], ICL(R).

[20] Elles 1922: 134.

nomenclature. With regard to issues involving taxonomic method, an individual geologist could still disagree—even if he did so at a disadvantage.

Sedgwick Returns to the Attack

Since Sedgwick was almost alone in advocating the revival of a fossiliferous Cambrian, any continuing challenge to the expanded Silurian depended primarily on his own activity in the field and at the Geological Society. Certainly no one else at Somerset House was prepared to take up the cudgels for his classification. Yet rather than following his closely argued paper of December 1846 with the supporting communications he had so sanguinely promised, Sedgwick did almost no new fieldwork until the summer of 1851. As usual this was with good reason. Additions and appendices to the fifth edition of his little sermon on university studies swelled it into eight hundred pages of "Geology, Psychology, Theology, Deism, Atheism, Pantheism, Procreation, Transmutation, Parthenogenesis, Academic training, Popery & Tomfoolery."[21] For all the oddities of its organization, the result was one of the most important statements of the anti-evolutionary argument in the years before Darwin. Sedgwick also became busier than ever with university affairs after his appointment to a parliamentary commission of inquiry in 1850. On several occasions illness or injury, including a serious fall and broken arm in 1849, precluded the possibility of work in the field.[22]

But Sedgwick continued sporadically to gather materials for his book on the older rocks. In 1848 he read a paper on the Skiddaw Slate of the Lake District, and two years later the British Association heard a discourse on the geology of the Scottish Borders. These papers provided brief opportunities for reasserting the validity of the Cambrian, as did discussions at Somerset House on topics relating to the older rocks. A particularly important dispute followed a paper by Murchison in February 1851 on Silurian rocks in the south of Scotland. In company with James Nicol, Murchison argued that his Llandeilo and Caradoc formations could be distinguished in the region, providing what his biographer called "a notable step in the progress of the extension of his Silurian domain over Britain."[23]

[21] Sedgwick to Jukes, 11 July 1850, CUL: 7652IIIE13 (copy); Sedgwick 1850.

[22] Clark and Hughes 1890, 2: 108-200.

[23] Geikie 1875, 2: 115-116. The discussion that followed is briefly recorded in Ramsay, Diary, entries for 4 and 5 Feb. 1851, ICL(R): KGA Ramsay 1/12, ff. 24v, 25r. In 1858 Murchison carried this extension even further, putting all the older gneisses

The reopening of full-scale debate a year later had its roots in a paper on Devon and Cornwall read by Sedgwick to the Geological Society in November 1851. In the ensuing discussion Murchison stood up to deliver his customary objections to the use of Cambrian in preference to Lower Silurian. However, William Hopkins—Sedgwick's close associate at Cambridge and the current president of the Society—promptly brought the meeting to order and postponed all further discussion until a special session could be held.[24] Presumably everyone hoped that an organized confrontation would end interminable repetitions of the 1846-1847 exchange. In later years Sedgwick complained bitterly about being officially prohibited from raising the question at subsequent Geological Society meetings. Such a proscription is not entered in the minute books, but it seems to have been generally accepted that an extended debate would close the question. Certainly Sedgwick entered the lists with the understanding that this would be "my final battle with Murchison," as he called it beforehand.[25]

The anticipated resolution of the controversy generated intense interest, and an audience that included almost all the specialists on the older rocks in Britain attended the meeting of 25 February 1852. As the geologist Gideon Mantell told his American friend Benjamin Silliman, the debate was "very warm" and lasted until after midnight.[26] The main event of the evening was a lengthy paper from Sedgwick, which was divided into two parts: the first half announced a major change in his correlation of the older rocks of the north of England with their counterparts in Wales, and the second delivered an elaborate attack on the Silurian system and its author.[27]

Although Sedgwick's new paper ostensibly focused on the Lake District, his changes there were predicated upon a far more momentous shift in his view of the structure of Wales. After a decade of resistance, he had finally accepted the succession elaborated by the Survey. Certainly Sedgwick would later criticize the accuracy of the official maps and sections in important details, and he consis-

---

and schists of the Scottish Highlands into the Silurian; see Geikie 1875, 2: 220-221. For Sedgwick's work, see his papers of 1848 and 1851.

[24] The paper was published in full as Sedgwick 1852a; Hopkins's action is described in Sedgwick to McCoy, 9 Nov. 1851, ML: frames 411-413.

[25] Sedgwick to McCoy, 6 Feb. [1852], ML: frames 483-485.

[26] Mantell to Silliman, 26 Feb. 1852, YUL: Silliman Family Papers, group 450, ser. II, box 22, folder 88. Sedgwick's own report of the discussion is in Sedgwick to McCoy, 8 Mar. 1852, ML: frames 181-183.

[27] Sedgwick 1852d.

tently repudiated their nomenclature and classification for the rest of his life. But by and large, the folly of any sweeping challenge to the government work (such as his questioning of the South Wales sections in 1846) had become all too apparent. De la Beche and his corps had cast a long shadow to the north ever since their entry into the Principality in 1841, and much of Sedgwick's work during the following five years can be seen as a last-ditch effort to record his status as a pioneer. Underlying all the reasons for his inactivity in research towards the end of the decade, one senses a certain hesitancy to speak out again until the completion of the official map. In this respect it is significant that his acquiescence in the Survey results should appear, not in a paper devoted to Wales itself, but rather as the assumed starting point for new correlations in the Lake District, an area not yet under the government hammers. Sedgwick had finally realized that his own knowledge of Welsh geology, however extensive, could never match the Survey's corporate expertise. His plight foreshadowed the fate of the tradition of gentlemen specialists in British science.

A comparison of the February 1852 paper with his earlier publications illustrates the changes that Sedgwick needed to make in order to coordinate his own ideas with the official findings. The principal alterations, already discussed in connection with the Survey, are summarized in Figure 7.4. As pointed out earlier, a thick sequence of rocks on separate horizons had proved in many cases nothing more than a series of the same beds repeated by folding and faulting. This was especially evident in the limestones of the Berwyn and Bala Lake region. Even as late as 1850 Sedgwick had placed these on separate horizons many thousands of feet apart. In his new paper, however, he joined the Survey in uniting them on a single level.[28] Since Sedgwick nowhere gave a convincing explanation of his reasons for this change of views, we are left to assume that he did so because of the authoritative government map. Sedgwick naturally hesitated to admit past errors; just as the informal etiquette of contemporary science condoned withholding the name of a mistaken predecessor from mention in print, so too did it allow glossing over one's own blunders after they had been corrected by others.

The placement of the Bala beds on a single horizon reopened the

[28] Sedgwick 1852d: 149. For his views in 1850, see Sedgwick to Jukes, 6 Apr. 1850, CUL: Add. ms 7652IIIE14 (copy); also Sedgwick 1848. Sedgwick and the Survey continued to differ in their interpretations of the igneous sequence even after the Survey maps.

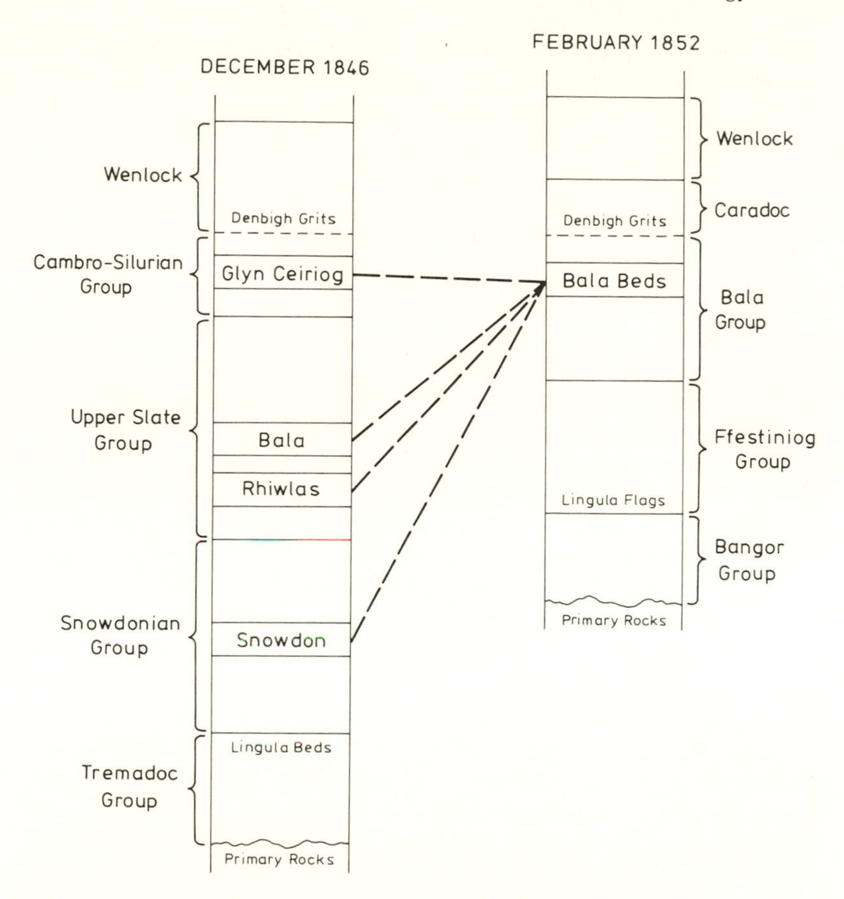

Fig. 7.4. Comparative columnar sections indicating changes to Sedgwick's view of Wales necessitated by the work of the Geological Survey.

old problem of the correct dovetailing between Cambria and Siluria. Despite frequent internal changes to his correlations in North Wales, Sedgwick had previously accepted the basic relationship between these two groups as set out during the joint tour of 1834. But now, in the wake of the Survey publications, he abandoned this critically important view for the first time. As a preliminary step, he silently withdrew his controversial positioning of the *Asaphus Buchii* beds above the Caradoc; the entire progressionist scheme related so closely to it, and of such importance for his classifications in the 1840s, was never mentioned again. Then, like the Survey, he took the all-important step of correlating Murchison's Llandeilo Flags in

South Wales directly with his own Bala beds in the north.[29] The stratigraphical separation maintained between these two formations ever since the early 1830s had suddenly disappeared. Even Sedgwick, who had tenaciously fought a link between Bala and any part of the Silurian for almost two decades, could not but accept the determinations of Jukes, Ramsay, Aveline, and the other Survey men responsible for the revised interpretation.

In the second half of his paper Sedgwick went on to pay De la Beche and his colleagues the sincerest form of compliment, by maintaining that they had corroborated point by point his views of the structure of Wales as elaborated in 1832. As far as this involved the fossiliferous rocks below the Bala Limestone, he could speak with some justification, for with the significant exception of the Snowdon beds, his main divisions had remained constant for twenty years and required little modification.[30] But when it came to the higher parts of the Cambrian sections, a far more extensive revision of the history of his research was required, one that led Sedgwick into a highly personal attack on his former collaborator. In previous chapters I have uncovered the intricate chain of reasoning that guided him in unravelling the structure of Wales during the 1830s and 1840s. From his own perspective in 1852, his work had a much simpler history: He had begun with a correct conception of the place of the Bala beds in the succession in 1832, but had subsequently been driven into error after error through misguided attempts to relate his sections to those of the Silurian. For the first time in the debate Sedgwick blamed Murchison for every failure, every wrong step, every deviance from the pristine truth revealed by nature. "His nomenclature was premature," Sedgwick wrote, "and his base-line was sectionally wrong; and, so far from leading to discovery, it retarded the progress of palaeozoic geology for, I believe, not less than ten or twelve years."[31]

Sedgwick's sensitivity to accusations of blame has already been illustrated by his vehement denunciation of Sharpe's historical statement. Up to 1852, however, he had always limited his complaints to a single valid point: Murchison was fully responsible for the geographical boundary that separated Cambria from Siluria in South Wales. But after Sedgwick accepted the Survey correlations,

[29] Sedgwick 1852d: 148-149, 167.

[30] Compare Sedgwick 1852d: 145-151, and 1847: 157-158. Sedgwick refused to admit that he had been wrong about Snowdon; see Sedgwick 1855: xli for his fullest discussion of the question.

[31] Sedgwick 1852d: 164.

his accusations concerning South Wales became much more serious. He had previously charged Murchison with errors of classification and nomenclature—of important mistakes in boundaries—but the new claim involved fundamental responsibility for ignoring a massive stratigraphical overlap. Murchison alone, he asserted, should be blamed for putting the South Wales slates below the Llandeilo formation during the preliminary dovetailing.[32] For a man who had contested the Survey roly-poly as late as 1846, this was a remarkable accusation indeed.

But Sedgwick's attack went still further. Murchison, he claimed, had not only been responsible for the mistakes in South Wales, but had also perpetrated the same error in the north. In previous papers Sedgwick had always assumed blame for any false correlations in North Wales, most notably his placement in 1834 of a band of supposed Caradoc strata at the base of the Denbigh Flags. His 1852 retrospective, in contrast, put the ultimate responsibility for every rejected view of the previous twenty years squarely on Murchison's shoulders. To show how Murchison could be guilty of so many mistakes in North Wales required a lengthy and ingenious explanation, for that region had always been regarded as Sedgwick's special province of investigation.

Since the two men had so effectively separated their scientific territories, their tour together in 1834 afforded the only opening for the shifting of blame that Sedgwick desired. From his perspective in 1852, the determinations Murchison had made during the traverse over the Berwyns accordingly became the critical events upon which the whole affair turned. Sedgwick began the composition of his account in the knowledge that in 1832 he had placed Bala, Meifod, and Glyn Ceiriog on a stratigraphical parallel, with the South Wales slates and the higher beds of the Berwyns immediately above. This arrangement bore strong (although ultimately superficial) similarities to the latest view shown on the Survey map, and it is hardly surprising that Sedgwick believed he had in some manner been led astray. However, rather than realizing that his "correct" grouping of the Bala beds had little meaning unless related to the overlying succession, he understandably grasped at it as his own original view of Wales, before Murchisonian errors had led him from the truth.[33]

According to Sedgwick, the great mistake had come in 1834 when Murchison broke the Meifod-Bala correlation and declared Meifod

[32] Sedgwick 1852d: 151, 158-160.

[33] Sedgwick 1852d: 151-152.

to be Lower Silurian Caradoc and Bala to be Cambrian. Sedgwick claimed that he had accepted this determination with great difficulty and only because of his "implicit faith" in the "perfection" of Murchison's "workmanship." The language of religious belief served an important function in his argument, for Sedgwick needed to explain why he had not abandoned the Meifod-Bala correlation before it supposedly stultified all his work on the older rocks for over a decade. In a complex and detailed retrospective he claimed that this determination had led to a remarkable series of mistakes, from 1834 when he packed a "hypothetical" band of Caradoc strata just below the Upper Silurian Denbigh Flags, to 1846 when he forced the South Wales slates into the Upper Silurian. Only during the summer of the latter year, he remembered, did the entire rotten edifice fall to the ground when the fossils of these slates proved unquestionably Lower rather than Upper Silurian. The Meifod Limestone just below (and the Bala Limestone with it) could not therefore be Caradoc as Murchison had said, but must occupy a lower horizon parallel to the Llandeilo. Sedgwick claimed that as a result he had returned to his original classification of 1832; from the end of 1846 he had defended views subsequently arrived at by the Survey.[34]

Anyone who has followed Sedgwick's actions as reconstructed in the preceding chapters will be in a position to admire his retrospective, even through this brief description, as one of the most remarkable technical myths in the history of science. Each of the errors in two decades of research was explained and every explanation pointed an accusing finger at Murchison. More surprising is just how little relation the entire account bears to the actual course of events, even granting the widest possible latitude for differences in interpretation. To cite only one example, the placement of the band of Caradoc strata at the base of the Denbigh Flags had not been "hypothetical," but was in fact based on extensive fieldwork in 1834; moreover, its positioning had always been largely independent of the Meifod beds. More important, Sedgwick had by no means defended the same view of the relation of his strata to those of Murchison in 1832, in 1846, and in 1852. When he stated any conclusions at all on the subject in 1832, it was only to separate the two groups by as thick a sequence as possible. In 1846 the Bala Limestone and Murchison's lowest formations were similarly kept apart

[34] Sedgwick 1852d. He had first developed this account in letters to Murchison; see Sedgwick to Murchison, 3 Feb. 1851, 17 July 1851, GSL: M/S11/257, 271a & b.

by vast successions both in North and South Wales. Only in 1852 did he unite the Bala beds with the Llandeilo, for reasons that had nothing to do with the rejection of an Upper Silurian age for the South Wales slates, but everything to do with the official Survey.

In later years Sedgwick and his biographers continuously altered the version of events recounted above so as to bring it into line with the results of new research. For example, in 1854 the Survey revised the position of the Bala beds and correlated them with the Caradoc formation rather than the Llandeilo. Obviously this rendered the foundations of the 1852 retrospective embarrassingly obsolete. Murchison's placement of Meifod in the Caradoc—supposedly the fatal error that falsely linked Cambria and Siluria—proved to be "correct"! As a result, Sedgwick ended his subsequent accounts by claiming that he had discovered the errors in Murchison's sections as early as 1843.[35] Confusion multiplied as he began to suffer from what he called "the strange clouds of oblivion which too often trouble an old man's memory"; retrospectives written at various times and for varying purposes were combined into self-contradictory pastiches.[36] Sedgwick's supporters in the debate continued to write similar accounts after his death. The outstanding example of such a wholesale fabrication must certainly be the remarkable chapter by Thomas McKenny Hughes in the *Life and Letters* of his predecessor in the Woodwardian chair. This chapter, the only one in the biography actually written by Hughes, has previously served as an authoritative history of the controversy. In it, dates are altered, individual statements are quoted completely out of context, and texts and sections are changed so that they appear to foreshadow later discoveries. The result possesses considerable interest as a polemical document, but for purposes of reference or historical interpretation is utterly worthless.[37]

Did Sedgwick simply lie? Given what we know of his character it seems highly unlikely that he engaged in conscious deception while composing his history, and the available sources fail to exhibit the kind of discrepancies that outright dishonesty might produce: public speeches, private jottings, letters to friends and enemies all em-

[35] Sedgwick 1854c: 499-500; 1855: xlix-li. For the Survey discovery, compare Ramsay and Aveline 1848 with Salter and Aveline 1854; Salter to Sedgwick, 8 June 1853, CUL: Add. ms 7652IIX57c also sheds light on the matter.

[36] The three longest retrospectives can be found in Sedgwick 1853a, 1855, and 1873. The "clouds" are in Sedgwick 1873: x.

[37] Hughes, "Sedgwick's Geological Work," in Clark and Hughes 1890, 2: 502-563, at pp. 508-555.

ploy different strategies and emphasize different points, but at any given time display a consistency of argument that bears the stamp of conviction. Under these circumstances it is safest to assume that Sedgwick believed in the basic accuracy of his account. The most plausible view is that he simply became so convinced of Murchison's error that *any* interpretation establishing his guilt could achieve the status of "truth." Otherwise his willingness to chop and change details in order to put his opponent constantly in the wrong seems almost impossible to explain. Significantly, Sedgwick gave the *Origin of Species* and the *Vestiges of Creation* similar treatment, in a manner that struck even sympathetic readers as unfair.[38]

Historians are by now thoroughly familiar with cases in which eminent scientists revise their past to serve some current end. Many men of much greater stature than Sedgwick—Newton, Freud, Darwin, Pasteur, Claude Bernard—are known to have done so,[39] and similar examples from Murchison are not difficult to find. Given the contemporary canons of historical practice and the context of controversy that give birth to Sedgwick's account, its inaccuracy is almost inevitable. Yet the depth, detail, and thoroughness of the revisions in this case make the issue worth stressing. They lead one to suspect, for example, that those studying other scientific figures have sometimes been too circumspect in relating contemporary evidence of letters or working notebooks to autobiographical statements, particularly when these reconstruct technical developments in lengthy and convincing specificity. Of course, no source gives the historian a direct insight into the mind of his subject, but Sedgwick's retrospectives make it clear that later statements written in the heat of controversy can sometimes be entirely misleading.

Crusading for Cambria

The demonstration of the polemical underpinnings of Sedgwick's account still leaves a much larger question unanswered. Why should the Cambridge professor of geology, a warm-hearted figure famed for his defense of the moral basis of scientific inquiry, launch such a vindictive attack on a friend of over twenty years standing?

[38] The somewhat similar case of H. E. Armstrong in debates about physical chemistry is analyzed in Dolby 1976: 385-391. Porter 1982b has an interesting discussion of the problem of scientific "truth" in connection with the work of Charles Lyell.

[39] For Newton: A. R. Hall 1980; for Freud: Sulloway 1979, esp. the valuable discussion on pp. 445-495; for Darwin: Herbert 1977: 191-194; for Pasteur: Mauskopf 1976, esp. 68-80, and a forthcoming book by Gerald L. Geison. The case of Bernard has been discussed by Holmes 1974 and several other authors.

On several occasions Murchison and his supporters gave a simple answer to this question: Sedgwick had quite simply taken leave of his senses and was senile or mad. As Forbes told Ramsay in 1853, "I had a very long talk with him & a very friendly one at the Athenaeum two days ago, but found him so vague & dreamy & so ignorant of the real natural-history question that I did not press any discussion."[40] Such comments, like Murchison's earlier accusation that Sedgwick was *"in love,"* cannot be taken at face value. Of all the weapons in the armory of scientific controversy, accusations of unreason were perhaps the most potent.

Subsequent authors favorable to Sedgwick's cause, but embarrassed by the personal tone of his remarks, have attributed his bitterness to a number of specific events. These certainly go some way towards explaining Sedgwick's increasing tendency to see Murchisonian machinations behind every setback for the Cambrian. Several of these grievances had already festered for several years, especially his anger at the supposed failure of Murchison to correct Sharpe's assignment of blame for the boundary drawing in South Wales. A more serious complaint had been added in 1846, when Sedgwick learned from the itinerant map-seller James Knipe that Murchison had brushed out the Cambrian colors on the 1843 map for the Society for the Diffusion of Useful Knowledge.[41] This map, which had been circulating for almost three years in thousands of copies, showed the Lower Silurian extending to the west coast of Wales for the first time (Fig 4.5). From Murchison's point of view such a coloring scheme was an obvious corollary of the loss of the Cambrian fauna, but for Sedgwick (who had actually helped construct the map) it seemed an unwarranted trespass made without his knowledge or consent.[42] For all his faults, however, Murchison certainly never planned to "steal" Cambria in secret, and his

[40] Forbes to Ramsay, 28 [Nov.] 1853, ICL(R). Lonsdale was of the opinion that Sedgwick was "not sane 'omnibus horis,' " a remark reported in Murchison's autobiographical journals: GSL: M/J18, pp. 38-39. I owe this and other references to Murchison's later journals to Robert Stafford.

[41] Sedgwick to Phillips, 10 Oct. 1846, ff. 120-129, UMO; Murchison 1843b.

[42] The SDUK map is often confused with the Survey maps of the late 1840s and early 1850s, thus making Murchison's action vastly more significant than it actually was: e.g. Clark and Hughes 1890, *2*: 214-215, Bowler 1976: 113, Speakman 1982: 80. For Sedgwick's help with the coloring of the SDUK map (although not its nomenclature), see Murchison to Sedgwick, 24 Feb. 1843, 27 Feb. 1843, 23 Mar. 1843, CUL: Add. mss 7672IIID41, 43, 44 (copies). Murchison specified in his letter of 27 Feb. 1843 the limited form of "Cambrian" he planned to use. By 1846 Sedgwick had completely forgotten this early correspondence, as shown by Sedgwick to Murchison, 1 Nov. [1846], GSL: M/S11/236a & b.

friend's alarm is best seen as the result of an unintended failure in communication.

But the real blow had come in mid-February 1852, only a week before Sedgwick was scheduled to read his paper at the Geological Society. Presumably while writing his historical account, he examined all his old publications, including the long paper of November 1843 on North Wales. Immediately after its publication Sedgwick had been so angered by Warburton's refusal to allow any checking of proof that he had put his offprints of "that ugly abstract" aside, unopened and unread.[43] What he found there in 1852 gave him a lasting shock, the bitter aftereffects of which were felt throughout the remainder of the controversy. Warburton had not confined his alterations to the minor technical slips outlined earlier. Rather, the entire purpose of the paper, which had been to support the temporary "Protozoic" designation for all the fossiliferous rocks below the Upper Silurian, had been subverted by a critical change of nomenclature on the accompanying map. Although no contemporary evidence shows precisely what the legend was to have said, the text of the paper—like Sedgwick's other publications from 1843 to 1846—uses "Protozoic" consistently. (For simplicity the redrawn version of this map included here as Figure 4.8 was brought into line with this intention.) In Warburton's published redaction, however, the legend read as follows:

—Carboniferous
—Upper Silurian
—Lower Silurian (Protozoic)

"Protozoic" is relegated to parentheses, and Sedgwick appears to be a supporter of the expanded Murchisonian classification. As he wrote in 1852, "Had the published map been allowed to pass, *in its present form*, after a revision by myself, I should virtually have surrendered the whole question now in debate."[44]

Since the text of the 1843 paper used "Protozoic," Warburton's equation of the term with "Lower Silurian" was probably made in

[43] For indications that Sedgwick did not in fact notice the legend on his own map for over eight years, see Jukes to Ramsay, 3 Mar. 1848, ICL(R); Sedgwick to Phillips, 10 Oct. 1846, ff. 120-129, UMO; and Sedgwick 1852d: 155.

[44] Sedgwick 1852d: 155. The explanation given in this paper, that he had written "Lower Silurian and Protozoic," is almost certainly not correct, for the Protozoic had included the Lower Silurian by definition. In later discussions, Sedgwick (e.g. 1855: l) often claimed that the legend was to have read Cambro-Silurian + Cambrian = Protozoic, but we have seen (Chapter Six) that he did not use "Cambro-Silurian" until Dec. 1846. The original map was published with Sedgwick 1844 and 1845a.

ignorance of the fundamental issues at stake. As a nonspecialist in Lower Palaeozoic geology Warburton undoubtedly wished to bring the nomenclature of Sedgwick's map into line with that used by the majority of his contemporaries. Although it is impossible to know if Murchison himself had any part in the alteration, he certainly knew about it, for his January 1847 reply to the revived Cambrian used the map as evidence for Sedgwick's supposed acquiescence in the expanded Silurian.[45] Unquestionably this altered map contributed to misunderstandings of Sedgwick's position, for it made the revival of the Cambrian seem a tardy affair conceived after initial acceptance of the Murchisonian usages. On the other hand, the importance of this single change is easily exaggerated. Even if the map had been correctly labelled with the "Protozoic" name, the fortunes of Sedgwick's nomenclature would doubtless have been equally dismal. The real importance of the alteration came, not when the map was published, but rather in 1852 when Sedgwick discovered that the mistake had been made. The inadvertent change (a unique occurrence in his *oeuvre*) multiplied in his mind into one of many, and he eventually even accused Murchison of secretly submitting altered abstracts of his papers, which were supposedly published in preference to the originals.[46]

This gradual accumulation of real and imagined grievances explains in part the anger with which Sedgwick exposed the supposed errors of his former collaborator. However, we must look deeper to find the sources of his extraordinary sensitivity to these provocations. Throughout his life Sedgwick was a sympathetic figure, a skilled geologist, and a great teacher, but by 1852 he was also a disappointed old man who had failed to write the big book that was to have been his major contribution to scientific research. Even as early as 1843 he had spoken of first publishing it in the abbreviated form of a pamphlet, to be "an introduction to *my larger work,* should I ever have health and courage to go through it."[47] By 1850 the unwritten volume weighed heavily on his mind, and little further progress had been made in the laborious process of composition. Looking ahead six years we find all thought of the *magnum opus* abandoned. "I am buried in the enormous mass of my materials,"

[45] Murchison 1847b: 170.

[46] For the multiplication of this single instance, see Sedgwick 1873: xxvii; Clark and Hughes 1890, 2: 201-202. The secretly altered abstracts are mentioned in Hunt 1878: 365, a claim "indignantly" repudiated in Geikie 1875, 2: 61-62.

[47] Sedgwick to Murchison, 31 Jan. [1843], GSL: M/S11/208, printed in Clark and Hughes 1890, 2: 54.

Sedgwick told Jukes in 1856. "I have neither time nor strength for the big Palaeozoic work I have long contemplated, & if my Synopsis were done, I should retire with a good conscience & I might then amuse my declining years . . . with some subjects half moral, half religious, half political, which I have been thinking about." His inability to write even this little summary, let alone a major book, provides an instance of what Jerome Ravetz has called "tragedy" in the history of science, in this case a research project overcome by events and incapable of completion.[48]

Sedgwick's failure had become increasingly apparent during the late 1840s as the Survey papered over his knowledge of North Wales with their maps and sections. Throughout the previous decade, De la Beche, Phillips, and other members of the Survey team had recognized Sedgwick's difficulties in completing his volume and generously offered full access to their unpublished results. Jukes especially urged his old teacher to gather his materials into final form before the official work totally obscured his priority:

> I am anxious that you should put on deliberate record some abstract of your labours on the palaeozoic rocks. Your contemporaries know how much you have done as a pioneer in opening up this department of science but the rising generation of geologists do not. . . . you should publish some condensed, well considered, work that may be appealed to by *name* hereafter in addition to the scattered works and oral information and instruction upon which a good many of us have built much of what little reputation we possess.[49]

But even as Jukes wrote these words it was already becoming too late for the proposed book to assume an important place in the literature of British geology, even had Sedgwick found the time and perseverance to complete it. Largely as a result of the work of men like Jukes himself, the standards of the stratigraphical enterprise had risen enormously since 1828, when Conybeare had outlined a few months of vigorous hammering to complete the geology of the older rocks. The planned work of a few summers had proved too much for a lifetime, and the individualistic tradition of the gentleman geologist—as represented by one of its ablest exponents—had simply failed when confronted by the complexity of the older rocks

[48] Sedgwick to Jukes, 26 Dec. 1856, CUL: Add. ms 7652IIIE9 (copy); Ravetz 1973, also Kubie 1953-1954.

[49] Jukes to Sedgwick, 10 June 1851, CUL: Add. ms 7652IIQ38b. Offers of Survey help are in Phillips to Sedgwick, 14 Oct. 1843, CUL: 7652IE58.

of Britain. North Wales, Sedgwick had despairingly told Phillips in 1843, "is too much for any man single-handed."[50] In the face of such disappointment, his wish for priority becomes easier to understand. As it became clear that he would not go down in history as the author of a great geological book, Sedgwick abandoned his long-standing flexibility and urged recognition of the Cambrian as a signal of his scientific achievement.

Larger issues than simple jealousy lay behind this desire for some acknowledgement of a lifetime's work in the older rocks. If Sedgwick had wanted only a name to his credit, he could have easily adopted "Cambrian" in the sense of the Survey, or in any one of the many compromise versions proposed during the last two decades of his life. But he did not do so. In the particular circumstances of the controversy, and of his life and career at Cambridge, the word "Cambrian" had achieved a value as scientific property far beyond its immediate purpose as a place-holder in the geological column or a color on geological maps. By the mid-1850s Sedgwick may have become an aging and frustrated old man, but he remained one of Victorian England's leading interpreters of the relationship between science, religion, and philosophy. Only in the light of his concern with the wider bearings of scientific research can the vehement persistence of his defense of Cambria be fully understood.

Out of the hundreds who heard the Woodwardian professor during his distinguished teaching career, only a handful—notably Ansted, Jukes, and Hughes—became engaged in scientific research. Sedgwick, unlike De la Beche at the Survey, had no ambition to train large numbers of full-time research geologists. This would have been inconceivable at Cambridge during this period in any case, for most of the students in his classes were intending clergymen. Instead, Sedgwick (who had taken orders himself) taught geology as part of the cultural equipage of the Christian gentleman. "It was my delightful task," he wrote at the end of his life, "to point out year by year to my Geological Class, the wonderful manner in which the materials of the Universe were knit together, by laws which proved to the understanding and heart of man, that a great, living, intellectual, and active Power must be the creative Head of the sublime and beautiful adjustments and harmonies of the Universe."[51] Imaginative and impressionistic, Sedgwick stressed the great facts of geology rather than its little ones and used his subject

[50] Sedgwick to Phillips, 10 Oct. [1843], ff. 46-53, UMO.

[51] Sedgwick 1873: xxxi. Sedgwick's religious views are fully detailed in Marston 1984.

to inculcate the student's powers of inductive thought. "I cannot promise to teach you all geology," he once told the class; "I can only fire your imaginations."[52]

Criticisms of the evolutionary theories of the *Vestiges of Creation* of 1844 provided a consistent theme for Sedgwick's later lectures. This work, like Darwin's equally objectionable *Origin of Species* of 1859, viewed the natural world from a perspective as wide-ranging as Sedgwick's own, but with a message and a method antithetical to those he hoped his students would embrace. Sedgwick's violent critiques of species transmutation have never been linked with his attacks on the expanded Silurian, and in one respect this is entirely sensible. The threat posed by Siluria pales to insignificance when compared with that of materialistic evolution. Whatever his failings, Murchison could hardly have been considered a materialist. Despite the comparatively secular cast of his science and private doubts about the revealed word, Murchison accepted without question the evidence for God's direct and constant involvement in the creation.[53]

But Sedgwick's second major criticism of the *Vestiges* and the *Origin* concerned scientific method, and it is here that parallels to the debate with Murchison become manifest. Darwin, Sedgwick believed, "had deserted utterly the inductive track—the narrow but sure track of physical truth,—and taken the broad way of hypothesis, which has led him (spite of his great knowledge) into great delusion."[54] The biblical image of the straight and narrow path unites the routes to salvation and to scientific understanding, the implication being that in losing one, Darwin had also turned from the other. Sedgwick's views on method, especially as expressed in the swollen fifth edition of his *Discourse,* bear an obvious similarity to those of his contemporary at Trinity, William Whewell. The *Discourse* discusses the physical sciences and their history at some length: Newton's work on gravitation, for example, is presented as the result of "observation upon observation . . . experiment upon experiment, gradually suggesting to the mind one of the forms of

[52] Clark and Hughes 1890, 2: 489. Sedgwick's teaching methods are well described in Porter 1982a.

[53] Sedgwick's criticisms of the *Origin* are conveniently available in Hull 1973: 155-170. For his views on the *Vestiges,* see Egerton 1970 and Sedgwick 1845c, 1850. A more general discussion is in Gillespie 1979. For Murchison's religious beliefs, see Murchison to Sedgwick, 19 Jan. 1838 [not sent], in Craig 1971: 498-500; Bowler (1976: 97) comments with insight on his attitude to *Vestiges.*

[54] Sedgwick to R. Owen, [28 Mar. 1860], in Clark and Hughes 1890, 2: 360-361.

general truth."[55] This "inductive effort of the mind," the process by which all science progressed, required far more than a mere massing of details—hypothesis had a role to play as well. The greatest violation of the inductive method, Sedgwick felt, was not the invention of theories but rather setting them up as objects of worship. This had been the fatal error of books like the *Vestiges*. Geology had been "the wildest field for speculation" and had suffered more than any other science from rashly promoted generalizations.[56]

Precisely the same warnings underpinned the polemics against Murchison. Indeed, Sedgwick used very similar language in pressing his methodological case against Murchison and the *Vestiges*, sometimes borrowing word for word.[57] In his view, Siluria had become an unalterable idol, impervious to correction even in the face of manifest mistakes. Sedgwick had always insisted that the inductive process of geology began with clearly determined sections; only after these were established could fossils play their part in classification and correlation. The principles at stake had not changed from the first stirrings of disagreement, although the battle had by 1852 shifted largely to the field of history. Needless to say, Sedgwick's retrospective accounts show him following these methodological precepts to the letter, making a connected sequence of decisions about the placement of particular strata—with the entire chain of reasoning ultimately vitiated by the single false link added by Murchison. The latter's triumph was all the more galling because it rested on false principles and a foundation of lies. "The personal question is indeed a paltry matter," Sedgwick wrote in the February 1852 paper,

> but it does involve a very important principle. Philosophical names are not to be given rashly; and premature names ought to be abolished; otherwise we barbarize our language, and retard the true progress of science. Scientific names are, or ought to be, the abstract representations of the highest conceptions of the human mind; which first dealing analytically with facts, then groups them together synthetically under their most general conception. The analysis of the phenomena comes *first*,—the philosophic names come, or ought to come *last*. Nor are philosophical names ever unimportant, even in mixed and progressive subjects like

[55] Sedgwick 1850: cc.

[56] Sedgwick 1850: cciii.

[57] For particularly clear examples, compare Sedgwick 1845c and 1850 with his 1852d and 1855.

> our own; for they are the very circulating medium of science; and if our coin be base, our scientific dealings can never prosper.[58]

Sedgwick wanted not only a name to record his labors, but also recognition of the principles that gave the name its value as scientific property.

The link between Sedgwick's concern for property and for philosophical method was not without its tensions. Robert Merton has suggested that priority disputes often bring institutionalized desires for originality and humility into direct conflict.[59] Although the strength of humility as a "norm" in Victorian science may be doubted (one searches the many volumes of Murchison's works in vain for a truly humble statement), such considerations did operate with some force in Sedgwick's case. An argument like his, based on the need for philosophical rigor and method in science, demanded that he appear a disinterested seeker after nature; but since the situation so clearly concerned the history of his own work, such a demand was virtually impossible to meet. Like most men of science, Sedgwick solved the problem by denying any egotistical desire for fame in long polemical accounts of the correctness and priority of his own views, all couched in terms of an overarching argument from first principles. Of course, the vehemence of his defense of the Cambrian—especially the serious distortions involved in his attacks upon Murchison—shows just how completely the "personal question" was bound up with the philosophical. But neither here nor in the analogous case of his anti-evolutionary crusade was Sedgwick merely hiding his "real" motives behind a figleaf of philosophy. From his viewpoint, what he believed to be an essential truth had to be defended, occasionally with a violence of language that more sober contemporaries found distasteful. "In my mind," he told Phillips in 1854, "plain truth is seldom spoken in a whisper."[60]

That this question of "truth" should ultimately resolve itself into assigning responsibility for a single (albeit important) mistake in correlation may sound as strange to modern ears as it did to many Victorian ones. But given the nature and strength of Sedgwick's views on scientific method, such a proceeding was almost inevitable. In the end the high value that Sedgwick placed on method made his debate with Murchison a moral and even a religious question. If scientific endeavor aimed to elucidate the work of the divine

[58] Sedgwick 1852d: 161.

[59] Merton 1973.

[60] Sedgwick to Phillips, 10 Feb. 1854, ff. 260-265, UMO. Hull 1973: 166-170, at p. 169, accuses Sedgwick of donning his philosophical cloak when it suits him.

Creator, then the baseless Silurian system merely attracted attention away from God's ordering of the materials of the earth's crust, a sequence that he firmly believed was expressed only by the Cambrian classification. His intention was not to literally match verses in Genesis with groups of strata in Wales, but rather to understand the order of the created world through a philosophically established classification. Ultimately Sedgwick thought that anything less than the strictest attention to method would produce a science that could not legitimately confront the forces of unbelief. In this sense, the Silurian threatened to poison the wellsprings of science at their source.

Rewriting the History of Siluria

Sedgwick's angry polemic in the February 1852 paper raised a storm of protest at the Geological Society. "Good scrimmage between Sedgwick and Murchison on the Lower Silurian and Cambrian question," wrote Ramsay in his diary after the meeting. "It was not an enlivening spectacle. Sedgwick used very hard words."[61] An abstract of the paper in the *Literary Gazette*, a general literary and scientific weekly, led to an exchange of published letters repeating all the old arguments about geographical propriety, palaeontological unity, and historical justice, but now in the full glare of the public gaze.[62] This change of venue was an extremely important development. Although the two men maintained cordial personal relations and used temperate language in their letters, they both felt that matters had reached a point where the opinions of a nongeological audience actually mattered. Murchison could not let a neutral abstract of Sedgwick's paper go unchallenged, while his opponent was equally determined to have the last word in any exchange. They were simultaneously attracted and appalled by the prospect of a public dispute; "no question can be settled by newspaper articles," wrote Sedgwick in April as he readied his pen to write yet another.[63]

Just how far Sedgwick's paper had transgressed the conventional bounds of scientific discourse is indicated by the events surrounding its full publication in May 1852. This remarkable incident gives

[61] Ramsay, Diary, entry for 25 Feb. 1852, in Geikie 1895: 197.

[62] For the abstract, Sedgwick 1852b; followed by Murchison 1852a, Sedgwick 1852c, Murchison 1852b, and Sedgwick 1852e. The letters were also printed in the *New Edinburgh Philosophical Journal*.

[63] Sedgwick to Murchison, [Apr. 1852], GSL: M/S11/276.

a unique insight into the inner workings of the Geological Society's governing Council. The crucial figure in the story was William Hopkins, then president of the Society. Interested in the application of mathematics to geology, Hopkins held what Forbes acidly called an "intense & Cantabridgian veneration for Sedgwick."[64] The organizational structure of the Society gave the president considerable power over the publication of the *Quarterly Journal*; in particular, he could ensure that a memoir passed from referee to print with only minimal interference from the rest of the Council. This is precisely what Hopkins did for Sedgwick. The paper was referred to Phillips, and although he noted its "unnecessarily controversial tone," he seems to have suggested only minor changes that the author accepted in full.[65] On 7 April the Council voted to publish the memoir "in the abstracted form recommended by the referee," but no one save Hopkins seems to have been aware how few changes that critical phrase implied. To their horror, the next thing most Council members saw was the full paper in print with its violent language virtually intact. Leonard Horner, a respected friend of all concerned, expressed to Murchison the alarm that almost everyone shared. After reading the paper carefully—for the first time—he concluded that "there are statements and expressions which Sedgwick ought not to have used towards you and which ought still less to have been printed in our journal. I speak of the manner and tone," he continued, "which go far beyond what is allowable in scientific controversy."[66]

In its defense the Council officially maintained that publication had been conditional on the omission of objectionable passages now printed "inadvertently."[67] But this was no more than a polite fiction. Contemporary evidence shows that the entire memoir had passed through all regular procedures. These were simply slack

[64] Forbes to Ramsay, 28 [Nov.] 1853, ICL(R). Hopkins's role in this complicated affair can be inferred from the letters in Craig 1971, esp. Hamilton to Murchison, 24 May 1852, at p. 487, and J. D. Hooker to Murchison, "Friday," at p. 496.

[65] Phillips to Murchison, 24 May 1852, in Craig 1971: 486-487. The original referee report has not been preserved; indications of what it said are available in Sedgwick to Phillips, 2 Apr. 1852, 27 May 1852, 29 May 1852, ff. 164-192 and 238-241, UMO; Sedgwick to president and Council of the Geological Society, 22 May 1852, GSL; and Phillips to Sedgwick, 16 Feb. 1856, CUL: 7652IIA4b.

[66] Horner to Murchison, 9 May 1852, in Craig 1971: 484-485. Italics in original removed. For the reaction to the full publication of Sedgwick's paper, see Craig 1971: passim. The "abstracted form" is mentioned in the Council minutes, 7 Apr. 1852, GSL: CM/1/7: 231.

[67] Council minutes, 19 May 1852, GSL: CM/1/7: 238-239.

enough at this time to give a friendly insider ample room to maneuver. Many Council members were furious at what Hopkins had done in their name. At their next meeting (19 May) they took the unprecedented step of cancelling the last fifteen pages of Sedgwick's article, even though five hundred copies of the issue had already been delivered to subscribers around the world. This attempted recall was unique in the annals of Victorian science, and Sedgwick understandably took it as a gross insult. Murchison, on the other hand, found it eminently acceptable and dropped his demand for a reply in the next number of the *Journal*.[68] Many leaders of the Society were relieved to thereby obliterate all traces of the affair (or as Sedgwick's biographers more graphically put it, "to remove the unclean thing from their *Journal*"[69]), particularly as a reply from Murchison would have drawn still more attention to the controversy. Other members of the Council, even those like Horner sympathetic to Murchison's wounded honor, condemned cancellation as overly harsh. After all, Sedgwick had been responsible only for writing the paper and not for its actual publication. Hoping to take advantage of this feeling, Hopkins and a small minority on the Council delayed implementation of the order and arranged for its reconsideration. Their strategy worked, and at the mid-June meeting competing factions compromised by leaving Sedgwick's outburst intact and permitting Murchison a calm reply.[70]

In constructing a retrospective favorable to the expanded Silurian, Murchison faced a far simpler task than Sedgwick had in justifying the Cambrian. Sedgwick referred to information in early notebooks, to brief abstracts in British Association reports, and to a complicated series of mistakes that had led him astray for many years. Murchison could send readers to published documents on the shelf of every geological library. For a man in his position such a quiet display of publications was in its own way as tendentious as Sedgwick's most virulent diatribes and came recommended as a highly effective strategy in controversy. Although everyone agreed that his old friend's bitter invective demanded a reply of some kind, many urged Murchison to make his responses as brief and reasoned as possible. "The Silurian nomenclature is now a *fait accompli* adopted everywhere," Sharpe told him in April 1852, "& it will

[68] Murchison to J. C. Moore, 1 June 1852, in Craig 1971: 487-489.

[69] Clark and Hughes 1890, 2: 216.

[70] Council minutes, 16 June 1852, GSL: CM/1/7: 243-247; Phillips to Murchison, 24 May 1852; Horner to Murchison, 3 June 1852, 11 June 1852; J. C. Moore to Murchison, "Friday Evening"; all in Craig 1971: 486-487, 490, 492-493, 495.

stand safe enough alone, but you raise doubts of it's soundness by not trusting it to itself."[71] Murchison had in fact been following this advice for some time. He made sure both in 1847 and 1852 that the Geological Society secretary read out only the titles of his controversial replies, so as to avoid unnecessary discussion of a matter settled in his favor; his papers on the subject were short and to the point. As part of this strategy, Murchison made his retrospective seem much less forced and argumentative than those of his opponent. Thus he generally exploited ambiguities in his past work, emphasizing aspects that had been confirmed and omitting those that had not. In contrast, Sedgwick's account contained many statements totally at odds with his published papers, let alone information in letters and notebooks. But for all the smooth gentility of Murchison's response, his anxiety about the attack on his system cannot be doubted and in fact largely explains the streamlining of the story leading to the expanded Silurian classification. Of the two men, Murchison is perhaps more likely to have constructed certain elements of his retrospective with an intent to mislead, although speculation on such matters is not very profitable. Certainly on the main points both Murchison and a host of supporters felt fully convinced of the justice of his case.

Although lacking the anger characteristic of his opponent's accusations, Murchison's narrative involved equally pervasive elements of myth-making. The essentials of his account have already been discussed in connection with earlier papers, but the main threads can be drawn together here. Unlike Sedgwick, who constructed his case against the Silurian on the basis of a fatal flaw in the correlations on the 1834 tour, Murchison elaborated the story as an aspect of the onward march of scientific progress, with the full Silurian classification foreseen almost from the beginning. He began by presenting his earliest work as though he had started in 1831 with the explicit intention of unravelling the Transition rocks. As we have seen, however, even at the end of his first tour through Wales he had not formulated any definite plans for such a study. This claim typifies Murchison's attempts to smooth the disjunctions on his path to the eventual expanded Silurian. For example, he maintained that his formations had been defined from the outset purely on fossil evidence, although his early notebooks show a reliance on lithological and structural criteria as well. Similarly, he quoted the *Silurian System*'s single reference to the possible need to

[71] Sharpe to Murchison, 17 Apr. [1852], GSL: M/S13/3.

extend the lower boundary of his system on palaeontological grounds, as if he had anticipated before 1839 the lack of a peculiarly Cambrian fauna. Murchison also reiterated his view that the line between the two systems was without geological meaning, merely a limit defining two separate palaeontological hunting grounds. Given this territorial view of the boundary, the critical separation between Bala and Llandeilo during the 1834 tour became entirely Sedgwick's own responsibility, for it had occurred on the western side of the border. "I am answerable," Murchison declared, "only for Silurian and Cambrian rocks described and drawn as such *within* my own region."[72] He explicitly blamed Sedgwick for the false dovetailing between the systems, with none of his previous polite ambiguity. To outside observers the determination of the collaborative boundary during the 1830s must have now appeared anything but the agreeable venture it had actually been.

Murchison refused to be pilloried for what he saw as minor mistakes, even had he been willing to admit that they were his doing in the first place. Consequently, while Sedgwick told a cautionary tale that pointed to the dangers of forgetting the true importance of sections, his opponent more straightforwardly recorded the progressive triumph of the palaeontological method. The errors so stressed by the one appeared to the other as mere nuisances to be ironed out without difficulty. Rather than attempting a frontal attack on the technical minutiae of Sedgwick's retrospective, Murchison characteristically expressed regret that his friend had not proceeded faster with the palaeontological aspects of his work. To the end of his life he argued that the "burial" of the Cambrian fossils in their original crates for so many years had automatically precluded any possibility of their priority. (As the bulk of these fossils had been called "Lower Silurian" by 1835, however, it is difficult to see how Sedgwick could possibly have satisfied this criterion even had they not been packed away.)[73] Just as Sedgwick looked upon the original elaboration of the Cambrian entirely in sectional terms and ignored his initial expectations of a characteristic fauna, so Murchison read his purely palaeontological definition of the Silurian back into the past by making the division between the two systems entirely dependent upon fossils. This same contrast colored their approaches to the history of geology as a whole. Although both hailed William Smith as the founder of the science in England, Murchison argued

[72] Murchison 1852c; the quoted passage is on p. 175.
[73] Craig 1971: 483; Murchison 1852c, esp. 176-177; Murchison 1854: 10.

that the humble engineer had mapped with fossils alone, while Sedgwick maintained that reliance on fossils had always followed the secure determination of the sections. Just what it was that Smith had "fathered" thus became clouded in obscurity.[74] Evidently they turned to the historical record as a rich resource to be exploited in controversy, much as they had previously employed the record of the rocks.

Murchison continued to rest his case on the fossils, but the completion of the Survey maps also enabled him to claim sectional grounds for the incorporation of Sedgwick's Cambrian into the lower half of his system. Taken at face value, this meant that the roly-poly found by Ramsay and De la Beche in South Wales now extended almost to the base of the known fossiliferous rocks. But as Sedgwick pointed out, this argument actually demanded relocating the type area used to define the lowest of the four original Silurian formations. In the *Silurian System* the horizon of the "Llandeilo" formation had been typified by the sequence found near the town of that name in South Wales, and Murchison had used it as a standard to place many other areas with similar rocks and fossils in the succession. After its own investigations, the Survey had decided that one of these areas, the district in Shropshire including Shelve and the Stiperstones, actually contained strata far older than any found near Llandeilo itself (Fig. 7.2). By moving his type area for the oldest Silurians from South Wales to Shropshire, Murchison could legitimate the incorporation of almost all the older fossil-bearing rocks within Siluria. By a fortunate chance (or a mistaken correlation, depending on one's point of view), Siluria had included good sections of these ancient strata after all.[75] In this way Murchison could finally meet Sedgwick's long-standing criterion of geographical propriety. His ability to do so was of no small importance, for "type sections" in stratigraphy had a significance comparable to "type specimens" in botany and zoology.

In many respects the professional government geologists had thereby brought the success story of Siluria full circle. Perhaps the most significant point, though, is that Murchison had possessed the practical good sense, extensive local assistance, and facility in composition to finish his *Silurian System* before they encroached on his

[74] Rudwick 1976c; for discussions of Smith's reputation, Laudan 1976: 223-224 and Rupke 1983: 191-193 are of interest, although neither gives an entirely balanced view of his approach to geology and its early diffusion.

[75] Murchison 1852c: 179-181, also 1854: 39; for the response, see Sedgwick 1855: lxxvii-lxxix.

efforts. In the mid-1830s he had even demanded official assurances that the Survey would not enter his territory until his own accomplishment was safely in print, and after 1838 he did almost all his geological work in Scotland, Russia, Scandinavia, and other regions beyond its immediate reach. The contrast with Sedgwick is evident.

Like everyone else, Murchison accepted the Survey view of the structure of Wales and used it as a standard to measure the accuracy of his own opinions as expressed in the *Silurian System*. In making this comparison he held a great advantage, for the Survey had colored North Wales with the shades of the Silurian classification; in reply, Sedgwick could only rather feebly suggest that the omission of "Protozoic" on his 1843 map had led De la Beche to assume his acquiescence in the expanded Silurian.[76] Rather than having to explain away the Survey's failure to adopt his views, Murchison could simply point his readers in the direction of the imposing new Museum of Practical Geology for proofs of his priority:

> There they will see that all the inferior slates and limestones of North Wales which contain fossils are named lower Silurian. . . . Why, then, did Sir Henry De la Beche, and his followers in the field, Ramsay, Aveline, Selwyn, and others, adopt this classification? and why has it been confirmed by Edward Forbes, Phillips, Salter, and the palaeontologists of the survey? Simply, that after a long and careful scrutiny these observers have satisfied themselves that the region called Cambria, at a time, it will be recollected, when none of its fossils were described, is made up of the same strata and contains the same organic remains as the lower Silurian rocks, whose contents were so long ago described by myself.[77]

As the impressive roster of names in this passage suggests, Murchison enjoyed reciting the roll call of those in his favor, particularly when they spoke with the collective power and expertise of the Survey.

Much as Darwin's appeal to the rising generation of naturalists received Sedgwick's rebuke, so too did Murchison find himself taken to task for continually stressing the number of authorities who supported him. In Sedgwick's view, popularity and widespread usage were irrelevant to the establishment of a correct no-

[76] This claim is first made in Sedgwick 1852c: 339, and most completely in Sedgwick 1854c: 500-502. Murchison's demand that the Survey keep out of Wales is evident from Buckland to Murchison, 12 June 1835, DRO: D138 M/F 221.

[77] Murchison 1852a: 278.

menclature; the case had to be decided on its own merits. This was not a small point, and Murchison noted that the rejection of the Lower Silurian would require revision of all Geological Survey maps, most geological and palaeontological memoirs, and every introductory textbook.[78] (There was no manual that Sedgwick could recommend to his lecture classes that used his Cambrian nomenclature.) But for Sedgwick, this wide diffusion of a mistake made its immediate rectification all the more essential. Science differed from politics, he wrote, precisely in that it appealed to an inductive method rather than the voice of the mob or the iron hand of dictatorial rule. It is important to emphasize that Murchison ostensibly shared many of these views on scientific method, and in a letter to Whewell even spoke of his "inductive process" in formulating the Silurian system.[79] Certainly his constant downward extension of the Silurian had been consistent with a methodological commitment to a purely palaeontological classification, given his broad definition of the Silurian fauna. With considerable justice, for example, Murchison argued that the little map drawn up for the Society for the Diffusion of Useful Knowledge in 1843 had not been issued "rashly," but only after a careful consideration of the fossil remains in the mountains of North Wales.[80] In his view, additions or corrections to the knowledge of the older rocks did not call for chaotic overturnings of the received classification. Science had grown by accretions, and nomenclature should reflect this historical process of development rather than being constantly restructured according to first principles. But even if Murchison occasionally emphasized the philosophical foundations of the palaeontological method, his main bent was almost entirely practical, and his defense of Silurian priority lacks the clarion call to inductive rigor so characteristic of his opponent. As Darwin once said, "How singular so great a geologist should have so unphilosophical a mind!"[81]

By 1852 the fortunes of the two former collaborators contrasted with all the drama of classical tragedy. While Sedgwick struggled to accept the failure of his plans to write a major book on the older rocks, Murchison geared up to produce a volume that described such strata not only in Britain but throughout the world. This volume, which appeared in 1854 under the appropriately expansive title *Siluria*, combined Murchison's own researches (as previously re-

[78] Murchison 1852c: 183.

[79] Murchison to Whewell, 19 Dec. 1846, O.15.48[53]; Sedgwick 1855.

[80] Murchison 1852b: 369; 1852c: 177.

[81] Darwin to Lyell, 2 Dec. [1859], in F. Darwin 1887, *2*: 236-237.

ported in the *Silurian System* and the *Geology of Russia*) with those of dozens of other investigators. Even by the mid-1840s the international reputations of the two men could no longer stand comparison, as is well illustrated by an 1844 poll for corresponding members of the Institute of France. Sedgwick obtained three votes; Murchison was elected with twenty-seven, more than all the other candidates combined.[82] By the early 1850s Sedgwick's program of research was on the verge of failure, impeded by his inability to undertake sustained literary composition, delayed by the complexity of the project itself, and finally superseded by the emergence of the professional Survey in Britain. Murchison's plans, on the other hand, had reached a remarkable fruition in the worldwide acceptance of his classification and the widespread distribution of his books. An apotheosis of sorts had been achieved at Dudley during an excursion of the British Association, when the Bishop of Oxford, Samuel Wilberforce, proclaimed him "King of Siluria" before an audience of cheering thousands. The international expansion of Siluria through the use of fossils rendered it a more precious possession to be defended by British geologists in its full integrity as a matter of national pride.[83] Although in many respects Murchison had less insight and intellectual power than Sedgwick, he wrote with a ready pen, possessed a clear sense of his limits and aims, and hammered in areas safely distant from the advancing march of the Survey corps. For both men, scientific nomenclature signalled many things, not least a natural and harmonious classification of the rocks. But in fighting for priority between Silurian and Cambrian, they effectively asked fellow geologists to judge the worth of their research methods, their reputations, and their scientific ideals.

In a pioneering article, Robert Merton—who more than anyone else has directed attention to the study of priority controversies—has argued that they shed light on scientific "norms" of humility and originality. As Sedgwick's case suggests, Merton's analysis usefully points to certain dilemmas that can arise in publicly claiming an original contribution to knowledge. But for all his insight, Merton assumes that priority disputants generally agree about what they have discovered, that a discovery is typically an identifiable and discrete unit for all participants whatever their situation and background. Hence his claim that all scientific findings are in

[82] Geikie 1875, 2: 31; for *Siluria*, see Murchison 1854 and J. C. Thackray 1981.

[83] Secord 1982, esp. 439-441.

essence "multiples"—that is, made more than once.[84] In thus reducing discoveries to their lowest common denominator, however, this approach strips individual controversies about priority of their conceptual substance. Such a view is a remnant of a sociology that places the contents of science outside its domain, what has often been criticized as a sociology of scientists rather than of science itself. As the present instance indicates, participants in a priority dispute often construe the meaning of their findings in ways so radically different that their seeming equivalence becomes extremely problematical in certain contexts. Because of this wealth of disagreement, the historical study of such controversies can provide an unusually explicit picture of the assumptions, values, and interests bound up in a science in a particular place and time.

Thus the choice between Cambria and Siluria demanded not only a knowledge of trilobites, brachiopods, and limestones, but also a series of social and philosophical choices as well. As so often in the history of science, the dispute pitted a major (and perhaps unnecessary) change against an established consensus; the righting of a perceived injustice against the equally worthy claims of prior possession; individual choice and constant flux against authority, order, and permanence. In their histories of discovery, both men drew upon political languages with deep roots in the English past. Sedgwick recalled a republican commonwealth of freeborn Britons, while Murchison employed images of kingship, international ascendancy, and imperial grandeur.[85] That one was a reforming Whig and the other a conservative Tory should come as no surprise, although as an overtly political contrast this point should not be pressed too hard. In any event, throughout most of the debate there could be little doubt as to the victor in the contest between "outsider" and "insider": fellow naturalists might express great affection and sympathy for Sedgwick and his views, but when putting their own reputations on the line, they almost always followed the dictates of Murchison. Despite accusations to the contrary, this was hardly ever through direct coercion or illegitimate use of patronage, but rather because utility and stability were seen as substantial virtues in their own right, in science as in other realms of human endeavor. As one commentator observed after the deaths of Sedgwick

[84] See Merton 1973: 286-324, 343-382. Criticisms of this view paralleling those below are available in Brannigan 1981 and (as part of a specific case study) in Kragh 1980.

[85] The Victorians' use of their political history is discussed more generally in Burrow 1981.

and Murchison, "The man of genius had been, as is generally the case, outstripped in the race by the practical man of method."[86] In the course of the debate, Cambria and Siluria became potent symbols of distinctive values in science, manifestations not only of different ways of seeing the rocks, but also of different ways of perceiving and obtaining scientific knowledge.

[86] [Dawkins] 1875: 190.

CHAPTER EIGHT

The Battle of May Hill

THE MOVEMENT of the debate into the historical sphere might suggest that the dispute held little further importance for the creation of new scientific knowledge. The Survey's work was done, and the principals seem engrossed by efforts to twist their findings to match the official results. But the pen was by no means replaced by the hammer in the controversy over Cambria and Siluria. In fact, the most important developments in Lower Palaeozoic geology during the 1850s grew directly out of the debate. During the summer of 1852 Sedgwick and his assistant at Cambridge confirmed their suspicions that Murchison and the Survey had confounded two separate sandstones under the designation "Caradoc." Although only a few geologists went directly over to Sedgwick's side, the entire question of the classification had been effectively reopened for the first time in over a decade. As this episode shows, the link between the defense of scientific property and the creation of new scientific knowledge can continue to operate even as controversy becomes increasingly bitter.

SPLITTING THE SILURIAN

At the end of her life, Cornelia Crosse remembered "the great battle between Murchison and Sedgwick" at the Liverpool meeting of the British Association in 1854. "But in the end," she wrote, "the 'King of the Silures,' as Murchison was called, seemed to claim too much for his kingdom under the soil; the ancestral blood of the Dalesman rose in bold Sedgwick's breast, and he on his side claimed the Caradoc sandstone for his own."[1] Although largely untutored in geology, she could scarcely miss the importance of the Caradoc, for by the end of 1852 it had already emerged as a focal point of the debate between Murchison and Sedgwick. To the end of their lives, the two principals in the controversy continued to center their attention on this formation and the widespread break at its summit, a gap that Sedgwick first traced during the summer of 1852.

[1] Crosse 1892, *1*: 46.

Most geologists, including Sedgwick, had always accepted the Caradoc formation of the *Silurian System* as a useful and widely understood natural grouping. For all their striving after novelty, the men of the Survey unavoidably saw the Welsh strata partly from the perspective of the pages of Murchison's great book, and they never questioned the unity of the Caradoc even after years of detailed mapping and section drawing. But during Survey fieldwork in the Malvern Hills (1843-1845), Phillips had noted that certain strata high in the Upper Silurian parts of his sections unexpectedly teemed with Lower Silurian Caradoc fossils. His official memoir of 1848 interpreted these as local exceptions to a general rule, the last gasps of a fauna on the verge of extinction. His explanation of the anomaly hints at connections between the Survey's environment-oriented stratigraphy and contemporary inquiries into the history of society:

> This is a difficulty of classification merely; a difficulty which occurs in almost every part of the series of strata, and everywhere troubles our systems, but which is perfectly in harmony with the well-understood operations of nature. Just as in the history of nations the germs of new social systems may be traced, and may even have periods of activity and influence, before older systems have died away; and just as a banished race may return and possess for a time its old domains; so in the Silurian Sea, during a pause or diversion of the currents which drifted the Caradoc sands, the [Upper Silurian] Woolhope limestones, which are the effect of the growth of shells and corals, began to appear; but the sand-drifts once again returned, before natural causes excluded them from the basin for ever.[2]

Sedgwick, of course, was far from accepting that classifications were such arbitrary affairs, but he did share Phillips's view of the position of the Caradoc. Even as late as February 1852 he described it as a single group of passage beds common to the Cambrian and Silurian rocks, intermediate both in geological horizon and in organic remains. He had by then withdrawn the compromise term "Cambro-Silurian" offered in December 1846 for these strata.[3] But

[2] Phillips 1848: 75. Similar parallels in Lyell's work are brought out in Rudwick 1979b. The Survey interpretation of the Caradoc in North Wales is briefly presented in Ramsay and Aveline 1848. For more on social analogies in the Survey's work, see Secord 1986.

[3] Compare Sedgwick 1852d: 142-143, 147, with Sedgwick 1847: 157-158.

the effect of the passage for the debate in practical terms continued to be the same, for Murchison and his followers could justifiably claim that the Caradoc welded Upper and Lower Silurian into a single unit with a single persistent fauna.

The idea that this passage might be mistaken arose through the palaeontological work of Frederick McCoy in Cambridge. Born in 1823, the son of a Dublin physician, he had previously catalogued the Silurian collections of Richard Griffith's Irish survey. Possessed of remarkable energy (he needed only five hours of sleep each night), McCoy had assisted Sedgwick ever since Salter's departure for the Survey in 1846, his chief task being the preparation of a complete catalogue of the Palaeozoic fossils in the Woodwardian Museum. Possibly his doubts about the soundness of the Caradoc were raised by the comments in Phillips's memoir; in any case, he began his work suspecting that it might include two separate groups of rocks on separate horizons, lithologically similar but palaeontologically very different. While changing the organization of the fossils from a zoological to a purely stratigraphical arrangement (the merits of both systems were hotly debated throughout the century), McCoy soon found that the standard localities for the Caradoc did fall into two distinct groups. One had fossils like those of the Lower Silurian, and the other had affinities with the Upper Silurian Wenlock formation. This separation is illustrated diagrammatically in Figure 8.1. McCoy immediately realized that if his view proved correct, Sedgwick would possess a marvelous wedge for splitting Murchison's enlarged Silurian palaeontologically in half. Rather than sharing a large percentage of their fossils in common, Sedgwick's Cambrian and his Silurian would be almost entirely distinct.[4]

There were difficulties with this view, however, and Sedgwick refrained from accepting it before he could reexamine at least some of the evidence in the field. In particular, Phillips had based his interpretation on sections in which the two different sandstones appeared to alternate with one another, thus obscuring the simple picture suggested by McCoy. The unwillingness of Sedgwick to accept a favorable innovation on purely palaeontological grounds shows his continuing emphasis on structural and sectional criteria, even when the issue had obvious implications for the validity of the Cambrian. At the same time, there can be no doubt that Sedgwick wished to see Murchison's Silurian divided in two. As he told McCoy late in 1851:

[4] McCoy, "Reminiscences of Professor Sedgwick," CUL: Add. ms 7652VJ8; for biographical information on McCoy, see *DNB*.

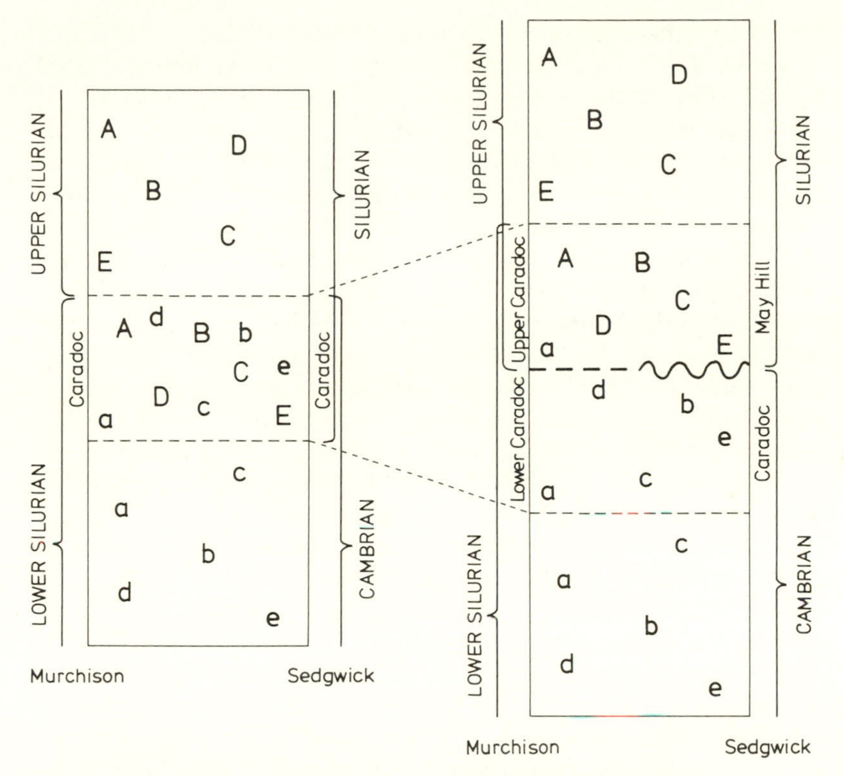

Fig. 8.1. Diagrammatic columnar sections showing the effect of the splitting of the Caradoc on the stratigraphical and palaeontological relations of the Lower Palaeozoic. Capital letters represent Upper Silurian species; lower case letters represent Lower Silurian (or "Cambrian") species.

You will see what I am driving at. The Caradoc is between two *systems*, if we use the word *system* in the sense of the *Government map*. But I trust that next summer I shall have the happiness of rapping it on the head with you to help me, & that then we shall be able to split the so-called Caradoc into two parts—one = upper Bala—the other = Wenlock grits. So that Wenlock, thus explained, may form the true Silurian base; all else being Cambrian, both zoologically & geographically.[5]

[5] Sedgwick to McCoy, 14 Nov. 1851, ML: frames 167-169. See also Sedgwick 1852d: 149-150. His hesitation is expressed in Sedgwick to McCoy, 6 Feb. [1852], 12 Mar. 1852, ML: frames 483-485, 190-198.

With their expectations thus primed by the evidence of the fossils, Sedgwick and McCoy planned a tour for the summer of 1852 through some of the best sections across the Caradoc in Wales and the Welsh Borders. Torrential downpours and previous engagements cut their expedition down to less than a week, however, and they only had time to thoroughly examine sections at May Hill and the Malverns. On this slender evidence, Sedgwick came before the Geological Society in November with a paper dividing the Caradoc into two groups. The first of these, at the summit of his Cambrian, continued as the "Caradoc" formation, and the second, at the base of the Silurian proper (Murchison's Upper Silurian), became a new group called the May Hill Sandstone.[6]

Within the specialized context of British Lower Palaeozoic geology, the implications of this revision were nothing short of sensational. A large number of beds that the Survey had placed in the Caradoc would have to be moved back into the Upper Silurian, where Sedgwick had consistently placed them prior to accepting the Survey results. If these recorrelations proved necessary, the number of species common to the Cambrian and Silurian in Britain would fall considerably. After the creation of the May Hill Sandstone, Sedgwick could argue for the first time since June 1843 that the fossil break at the base of Murchison's Upper Silurian would prove greater than that found at its summit, thus making the two halves of the Silurian more distinct than the Silurian and the Devonian. Moreover, this gap brought the British sequence into a closer parallel with the situation known to prevail in other important Palaeozoic districts throughout the world, especially in America and central Europe. These were large claims to make on the basis of a few days in the field, and Sedgwick's listeners at the Geological Society were understandably sceptical. They viewed McCoy's palaeontological work with particular suspicion, and the fossil experts on the Survey led the attack. Forbes, the chief palaeontologist, reported the discussion in a letter to Ramsay:

> At the society Sedgwick came out with his story. . . . The argument was of course founded on fossils—the data supplied with McCoy. The impression left not only on my mind, but also on Strickland's, Austen, Sharpe, Morris & all I "conversed" with was that McCoy (who was there & spoke) had ["got up" *del*] cooked the fossil evidence to please Sedgwick & misled him. Salter spoke

[6] Sedgwick 1853b. The tour is described in detail in Sedgwick to Phillips, 1 Oct. 1852, ff. 204-218, UMO.

> *very well* & from good data. His speech was very convincing & went to show that the Survey was justified in both cases. There was the usual spar between Murchy & Sedgwick about Cambrian. The discussion was lively & good.[7]

Given the large number of important sections that Sedgwick and McCoy had not examined, there was plenty of room for doubt. For the first time in his life, Sedgwick was even accused (by Phillips) of being an "*ultra* palaeontologist" for relying so heavily on fossil evidence in local districts. He immediately denied the charge; however, additional sections clearly required reinterpretation before the splitting of the Caradoc could be considered as established.[8]

Although the men of the Survey had begun by opposing the new theory with considerable vigor and even a certain arrogance, they soon suspected that Sedgwick and McCoy really had found an important error in the official sections. Salter, citing the fossil evidence, partially recanted to Sedgwick in February 1853, and in April Ramsay and Jukes read papers at the Geological Society which concluded that certain beds formerly called Caradoc were actually sandy parts of the Wenlock. McCoy, who attended this meeting, was understandably annoyed at their failure to acknowledge Sedgwick's previous paper on the Caradoc question. In a process of historical revisionism that will have a familiar ring, the Survey now claimed that McCoy's discovery merely repeated work done by Phillips, Ramsay, and Talbot Aveline in the 1840s.[9]

In this case, however, actions spoke louder than words. From this point onwards it was a race to the field, with the Survey as eager to correct its own errors as Sedgwick was to find them. That the official organization should be compelled to return to a region already mapped demonstrates the potential importance of the separation of the May Hill Sandstone (or "Pentamerus beds" as the Survey initially called them) from the true beds of the Caradoc, especially as De la Beche felt considerable pressure from his superiors to complete the mapping of England and Wales. Although the men of the Survey had a strong stake in the geological lines already printed on

[7] Forbes to Ramsay, 4 Nov. [1852], ICL(R).

[8] Phillips to Sedgwick, 24 Nov. 1852, CUL: Add. ms 7652IIX30. See also Sedgwick to Phillips, 27 Nov. 1852, ff. 222-225, 249-252, UMO; and McCoy to Sedgwick, 29 Nov. 1852, CUL: 7652IIX43b.

[9] Ramsay 1853, Jukes 1853; Salter to Sedgwick, 25 Feb. [1853], CUL: Add. ms 7652IIX40; for McCoy's reaction to the Survey claim, and a report of the discussion at the Geological Society, see McCoy to Sedgwick, 22 April 1853, CUL: Add. ms 7652IIX55b.

the official map, they had an even stronger commitment to its accuracy. Ramsay, as local director for Great Britain, deputed Aveline and Salter to examine some of the critical sections. Aveline (in Geikie's words, a "tall, dark, silent, big-booted man who strode with gigantic steps over the hills") and Salter (a nervous excitable character) must have made an odd pair in their excursion during the spring and summer of 1853.[10] In any event, they immediately found evidence for the accuracy of McCoy's surmise at many sections that Sedgwick and his assistant had been unable to check.

One of these sections, along the River Onny in Shropshire, typifies the most interesting feature of the battle over the May Hill Sandstone: the importance of theory-dominated observation in a classificatory and descriptive science like geology. Previously, the dozens of geologists who had observed the classic section along the Onny had always seen a perfectly conformable sequence ascending from Caradoc into Upper Silurian without a break. Because Sedgwick believed that the fossils from this locality were "true" Caradoc species, he immediately realized that if this succession really was conformable, it afforded no room for the May Hill formation—a serious objection to his proposed attempt to separate the Caradoc into two groups. Thus if the new theory was correct, the Onny section should show an *un*conformity on the horizon of the May Hill Sandstone, a gap in the succession representing a period of nondeposition or subsequent erosion. Rechecking this section in the light of Sedgwick's prediction, Aveline and Salter immediately found a somewhat smaller but clearly marked stratigraphical overlap as illustrated in Figure 8.2. Despite their initial hostility to the splitting of the Caradoc and their vested interest in the published documents of the Survey, the two men could do little else than note the fossils and recolor the map. Once pointed out, the phenomenon was there for anyone trained in geology to see, and hundreds of students continue to do so every year.[11]

Murchison and all those who had followed him in visiting the locality for over thirty years had failed to observe the Onny uncon-

[10] Aveline is described in Geikie 1904: lxvii. For the pressure felt by De la Beche, see De la Beche to Ramsay, 17 May 1850, 31 July 1854, ICL(R). Ramsay's role is described in Geikie 1895: 204-206. Salter 1867 provides an entertainingly egocentric account of the tour.

[11] Salter and Aveline 1854: 70. Sedgwick 1853b: 228 explains his position. A similar case involving the perception of an unconformity is given by Oldroyd 1972a; also relevant is Rudwick 1974, on Darwin and the Glen Roy problem.

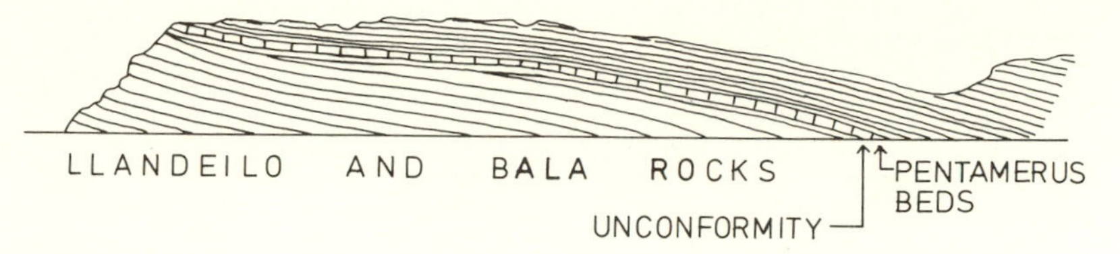

Fig. 8.2. Traverse section along the banks of the River Onny, 1853.

formity, not through any gross defect in their scientific methods, but simply because they had no reason to expect it. Guided by appropriate expectations, a properly instructed observer would henceforth immediately find the unconformity; moreover, anyone who had accepted the rules and social conventions of geology would have to admit its presence or risk being dismissed as a blind man or a crank. In a sense, then, the gap had been "waiting" to be found, but only by an observer whose vision had been properly primed. The priming in the present case, as we have seen, originated in McCoy's loyal wish to provide Sedgwick with palaeontological grounds for two separate groups among the older fossiliferous rocks. Although the Survey did not use the Onny discovery to support the Cambrian classification, there can be little doubt that Aveline and Salter would never have returned to the Welsh Borders if Sedgwick had not read his paper the previous November.

Many similar examples could be cited, for during the next three years Aveline, Salter, Ramsay, Sedgwick, and McCoy ranged across Wales splitting the Caradoc in half and tracing previously undetected unconformities. The "errors" now revealed had not been confined to Murchison alone; Sedgwick found stratigraphical gaps in North Wales as well, often in places where he would least have predicted them ten years before. In a paper read to the Geological Society in May 1854, Sedgwick wrote that the beds at Glyn Ceiriog and Mathrafal, which he had consistently viewed as offering a conformable transition into the Denbigh Flags, were in fact separated from them by a break of some consequence (Fig. 8.3). As in the Onny section (which he described in the same paper), important strata were missing. For almost thirty years, the Glyn Ceiriog strata had held their high place in his sections, the passage there serving as one of the cardinal points in his interpretation of the structure of North Wales. Now, in the wake of the discovery of the May Hill Sandstone, that conformable succession disappeared, and Sedg-

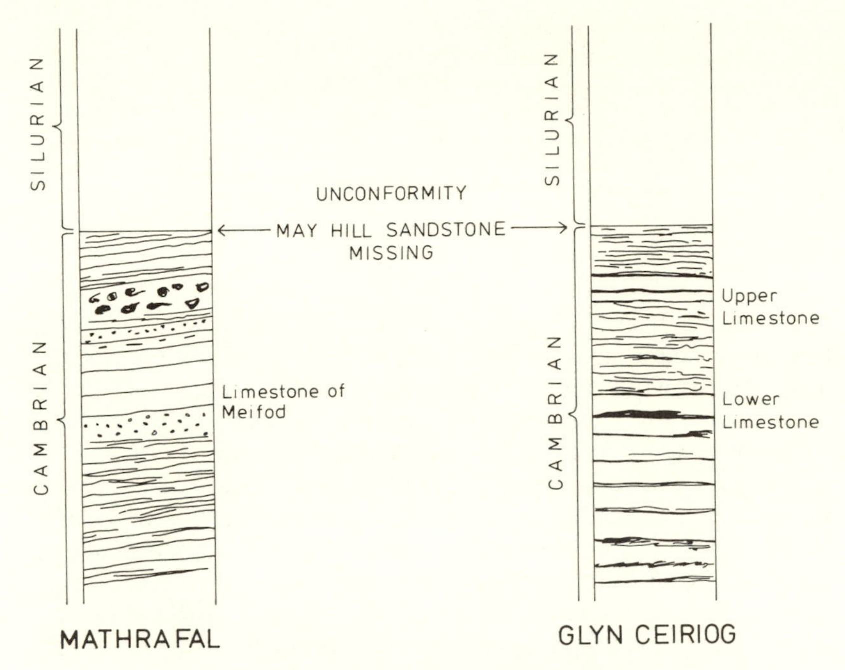

Fig. 8.3. Columnar sections by Sedgwick showing the junction between the Cambrian and Silurian in North Wales, 1854.

wick was forced to explain away the evidence that had led him to see a passage at Glyn Ceiriog in the first place.[12] Even better than Murchison's error at Onny, this example shows that the changed perspective did not reflect any defect in methods, but represented the result of expectations altered by the discovery of the May Hill Sandstone.

The failure to observe the unconformities like those at Glyn Ceiriog, Mathrafal, and Onny is easy enough to understand, for in many cases the exposures were obscure and the exact points of contact difficult to see on rapid examination. Occasionally, however, the confusion between the May Hill Sandstone and the true Caradoc resulted from more serious mistakes due to carelessness and haste. In one of their earliest boundary-tracing tours in 1853, Aveline and Salter visited a lime kiln on the Hollies Farm near Acton Scott in Shropshire, which had served Murchison as one of the richest and most typical sources for the Caradoc fossils illustrated in the

[12] Sedgwick 1854b: 304-306. In Sedgwick 1853b: 228 he pointed out the need to postulate an unconformity.

Silurian System. The intermingling of Lower and Upper Silurian fossil types was here particularly striking, and the two Survey geologists soon discovered why: the kiln itself was made out of Lower Silurian Caradoc, but the material burnt inside originated from a higher quality calcareous band with fossils of the Upper Silurian May Hill or Pentamerus beds. In the early 1830s Murchison had gathered specimens from the kiln and its contents indiscriminately, not realizing (as he might have done, had his fossils been collected *in situ*) that specimens from more than one stratigraphical horizon were present.[13] The mistake was an elementary one, but easily made—indeed, Sedgwick himself became tangled up in a similar error in 1853 by confusing the names of two important bridges across the Onny.[14] The fundamental difficulty is suggested by a passage from Darwin's instructions to the beginning geologist in the 1849 Admiralty *Manual of Scientific Enquiry*:

> It is highly necessary most carefully to keep the fossils found in different strata separate; it will often occur in passing upwards from one bed to another, and occasionally even without any great change in the character of the rock, that the fossils will be wholly different; and if such distinct sets of fossils are mingled together, as if found together, undoubtedly it would have been better for the progress of science that they had never been collected.[15]

Certainly De la Beche must soon have wished that he had never heard the word "Caradoc," for the next few years required extensive tours for the resurveying of that formation throughout Wales. He permitted these reexaminations only with the greatest reluctance and after repeated urgings from his subordinates. As he told Ramsay in 1854, "the very sound of such matters sets me adrift."[16] In North Wales (along the base of the Denbigh Flags), South Wales, Pembrokeshire, the Malverns, and the rest of the Border country, important readjustments were shown to be necessary, and revised editions of the one-inch maps began to appear in the mid-1850s.

The entire incident was evidently a great embarrassment to the men of the Survey. Some of the younger field geologists, especially Ramsay, Jukes, and Aveline, began to look on the early maps of the Survey with disdain, for patent errors on the published documents

[13] Salter 1867; Salter and Aveline 1854: 66. See also Murchison 1839: 217, pl. 19.

[14] Sedgwick 1853b: 228, corrected in Sedgwick 1854b: 302-303. Also Salter to Sedgwick, 30 Apr. [1853], 9 May 1853, CUL: Add. mss 7652IIX57a, 57b.

[15] Darwin 1977b: 232.

[16] De la Beche to Ramsay, 31 July 1854, ICL(R).

of a supposedly authoritative source reflected on their personal reputations and professional status. "I cannot but think," Ramsay told De la Beche after the first summer of revisions, "that when by new lights shining out, omissions or errors are discovered it is better to mend them as soon as we know the way, than to leave them open to amateur carpers."[17] Whether or not Ramsay here referred specifically to Sedgwick (who scarcely deserved such an epithet) the threat to the career geologists posed by his work should be immediately evident. However, once De la Beche mobilized the resources of the Survey, an individual geologist—even one of Sedgwick's ability—could not really hope to compete with its work from the standpoint of comprehensiveness. Sedgwick's three papers on the May Hill question described single traverses at critical localities, while the government geologists conducted mapping on a much wider scale, along the boundaries of individual formations.[18] Thus even if the impetus for splitting the Caradoc came from Sedgwick and McCoy, primary credit for the actual remapping in the field belonged to Aveline, Ramsay, and their associates on the Survey.

By the end of the summer of 1853, with Sedgwick preparing long papers for the Geological Society and the British Association, and the Survey deep into its revisions of the one-inch maps of Wales, there could be little doubt that McCoy's insight had been prophetic. Throughout Wales, the May Hill Sandstone was proving to be distinct from the Caradoc, and although the precise extent of the gap remained uncertain, an unconformity seemed to extend with unusual persistence between the two halves of Murchison's Silurian. For the first time since the naming of the two systems in 1835, the progress of research in Wales had uncovered a result fundamentally unfavorable to the palaeontological unity of the Silurian system. British geologists, almost all advocates of the expanded Silurian classification, were suddenly forced to incorporate this uncomfortable but incontrovertible phenomenon into their picture of the older rocks.

Since the discovery of the twofold nature of Murchison's Caradoc had originated in McCoy's desire to support the Cambrian, it is scarcely surprising that Sedgwick immediately used the May Hill Sandstone as ammunition in the battle with Murchison. We have already seen that he argued from the new palaeontological data for an

[17] Ramsay to De la Beche, 21 Nov. 1853, NMW, printed in Geikie 1895: 205.

[18] Sedgwick 1853b, 1854b, 1854c, should be compared with the revised editions of the Survey maps and sections which appeared during the mid-1850s.

almost complete separation of the Cambrian and Silurian faunas, not only in Britain but throughout the world. At the same time, the discovery of the May Hill Sandstone also became the key to his historical and personal arguments against Murchison. The mistake introduced by the false sections across the Caradoc, Sedgwick believed, demonstrated "the great hindrance to good classification and real progress which arises out of the premature adoption of definite scientific names."[19] Although the basic import of Sedgwick's attack remained unchanged, the discovery of the May Hill Sandstone gave his historical reconstructions a more plausible basis: most geologists held both men responsible for the determinations of the 1834 joint tour on which Sedgwick had previously rested his case, but they attributed the error in the Caradoc to Murchison alone.

For Sedgwick's tiny but growing band of outright supporters, confirmation of the new findings came as welcome news. The American geologist Henry Darwin Rogers, the first advocate of any importance from outside the British Isles, had already been convinced by the arguments of the February 1852 paper. "Sedgwick's beautiful classification and nomenclature of the British rocks is infinitely better in harmony with our American Palaeozoic Geology than Murchison's," he had told his brother William in December of that year. "He calls all the Palaeozoic *one system*, and terminates the Cambrian with the Caradoc, just where we would draw our strongest equivalent line. . . ."[20] The discovery of the May Hill break only confirmed his conviction of the merits of a distinct Cambrian. Previously the type sections in Wales had presented an unusually gradual transition between the Upper and Lower Silurian; now the abrupt break Rogers recognized in America had found its match in the standard sequence in Britain. It is important to note that Rogers doubted the utility of palaeontological correlation over long distances and shared Sedgwick's faith in the primacy of sections and structure. His compatriot, the New York Survey geologist James Hall, took a much more sanguine view of the use of fossils and was correspondingly more favorably inclined towards Murchison's classification, although he too had some reservations about the inter-

[19] Sedgwick 1854b: 367. The change in emphasis is best evidenced in Sedgwick 1855 and 1873.

[20] H. D. Rogers to W. B. Rogers, 16 Dec. 1852, in W. B. Rogers 1869, *1*: 328-329. Gerstner 1979 offers a helpful exposition of the classification used by the Rogers brothers.

national applicability of the Silurian and generally employed local names in his stratigraphical descriptions.[21] However, suspicion of the use of fossils was by no means the only factor in determining support for the Cambrian. The case of McCoy shows this clearly, as does that of James de Carle Sowerby, another palaeontologist who (privately) advocated Sedgwick's views even before the May Hill affair. Besides Rogers, Hopkins, Sowerby, and McCoy, most of Sedgwick's supporters were local amateurs, frequently men who in earlier years had sent him specimens and helped in his fieldwork among the older rocks; their specialized expertise on such issues was often unrivalled, although their views on the great questions of classification counted for little among the specialists of the metropolis. Thomas Gough of Kendal, Samuel R. Pattison of Launceston, John D. Pring of Taunton, and John Giles of Liskeard all sent strong letters of support; but as Giles admitted, they were just "small folk."[22]

Sedgwick and McCoy's discovery, as extended and confirmed by the Survey, did not result in any sudden rush to support the full Cambrian classification. But even though almost all geologists continued to call rocks on both sides of the May Hill Sandstone by the name "Silurian," one can easily detect a quickening of interest in Sedgwick's proposals. At the Hull meeting of the British Association in September 1853, for example, he addressed a large audience on the subject of Lower Palaeozoic classification, making great play of a "glorious letter" from Henry Darwin Rogers in favor of the Cambrian—and with good reason. Although Rogers suffered the considerable disadvantage of being an American in a field patriotically viewed by the British as their own preserve, his support potentially placed the geology of an entire continent at Sedgwick's disposal.[23] In the ensuing discussion Phillips refused to commit himself one way or another on the question of nomenclature, but he evidently believed (as his private letters to Sedgwick also show) that the new discoveries had substantially buttressed the case for a ma-

[21] Hall's views are summarized in Schneer 1978 and Clarke 1923.

[22] Giles to Sedgwick, 22 Apr. 1854, CUL: Add. ms 7652IIM40; also Gough to Sedgwick, 30 July 1852, CUL: Add. ms 7652IIX21a; Pattison to Sedgwick, 22 Sept. 1852, CUL: Add. ms 7652IIX25; Pring to Sedgwick, 25 Oct. 1854, 30 Oct. 1854, CUL: Add. mss 7652IIL14a, L5a; and Sowerby to Sedgwick, 1 Sept. 1852, CUL: Add. ms 7652IIX24.

[23] Sedgwick 1853a; for the "glorious letter," see Sedgwick to McCoy, 8 Mar. 1853, ML: frames 134-145, and the letter itself: H. D. Rogers to Sedgwick, 15 Feb. 1853, CUL: Add. ms 7652IIIB73.

jor classificatory break at the base of the Upper Silurian. In the absence of Murchison, debate was rather slow, although his old travelling companion Hugh Strickland set forth the continuing argument for the unity of the Silurian. But even if victory thus came partly by default and before a popular audience, this was the first occasion since the late 1830s that Sedgwick's proposals were not rejected out of hand. "The Cambrian monarch declared that he had now found a base line to the territory of the Silures which his Silurian majesty has missed," wrote a hostile commentator in the *Literary Gazette*, "and as there was no one present to dispute it, the debateable ground on the confines of these two great geological systems is, for the present, to be assigned to Cambria."[24]

As the May Hill discovery carried opinion towards Sedgwick, how did Murchison react to the threat? By this late date there could be no question of abandoning the fully extended Silurian classification. He had missed the Hull meeting, not from any fear of confronting Sedgwick, but rather because he was busy gathering materials in Germany for his new book, *Siluria*. The Survey's confirmation of the need for the reclassification of the Caradoc came at a highly inconvenient time, just as the pages describing that formation were passing through the press. Trying to cut his losses to a minimum, Murchison called Sedgwick's May Hill Sandstone the "Upper Caradoc" and maintained that its palaeontological affinities were with the Lower rather than the Upper Silurian. However, the chapter remained in something of a muddle. For example, a woodcut (Fig. 8.4) had been prepared showing the characteristic Upper Caradoc fossils, but Murchison was forced to explain in a footnote that at least one of them, *Orthis alternata*, actually belonged in the underlying beds.[25] Moreover, by the time he had finished the book, his view of the affinities of the Upper Caradoc had become untenable. Rather than continuing to bracket this formation with the Lower Silurian, he presented it in an appendix as a passage bed uniting the two halves of his system. In this opinion, so totally at odds with that put forward by McCoy, Murchison had the important support of Salter, who had undertaken the fossil work for *Siluria* on a freelance basis. Salter argued that certain fossils (notably the brachiopods *Pentamerus oblongus* and *Atrypa hemispherica*) rendered the affinities of the group still sufficiently close to the Lower

[24] *Literary Gazette*, 17 Sept. 1853, no. 1913: 913; also *Athenaeum*, 24 Sept. 1853, no. 1352: 1134-1135.

[25] Murchison 1854: 86-87, 98.

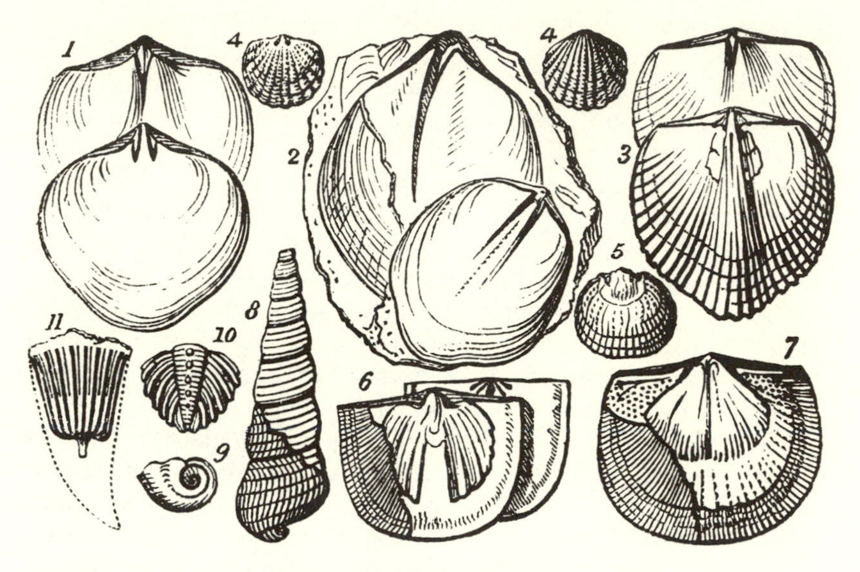

Fig. 8.4. "Upper Caradoc" fossils pictured in the first edition of Murchison's *Siluria* (1854), including *Pentamerus oblongus* (no. 2), *Atrypa hemispherica* (no. 4), and *Orthis alternata* (no. 6).

Silurian for it to be retained as a bed of passage.[26] In the appendix added just before his book entered the shops in 1854 Murchison explained this blending and justified using "Upper Caradoc" as a connecting link. The fact remained that this effectively retracted the view offered in the main body of his text and was an admission of error accomplished with great pain. For the first time since the publication of the *Silurian System*, new data from the field had proved unfavorable to Murchison's extended classification. At best he might stress the geographical limitations of the unconformities being found at Onny and elsewhere in Wales and follow Salter in stressing the transitional aspects of the so-called Upper Caradoc.[27]

But the need for concessions on Murchison's part continued even after the publication of the first edition of *Siluria* in 1854. In the fol-

[26] Salter and Aveline 1854: 72; also Salter to Ramsay, 14 June 1853, 12 July [1853], ICL(R). Aveline did not agree with this interpretation, as is evident from internal evidence (e.g. p. 66) in their joint paper, and from slightly later correspondence: Aveline to Ramsay, 18 Oct. 1856, ICL(R); Ramsay to Aveline, 23 Dec. 1855, BGS: GSM 1/420(A), f. 307.

[27] Murchison 1854: 483-492. A clear summary of his views at this time can be found in Murchison to Lyell, 13 Feb. 1855, APS (D/L): B/D25L.

lowing year De la Beche died, and in order to prevent the Survey from falling into the hands of a non-geologist, Murchison assumed the directorship. This occurred just as Aveline and Salter were completing their remapping, and it only remained for Murchison to choose a permanent name for the May Hill Sandstone or Pentamerus beds. At first the new director tried to get the Survey geologists to accept his "Upper Caradoc," a proposal which drew a protest even from the usually taciturn Aveline.

> What an extraordinary man Sir R. is, I think he would sacrifice truth and every thing else rather than his first ideas of the Silurian systeme should be in the least controverted, one would fancy that any man would be satisfied with the establishing such a great system and would be glad and not jealous with any working out and improving the interior of that systeme. The Survey has done every thing for him. All they have done has only added more value and importance to Siluria and it is very poor for him to make to say that all his original ideas were quite correct in every subdivision in spite of our labours showing the contrary. If not to ignore the work of our hands certainly the work of our brains our thoughts our conclusions. He would render us to mere labours, collectors for his Siluria. I hope we will never give in to him with the Pentamerus beds, if we give that we give all.[28]

Jukes, Ramsay, and Salter all agreed that Murchison's avoidance of any changes in the received Silurian classification was unfair not only to the evidence, but to their reputations and even to their old opponent at the Geological Society, Sedgwick himself. "It would be but a small act of grace in him to Sedgwick & the survey to yield the term upper caradoc," wrote Ramsay to Aveline, "but he is not unlike the Emperor Nicholas in that respect he hates giving up anything—even a name—that he has ever occupied."[29]

In the end Murchison reluctantly backed down and accepted the need for at least a minimal change in the formations of the Silurian. As matters stood, the text of *Siluria* and the statement in the appendix contradicted one another.[30] Moreover, Murchison's name for the May Hill Sandstone—the Upper Caradoc—though expressing his wish to keep the Silurian nomenclature intact, incorrectly im-

[28] Aveline to Ramsay, 18 Oct. 1856, ICL(R).

[29] Ramsay to Aveline, 10 Oct. 1856, BGS: GSM 1/420(A), f. 337. For Salter and Jukes, see Salter to Ramsay, 27 Oct. [1856], 15 Sept. [1857], ICL(R).

[30] The inconsistencies are pointed out in Salter to Murchison, 12 Jan. [1855], GSL: M/S3/1; see also the comments in Sedgwick 1855, esp. p. lxxxi.

plied that the group's strongest affinities were with the Lower Silurian, a view even Salter could not sanction. In order to take account of these difficulties and to placate his staff, Murchison wrote a new chapter in 1856 for the next edition of *Siluria*. He abandoned the objectionable term "Upper Caradoc," but did not accept either "May Hill Sandstone" or "Pentamerus beds." Instead, Murchison modelled his classification on the succession in South Wales near Llandovery, where a sequence traced by Aveline in 1855 offered a graceful way of avoiding the apparent break being found at the base of the Upper Silurian. In following the course of the May Hill Sandstone in South Wales, Aveline had discovered that still another bed intervened between it and the underlying or "true" Caradoc (Fig 8.5). The palaeontological affinities of this intervening stratum were with the Lower Silurian, although a number of fossil species appeared to be held in common with the May Hill Sandstone. In Murchison's 1856 classification, the troublesome May Hill Sandstone became the "Upper Llandovery," and Aveline's newly discovered stratum just below became the "Lower Llandovery." The Upper Llandovery was allied with the Upper Silurian, the Lower Llandovery with the Lower Silurian; but by uniting these two formations under a single name, Murchison patched over the yawning gap in his previously conformable succession, as shown in Figure 8.6. In 1856 almost all British geologists agreed that the break between the Upper and Lower Silurian was one of the most substantial in the entire geological record, and yet the rocks on either side were bracketed together under the Llandovery title. The use of Llandovery on the official maps and in *Siluria* eventually led to its acceptance by British geologists, but this should not be allowed to obscure its origins as a compromise device to save the Silurian as a single system of life.[31]

Although "Llandovery" gradually limped into use in geological textbooks and memoirs, there can be little doubt that the battle of May Hill represented a victory for Sedgwick and his supporters. For the first time in many years, Murchison himself had been forced to concede a point of considerable importance for the geology of the older rocks, one that made the unity of the Silurian a question for further research and inquiry. "It was sadly against his inclination," wrote Geikie in his biography, "that he was driven to admit that, in

[31] Murchison 1859: 94-114; also Murchison to Barrande, 21 Feb. 1857, in Geikie 1875, *2*: 308-310. Although described on the title page as the third edition, the 1859 publication of *Siluria* was in fact the first printing since the original of 1854. This is clearly explained in J. C. Thackray 1981.

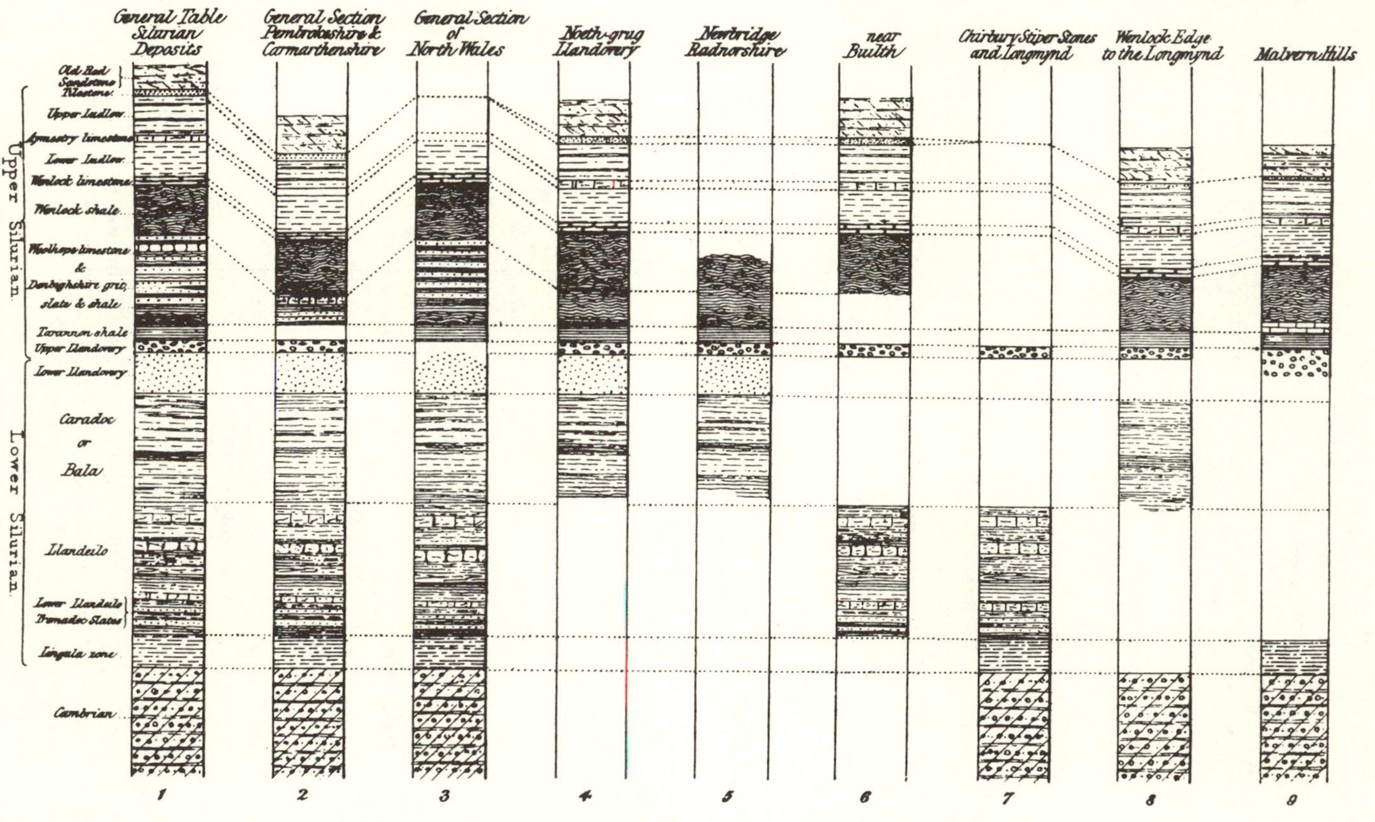

Fig. 8.5. Comparative sections of Silurian deposits in Wales prepared by Aveline for the 1859 edition of *Siluria*. Murchison's choice of this particular assemblage of sections, it should be noted, was designed to emphasize those localities that bridged the break at the base of his Upper Silurian.

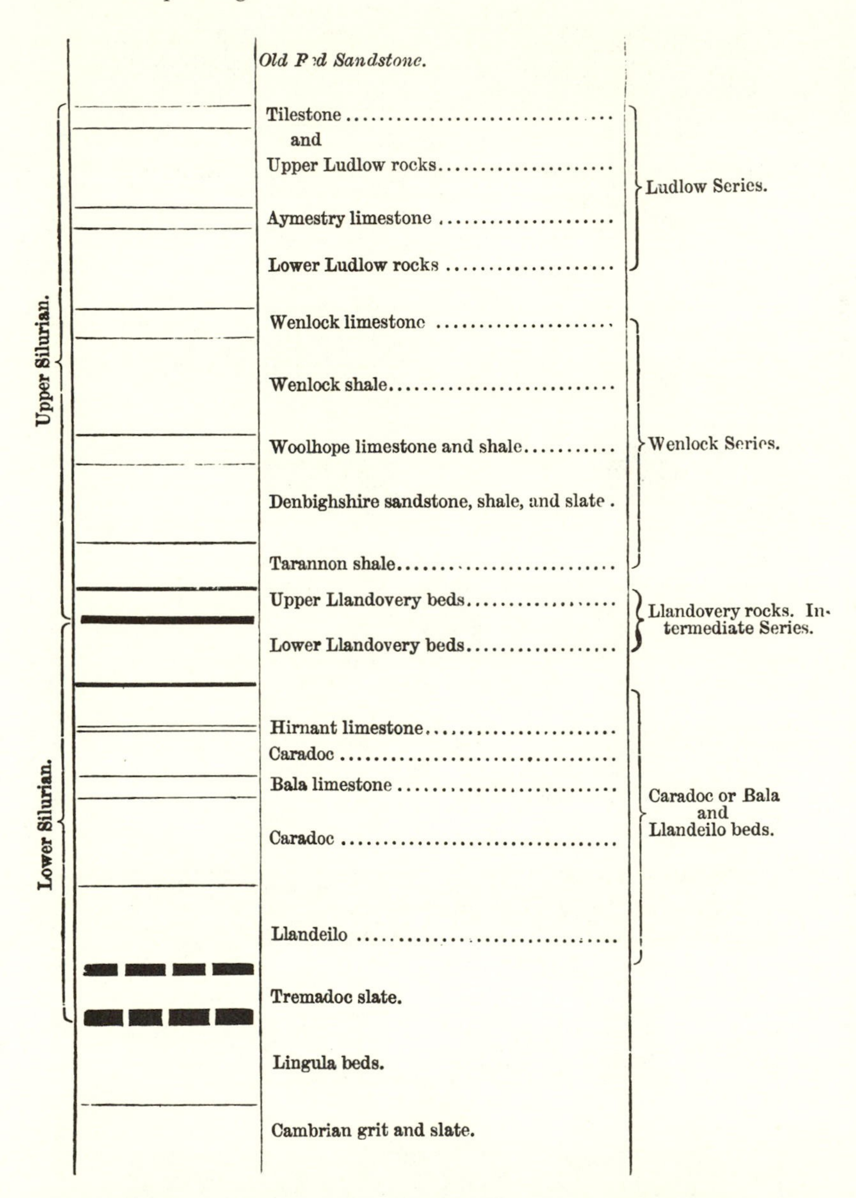

Fig. 8.6. Succession of the Lower Palaeozoic rocks in North Wales as shown by Ramsay in his Survey memoir of 1866. Thin lines between formations indicate the absence of any marked stratigraphical and palaeontological break, and the thicker lines are drawn in proportion to suggest the relative importance of the gaps in the sequence. Note the presence of such an unconformity between the Upper and Lower Llandovery.

this respect, his series was not the unbroken whole which he had represented it to be."[32] As the widespread importance of the gap became clear, most geologists did not embrace the version of the Cambrian favored by Sedgwick, Hopkins, Rogers, and McCoy. But they did begin to doubt the wisdom of classing all the older fossiliferous rocks as Silurian.

The Breaking of a Friendship

The May Hill episode demonstrates that the controversy continued to prove a productive source of new scientific knowledge. At the same time, though in a less direct way, it shows that any such discoveries were bound to have little effect on the respective positions of Sedgwick and Murchison themselves. However sharp the wedge ready to split the Silurian, Murchison would always find some way to deflect it; similarly, whatever proofs were offered for the presence of a single grouping in the older rocks, Sedgwick would always come up with grounds for maintaining the separate existence of the Cambrian. Although the controversy over Lower Palaeozoic classification continued as an important impetus to scientific discovery by other geologists—including the staunchest allies of Siluria and Cambria—Murchison and Sedgwick in their final years contributed more to the volume of the debate than to its resolution.

Mincing no words and missing no opportunity to raise the issue, Sedgwick throughout the 1850s repeated the litany of his former collaborator's errors until compromise became impossible. Even the publication of his papers through the usual channels became difficult, for his forthright innuendoes ranged far outside the usual boundaries of Geological Society discourse. The November 1852 notice of the splitting of the Caradoc was Sedgwick's last paper in the *Quarterly Journal*, for the Council—now out of Hopkins's sympathetic control—refused to print any further statements on either the Cambrian or the Silurian side of the question. (This prohibition was never entered in the minute books, but this omission was typical of many of the Society's important decisions.) Murchison warned Henry Darwin Rogers, for example, that a paper on the American Palaeozoics would be unwelcome if it trespassed upon the forbidden ground.[33] In consequence Sedgwick and his supporters took their case outside the narrow circle of specialists, particularly to popular audiences at the annual meetings of the British Associa-

[32] Geikie 1875, 2: 309.

[33] Rogers to Sedgwick, 25 June 1856, CUL: Add. ms 7652II.I24.

tion. Throughout the 1830s and 1840s the Association had served Murchison and Sedgwick as a forum for establishing claims to discoveries and for vendettas against mutual antagonists like De la Beche. But in Sedgwick's hands, it now became a prominent platform for pressing an anti-Murchisonian crusade. As Rogers said, the Association was "perhaps after all a fairer court of opinion than the Geological Society." Needless to add, their opponent scarcely shared this view.[34] The discussions at "Section C" at Hull (1853), Liverpool (1854), Cheltenham (1856), and Manchester (1861) offered the entertaining spectacle of Britain's leading geologists airing a dispute in public with a vehemence seldom seen in scientific debate. The Liverpool meeting, which featured papers by Sedgwick and Murchison on two separate days, was the scene of perhaps the most spectacular verbal pyrotechnics ever seen at an Association meeting. As Murchison told the relatively peaceful geographers, his old friend seemed intent on extending an ancient proverb:

> A spaniel, a wife, and a walnut tree,
> The more you beat them, the better they'll be.[35]

Not only Murchison, but most other men of science were shocked by Sedgwick's attack. "It is no wonder that you were pained by the tone & matter of the discussion on Palaeozoics at Liverpool," Phillips told him, "as who was not?"[36] Forbes, who had chaired the stormy sessions, expressed fears that the publicity being given to the issue would lead the uninformed to think that "all geology had come to a hitch," thereby opening the door to Scriptural geologists who could use the dispute to cast doubt on the entire geological enterprise, with its indefinite time span and denial of a universal deluge. In popular reviews and essays, Forbes and other authors took great pains to combat this impression. Many felt that Murchison's own *Siluria* was marred by a disputatious tone unsuitable for a permanent work of reference. "The controversies of naturalists,"

[34] Rogers to Sedgwick, 25 June 1856; Murchison's reaction is evident from Phillips to Murchison, 29 Oct. 1854, EUL: Gen. 525. For his own earlier uses of the Association as a debating arena, see Morrell and Thackray 1981.

[35] Quoted in the *Liverpool Courier, and Commercial Advertiser,* Supplement, 27 Sept. 1854, no. 2439, *47*: 354. The best report of the debate is in the *Athenaeum,* 14 Oct. 1854, no. 1407: 1243-1244.

[36] Phillips to Murchison, 29 Oct. 1854, EUL: Gen. 525. See also Phillips to Sedgwick, 29 Oct. 1854, CUL: Add. ms 7652IIL46, and Nicol to Murchison, 14 Oct. 1854, GSL: M/N8/21a & b. For Sedgwick's point of view, see Sedgwick to Phillips, 3 Nov. 1854, ff. 277-284, UMO, and Sedgwick 1854c: 493-494, 1855: lxviii-lxx.

Forbes admitted in a review of the volume, "are seldom creditable to the popular reputation of science."[37]

After the fracas at Liverpool, considerable pressure was exerted upon Sedgwick and Murchison to avoid further unseemly conflicts, although pulses quickened every time they both appeared on a platform. Sedgwick brought up the issue in discussion at the Association on several later occasions ("with great eloquence and his usual fervour," as William S. Symonds once remarked) but never gave another paper of his own on the subject. However, memoirs by other men, such as a controversial lecture by Rogers at the 1856 Cheltenham meeting, gave ample opportunity for comments.[38]

Although discussions at the Association remained relatively open affairs, its publications were not. Here controversy was avoided just as thoroughly as in the Geological Society *Quarterly Journal*. Thus Phillips, acting as assistant secretary to the Association, requested the right to excise all complaints against Murchison and the Survey in Sedgwick's paper from the Hull meeting.

> We must not, in our Volume, raise any discussion of such points—this is always enjoined on me; for this reason we omit one of the most interesting parts of our meetings, viz the discussion after papers. But if it were not so, still, as deeply interested in obtaining you a fair hearing, I should wish to omit all such references, & leave the case on its own good basis of facts & reasoning *ad rem*. It seems to me that by these last papers . . . you have acquired a position of too much value to be defended by shooting at clouds and bushes—Let me take out all that reflects on the judgment, dicta, & publications of others, and your paper will be an authority for many a day & year.

A lengthy correspondence produced at last a solution, and although preferring a more forthright approach, Sedgwick was reasonably satisfied. "Spite of your gelding knife," he wrote Phillips, "it contains, I think, sufficient masculine vigour to help in the propagation of truth."[39]

[37] [Forbes] 1854b: 388-389. Similar views are expressed in [anon.] 1855: 59.

[38] Symonds is quoted in C.J.F. Bunbury to Horner, 25 Aug. 1856, in Bunbury 1890-1893, *Middle Life*, 2: 429. The Cheltenham discussion is mentioned in *Athenaeum*, 23 Aug. 1856, no. 1504: 1060, and in Sedgwick to Lyell, 28 Apr. 1857, in Clark and Hughes 1890, 2: 324. Ramsay, who chaired the three hours of disputation, commented privately that "the whole thing was a bore"; Ramsay, Diary, entry for 11 Aug. 1856, ICL(R): KGA Ramsay/1/24, f. 117r.

[39] Sedgwick to Phillips, 22 Mar. [1854], ff. 270-273, UMO; Phillips to Sedgwick, 8

The Hull article was Sedgwick's last in the Association proceedings. With so many channels closed against his insistent linkage of moral outrage and scientific discovery, he considered trying the Cambridge Philosophical Society *Transactions*, but settled instead on the unvetted *Philosophical Magazine* and introductory notices to catalogues of the Cambridge fossil collection. Tourists relaxing in the Lake District with William Wordsworth's *Guide to the Lakes* must have been startled to find an angry account of the controversy appended to that work; Sedgwick had sent it to the publisher in 1853 in response to a request to bring his earlier discussions of the regional geology up to date.[40] Effectively silenced in many of the most important publications in Victorian science, Sedgwick proclaimed the case for the Cambrian wherever he could.

During the last twenty years of his life, Sedgwick assumed the character of a geological martyr, and he emphasized his heroic solitude in defending what he recognized as the true principles of reasoning in science. "I fought my way, single-handed, to a knowledge of the Cambrian series . . ." he wrote, "and I have defended my classification single-handed, and without any fear of consequences. . . ."[41] Sedgwick's struggle against the "suppression" of the Cambrian provided a precedent for other geologists who suffered from what they considered an unfair use of authority. Hugh Falconer explicitly followed his example in a controversy surrounding the publication of Lyell's *Antiquity of Man*, and Jukes referred to Warburton's infamous alterations of the 1843 paper on North Wales when he became entangled in his own dispute with Murchison and the Geological Society. Those outside the mainstream of North American geology, like T. Sterry Hunt and Jules Marcou, similarly looked to Sedgwick as a symbol of resistance to scientific oppression.[42]

While Murchison was characterized as an authoritarian or even dictatorial leader in science, with a classification representing the values of established order and consensus, Sedgwick came to be

---

Feb. 1854, CUL: Add. ms 7652IIM31. The paper in question appeared as Sedgwick 1854a.

[40] Sedgwick 1853a. Thoughts of publishing in the Philosophical Society *Transactions* are mentioned in Sedgwick to McCoy, 25 Jan. [1853], ML: frames 220-221. The catalogue prefaces are Sedgwick 1855 and 1873.

[41] Sedgwick 1855: xcvii.

[42] Falconer to Sedgwick, 29 Apr. [1863], CUL: Add. ms 7652IIT16; further background is available in Bynum 1984. For the other figures mentioned, see Jukes 1867: v; Hunt 1872a; Marcou to T. R. Jones, 20 May 1884, BL: Add. ms 42581, f. 108.

Fig. 8.7. Sedgwick in old age.

viewed as a lone figure of great integrity and passionate intensity: a Victorian sage. Jukes compared his literary style to that of Carlyle, and his speeches and addresses were justly famous. "He is indeed a grand living example of the truth of Wordsworth's philosophy," a student remarked after a lecture. "He has been schooled by Nature into a 'divine old man.' "[43] Affectionately called Adam, "first of

[43] The comparison with Carlyle is in Jukes to Ramsay, [July 1850], ICL(R). Sedgwick is called the "divine old man" in Richard Wilton to George Morine, 6 Nov. 1849, SM.

Fig. 8.8. Murchison in his late seventies, probably soon before a stroke deprived him of the use of his left side.

men,'' or ''the good old patriarch,'' Sedgwick became associated with an older pastoral ideal of England as a nation of independent yeoman farmers like those who still lived in the Yorkshire dales.[44] When Murchison was referred to as a ''knight of the hammer,'' the

[44] For Sedgwick's nicknames, see Geikie 1875, *1*: 222, and Clark and Hughes 1890: passim. Speakman 1982 emphasizes Sedgwick's lifelong connection with the Dales.

overtones were of romantic conquest and feudal might; when the phrase was used of Sedgwick, the moral qualities of the knight were emphasized. Salter—admittedly not quite in his right mind at the time—described this situation very vividly to Sedgwick, deriding Murchison as "the false knight who succeeded to De la Beche's patrimony." "I am your pupil, not your critic," he wrote in 1867, "the squire whom you are to knight before you die. But you were generous to fallen Murchison, & there, I love you for that. You've unhorsed him, & you spare his life." Many of Sedgwick's opponents granted that his quest after Cambria was well intentioned, however misguided or futile: Forbes compared him with Don Quixote, and made McCoy into Sancho Panza.[45]

In the end Sedgwick's righteous indignation led him into a self-imposed exile from the metropolitan geological community. His publication difficulties were partly responsible for his decision in 1856 to break off all dealings with the Geological Society, although he never formally resigned his Fellowship.[46] The last straw came when the Council refused to print a strongly worded conclusion to one of his May Hill papers. Sedgwick fully acknowledged their right to do this, but wished to have it appear in full and sent it off to the *Philosophical Magazine* instead. Outside publication of a paper previously delivered at the Society, without the express written permission of the Council, violated one of the organization's cardinal rules, and William Hamilton severely censured Sedgwick in his 1856 presidential address. This condemnation, though probably justified, confirmed Sedgwick's suspicions that the Society was under Murchison's control. Despite public apologies in subsequent addresses and repeated private interventions, he remained implacable.[47] By severing ties with the Geological Society, Sedgwick threatened not only its gentlemanly codes of behavior, but also one of the principal foundations of its success as a scientific institution—the relatively strict separation maintained between personal rancor and geological debate. This had been threatened on many previous

[45] Salter to Sedgwick, 8 Dec. 1867, CUL: Add. ms 7652IIFF20; Forbes to Ramsay, 28 [Nov.] 1853, ICL(R).

[46] See esp. the complaints in Sedgwick 1855.

[47] The paper in question was Sedgwick 1854b. Although not present at its reading, Sedgwick was perfectly aware that it had been officially received, despite the polite fiction maintained in the presidential apology (Portlock 1857: lxxvi). For the offending address, see W. J. Hamilton 1856: xliv-li, and the reply, Sedgwick 1857. Sedgwick's papers are filled with letters begging him to return to the Society; see esp. Horner to Sedgwick, 16 Dec. 1858, 6 Jan. 1859, CUL: Add. mss 7652IIG31a, b; Phillips to Sedgwick, 20 Jan. 1859, 24 Jan. 1859, CUL: Add mss 7652IIP8a, b.

occasions, but never before had a prominent Fellow become totally alienated from the Society.

At the same time Sedgwick also withdrew from any pretense of friendship with Murchison. It may come as some surprise to those familiar with scientific controversy in other historical circumstances that this estrangement had not occurred at the first signs of open disagreement many years before. In 1846-1847 there had certainly been some indications of strain in their relationship, and they felt the need to reassure the outside world of the unaltered continuation of mutual esteem and good feeling. Their correspondence continued on much as before, its lessening frequency only a natural outcome of the end of their active collaboration and their diverging paths of research. Only a few days after the charges and countercharges at the February 1852 meeting of the Geological Society, Sedgwick stayed overnight at Belgrave Square in order to attend one of Murchison's soirées.[48] From that point onwards, however, they rapidly drifted apart, reiterating their respective positions until reconciliation became impossible. By the end of 1856 "Dear Murchison" had become the coldly formal "Dear Sir Roderick," and their friendly interchanges were never resumed. Soon they were not even on speaking terms. The death of Lady Murchison in 1869 led Sedgwick to pen a heartfelt letter of condolence, but contrary to the statement in Geikie's biography it was not the last exchange between the two men. "This appears to me so like monomania," Murchison told Huxley of Sedgwick's final epistle, "that it is hopeless for me to make any concession which can satisfy such a man."[49] He had made many attempts to conciliate his former coadjutor, but understandably was unable to give way on the one critical point. He simply could not accept the full Cambrian classification and the version of history that such an acceptance would imply.

Patronage and Power

In many respects the battle over May Hill was the last time either Sedgwick or Murchison contributed new ammunition from field research to the dispute. But arguments over unconformities and fossils were far from the only controversial resources at their disposal.

[48] Murchison to Featherstonhaugh, 22 Mar. 1852, CUL: Add. ms 7652IIKK74; for the earlier reaffirmations of friendship, see Murchison 1847b: 165, and Sedgwick to Murchison, [Jan.-Feb 1847], GSL: M/S11/238a & b.

[49] Murchison to Huxley, 21 May 1869, ICL(H): 23, f. 186. For Sedgwick's feelings, see Sedgwick to McCoy, 10 July 1856, ML: frames 526-529.

Posts were few and salaried positions hard to come by in Victorian science, and a strong recommendation from Murchison or Sedgwick was one of the surest aids a young aspirant to a job in geology or natural history could possess. Their part in the later stages of the dispute thus affords an opportunity to assess the role of patronage and personal power in maintaining a scientific consensus. Such matters, for all their importance, have as yet received little systematic study from historians.[50]

In the eyes of his opponents and a number of modern commentators, Murchison spent his last years trying to do for Siluria what the Czars had done for Russia. Together with his proven wealth of geological expertise, Murchison's social standing in science and the world at large gave him unparalleled power within the natural history community.[51] We have already seen that Ramsay obtained his Survey post in the 1840s through Murchison's intervention, and many other candidates owed professorial chairs, museum posts, or foreign surveyorships to the same good offices. Murchison helped bring Phillips down to London in 1834, took Forbes from Edinburgh in 1841 and sent him back thirteen years later as Regius Professor of Natural History, found the struggling Archibald Geikie a place on the Survey in 1855, publicized Thomas Davidson's work on brachiopods and offered to finance its publication—the list could be extended for many pages.[52] In contrast, Sedgwick possessed little power on the London scientific scene. His assistance could sometimes prove useful in obtaining an academic post, but by the 1850s his influence with the British Museum, the Survey, and the Geological Society seems to have been slight. Although Sedgwick's status as a referee cannot be ignored—Jukes, Rogers, McCoy, Harry Seeley, and many others relied upon his testimonials—his aid could not bear comparison with the rich network of patronage controlled much more directly by Murchison.[53]

[50] Exceptions include the superb analyses in Desmond 1982 and Morrell and Thackray 1981.

[51] Stafford 1984, Eyles 1971, and Gilbert and Goudie 1971, provide convenient summaries of Murchison's involvement in these activities, and Morrell and Thackray 1981 bring out his skill as an administrator in the British Association. For background on the "public men" of science, see Turner 1980.

[52] For Phillips: Morrell and Thackray 1981: 440-441; for Forbes: Wilson and Geikie 1861; for Geikie: Geikie 1924; for Davidson: Davidson to Murchison, 27 Jan. 1865, 10 Nov. 1867, GSL: M/D6/4, 6. The offer of help in financing the brachiopod monograph is mentioned in Ramsay, Diary, entry for 24 Jan. 1847, ICL(R): KGA Ramsay/1/8, f. 15v.

[53] Sedgwick's role in writing testimonials is evidenced by Sedgwick to S. P. Wood-

Precisely what effect did all this have on the debate? On occasion, Murchison was accused of using his position unfairly, "by the length of his purse buying the brains of other men" as Jukes once put it. George Poulett Scrope expressed a similar opinion to Geikie, Murchison's chief protégé. "But M.," he wrote in reference to the dispute, "with his house in Belgrave Square, his fine friends and the Presidentship of the R. G. S. and the G. Societies was of course the victor in the struggle."[54] Needless to say, both these opinions are suspect. Not only had Jukes studied under Sedgwick, but he was also embroiled in his own dispute with Murchison, and by his own admission Scrope knew almost nothing of the Lower Palaeozoic rocks. However, even if the crude reductionism that their comments imply is rejected out of hand, it remains highly unlikely that Murchison could have exercised his patronage without taking into account a subject as near to his heart as his Silurian classification.

Compared with the situation in France, where Élie de Beaumont openly dispensed favors to those who supported his controversial theories, Murchison wielded his immense influence with considerable discretion and tact. Of course, to his way of thinking Siluria was simply good science and its adoption one sign (although not a necessary one) of an intelligent man. He went to great lengths, for example, to obtain a professorship of geology in Cork for Robert Harkness, a warm advocate of his classification. "Do not make *me* appear to be *instrumental*," Murchison wrote after hearing of a successful outcome, "for I still have enemies elsewhere. But I am bound to adhere to so good a Silurian as yourself."[55] Precisely who these enemies were remains unclear; one would need to know more about the backstage politicking for the chair. But it does seem evident that Harkness—like Ramsay, Forbes, Geikie, Nicol, John Bigsby, and John Morris—was one of the highly able and, in many cases, younger geologists who found favor in Murchison's eyes in part because of their strong support for the continued extensions of Siluria.[56] Such considerations almost always came into play when

---

ward, 29 Nov. 1847, APS(W); Seeley to Sedgwick, 10 July 1866, CUL: Add. ms 7652IIEE12; Jukes to Ramsay, 8 Oct. 1850, ICL(R).

[54] Scrope to Geikie, 4 Feb. 1873, GSL; Jukes to T. M. Hughes, 27 May 1867, CUL: Add. ms 7652VF1.

[55] Murchison to Harkness, 20 Apr. 1853, in Goodchild 1882: 153; original at CUL: Add. ms 7652IVB6. For Élie de Beaumont, see many contemporary comments; e.g. Lyell to Murchison, 5 Jan. 1860, GSL: M/L17/34a & b.

[56] Most of these figures have already been discussed. For Bigsby, see Bigsby to Murchison, 5 July 1865, 5 Feb. 1869, GSL: M/B14/1,2, and Bigsby 1868; for Morris, see

choosing among several qualified candidates, or in dictating the tone of particular testimonials.

The ambivalence characteristic of the use of patronage is illustrated with special clarity in the case of Frederick McCoy, Sedgwick's assistant at the Woodwardian Museum and the man largely responsible for splitting the Caradoc in two. As an Irish Catholic, McCoy already suffered sufficient disadvantages in the competition for positions. Although he spoke up for Sedgwick at the Geological Society, he took pains at the outset to disassociate himself from the unpopular views of his employer, perhaps explaining in part why he did not appear as coauthor of the May Hill papers. A letter to Murchison in June 1852 shows that he pretended to know nothing of the controversial issue even while he was trying to convince Sedgwick to reexamine the Caradoc sections:

> I hope you and Professor Sedgwick have long before this settled to your mutual satisfaction the bounds of your grounds? I feared I should have come in for some knocks, although I have never intruded myself into the discussion but confined myself to identifying the fossils to the best of my ability and registering them faithfully. A smack from you would probably ruin my prospects, and I think undesirably—but I believe you spare the weak in as marked a manner as you grapple with the strong.[57]

Even after McCoy's efforts for the Cambrian could no longer be concealed, Murchison promised to remain unprejudiced, although the issue could not be avoided entirely. "One of the only real pleasures in this turmoil of life," he wrote in reference to a possible opening in Ireland, "is to have the satisfaction of zealously aiding a good & able man: and although it had been whispered to me that you were the prime agent in pulling down the Silurian Mansion I had built for myself, let me say, first that I never believed the story, & next that I am happy to give you the strongest proof of my esteem."[58]

The question of McCoy's future came to a head early in 1854, when Forbes's acceptance of the Edinburgh natural history chair vacated the star position in British palaeontology—the curatorship of the Survey's fossil collections in London. Sedgwick made every effort to obtain the post for McCoy, while Murchison quietly (and suc-

---

Geikie 1875, 2: 153, 172, and McCoy to Sedgwick, 15 May 1854, CUL: Add. ms 7652IIIB62.

[57] McCoy to Murchison, 12 June 1852, in Craig 1971: 494.

[58] Murchison to McCoy, 10 Jan. 1853, ML: frames 329-331.

cessfully) did his best to have it offered first to John Morris and then to Salter.[59] Both Morris and Salter had labored on *Siluria* and supplied anti-Sedgwickian ammunition in the May Hill disputes. All concerned viewed the curatorship as a potential reward for services rendered. Behind this seemingly naked struggle for power, however, were equally important questions about ability and merit. McCoy's early work had not been highly regarded, and in all honesty Forbes could offer little encouragement to the Irishman's quest to succeed him.[60] Moreover, Morris and Salter not only supported Murchison in the controversy, but they were both outstanding palaeontologists, certainly McCoy's equals in experience and expertise if not in industry.

A few weeks afterwards, McCoy was packing his bags for Australia to assume a post that everyone agreed he deserved—the geology chair at the newly founded University of Melbourne. Even Murchison had written to support his candidacy. As Murchison explained to John Herschel, if he had refrained from writing a testimonial, "I should not have acted fairly towards an able man. . . ." Sedgwick, in contrast, penned a letter of recommendation that was almost embarrassingly eulogistic.[61] In any event McCoy's exile to Australia proved a blessing (in disguise?) for Murchison, as it deprived Sedgwick of his chief supporter and principal source of palaeontological ammunition in the battle over the gap at the base of the Upper Silurian. This did not go unnoticed at the time: James Nicol, then currying favor with Murchison in hopes of obtaining a position at the University of Aberdeen, told him to "send McCoy to Australia where he might do some good work and not stir up mischief between old friends."[62]

What has always been regarded as the most noteworthy addition

[59] The offer to Morris is noted in McCoy to Sedgwick, 15 May 1854, CUL: Add. ms 7652IIIB62, and that to Salter in Secord 1985b. Sedgwick even went so far as to think of McCoy for the Edinburgh chair itself, in preference to Forbes: see McCoy to Sedgwick, 3 May 1854, CUL: Add. ms 7652IIIB72.

[60] Forbes to Sedgwick, 31 Dec. [1853], CUL: Add. ms 7652IIM10.

[61] Murchison to Herschel, 30 May 1854, ML: frames 272-274; Sedgwick to Herschel, 26 May 1854, ML: frames 439-444.

[62] Nicol to Murchison, 9 July 1854, GSL: M/N8/16 contains the quoted passage; see also Nicol to Murchison, 6 Nov. 1854, EUL: Gen. 1999/1. For the Survey post, see McCoy to Murchison, 9 May 1854, GSL: M/M1/5; McCoy to Sedgwick, 15 May 1854, CUL: Add. ms 7652IIIB62. McCoy certainly thought he had been unfairly dealt with; see McCoy to Sedgwick, 15 July 1872, CUL: Add. ms 7652IIHH63: "Sir Roderick never forgave me for my part in the triumph of your views, small as it was, and did me much mischief. . . ."

to Murchison's hold on the reins of scientific power came in 1855, when he succeeded De la Beche as director of the Geological Survey. But the importance of this event for the controversy has been much overrated. Even his greatest opponents admitted that Murchison took over the position largely as an act of courtesy, to keep the organization from being split up or taken over by a nongeologist. It was a critical moment in the history of the Survey, and Murchison cancelled his plans for "retirement" in the interests of the scientific community as a whole. Sedgwick recognized the desperate state of affairs and willingly added his name to the long list of supporters, even though he was drifting towards estrangement with his former collaborator and was well aware of the Survey's importance in determining all questions of nomenclature.[63] Faced by an outside threat to the integrity of one of their principal institutions, the scientific elite closed ranks and set lesser disagreements aside. In any case Murchison could initiate only a limited amount of change in the official memoirs and maps relevant to the controversy, for De la Beche had already established the basic groundwork of the Survey nomenclature. In certain matters the official maps and the private publications of Murchison were thus initially at odds. The Llandovery compromise on the May Hill issue demonstrates that the new director could by no means simply force his lieutenants to accept his views.

With the exception of one or two cases involving close family relations, I have not encountered any instance of Murchison or Sedgwick supporting an individual who was not generally recognized to be qualified for the position in question. In the competitive world of mid-Victorian science, there were usually knowledgeable, competent candidates in abundance, and for either of them to have promoted ill-prepared applicants merely because they adopted a particular position in controversy would have severely damaged their own reputation, let alone their unbending sense of propriety and honor. Moreover, like all leading scientists, Murchison and Sedgwick were engaged in a variety of other disputes. Only the most servile intellect could have accepted a suitable position on all of them, and the support of such a man would have been worth little. Thus Ramsay and Geikie publicly supported Murchison on the Silurian question, but strongly contested his views on glaciers, erosion, and the origin of species. Nicol wrote a laudatory notice of *Siluria* for the *Edinburgh Review*, and although this remained unpublished, he

[63] Murchison to Sedgwick, 30 Apr. 1855, in Geikie 1875, 2: 188-189.

aided in other ways as well; but the two men disagreed profoundly about the geological structure of the Northwest Highlands, an area of particular importance to Murchison's later research. Forbes gave the Silurian his prestigious support right up until his tragic death late in 1854, while simultaneously adopting many Lyellian views on the early history of life. In the late 1850s Jukes moved notably closer to accepting Sedgwick's Cambrian, but he was also a fond supporter of Darwinian evolution.[64]

These examples of divided support are not intended to cast doubt upon the immense significance of patronage in Victorian scientific life. Indeed, the personal bitterness of the Sedgwick-Murchison controversy brings to light aspects of its operation that typically remain hidden from view. As in the Huxley-Owen debate, protagonists ignored institutional power to their cost.[65] It is equally clear that naturalists did not line up on one side or another in any particular dispute simply to gain favor from a great name. Those who did so were typically outsiders, ranging from provincial autograph hunters to sycophantic job-seekers. Taxonomic palaeontologists were placed in perhaps the most difficult position, for theirs was in one sense a "service industry" to stratigraphy, and they were expected to acquiesce in the views of their employers. But even here, as the cases of McCoy and Salter suggest, the situation was not a simple story of blackballing or *force majeure*. Running roughshod over generally perceived claims of merit would have been just as irrational as shutting one's eyes to the existence of the May Hill Sandstone.[66]

Ironically, authors opposed to a social approach to scientific knowledge have sometimes been more than willing to adopt a "coercive model" to explain Murchison's success in the Cambrian-Silurian dispute. But consensus in Victorian science was not achieved through coercion. The situation was far more complicated. Murchison, who had successfully engineered the success of his classification for over a decade, was by no means able to do so indefinitely. In the end he came perilously close to suffering the fate of the Czars whose methods he had so admired. Notably, as he resorted to stronger measures when his ideas met with resistance, the

[64] For Jukes's support of evolution, see Jukes to Darwin, 27 Feb. 1860, transcript in Wilson 1970: 361; for Nicol, see Geikie 1875, 2: 212-239, and Nicol to Murchison, 14 Oct. 1854, GSL: M/N8/21a & b; for Ramsay and Geikie, see G. L. Davies 1969 and Geikie 1875, 2: 316-322. For Forbes and Lyell, see J. Browne 1983 and Rehbock 1983.

[65] Desmond 1982, Turner 1978.

[66] For this point more generally, see Barnes 1982.

strategy backfired, and his reputation rapidly declined after his death in 1871. John Judd, professor at the Royal School of Mines in London, expressed in a letter to Lyell the common belief that the fully extended Silurian classification would be abandoned without its author's constant vigilance: "I fear that a 'great rebellion' if not a 'grand revolution' is imminent in that ancient kingdom,—and that Autocratic government has become impossible within it; neither Ramsay nor Geikie being able to wield the iron sceptre and wear the silken glove of Sir Roderick—The reign of Elizabeth has been followed by that of the Stuarts!"[67] While patronage could be a powerful tool for rewarding supporters and crushing enemies, it demanded careful use lest it lose all effectiveness.

By the end of 1853 the findings of Sedgwick and McCoy had already been used to weaken Murchison's position. What would prove to be an even greater setback for the "old regime" was emerging out of work by one of Siluria's most ardent adherents.

[67] Judd to Lyell, 11 Oct. 1874, EUL: Lyell 1/3103. For criticisms of social coercion as an explanatory tool in writing the history of science, see Shapin 1982: 194-198.

CHAPTER NINE

The Creation of an Alternative

WHILE the debate between Murchison and Sedgwick focused on the break at the base of the Upper Silurian, other important developments took place in the study of fossiliferous rocks much lower in the geological column. First in Bohemia, and later in other countries throughout the northern hemisphere, geologists (some of them firmly committed to Murchison's views) found an important group of fossils *below* the Lower Silurian. This "Primordial" fauna, as it was generally called, played in many ways an even more important part in the eventual resolution of the Cambrian-Silurian question than did the discovery of the break at May Hill. Although the issue was far from being settled simply by these findings, Phillips, Lyell, and other leading naturalists felt that here at last was the long-awaited "Cambrian" fauna, distinct enough from the Lower Silurian to deserve a separate name. Like the May Hill dispute, the debates about this new fauna exemplify the role of controversy in stimulating scientific discovery, this time in hands other than those of the principals.

THE PRIMORDIAL FOSSILS

To understand the eventual resolution of the stalemate, we must first turn our attention abroad, for what eventually became accepted as the basis for a separate Cambrian was uncovered not in Britain, but rather in Bohemia by an expatriate Frenchman named Joachim Barrande. Educated at the École Polytechnique in Paris, Barrande became a tutor to the royal family and went into exile with them after the Revolution in 1830. He spent the remainder of his life in Prague, first as a tutor and later as an engineer with a railway company. During the 1830s Barrande became interested in the strata and fossils of the surrounding countryside. Like many foreign men of science, Barrande found that Murchison's *Silurian System* provided a palaeontological key of unprecedented utility for the correlation of these local rocks with an established succession. Barrande's work in the Prague basin paralleled similar studies throughout Europe, but was exceptional in its scale and quality. He went to unprecedented lengths to obtain the finest fossil specimens for his collection: opening special quarries, hiring large numbers of workmen, and spending most of a modest private fortune. A small monograph on trilobites appeared in 1846 as the first fruit of his labors, presaging the monumental *Systême Silurien du Centre de la Bohême* which began publication in 1852. This gigantic work, which eventually occupied twenty-nine large quarto volumes descriptive of several thousand fossil species, was published over a fifty-year period and included several volumes prepared posthumously from materials left by the author.[1]

Barrande's remarkable specimens all came from a relatively small area around Prague, which he described as a basin bounded on either side by crystalline rocks, as shown in Figure 9.1. Each stratigraphical group, or *Étage*, received a separate letter from *A* to *H*. As the title of his work implied, Barrande then correlated these groups with those of the standard British sequence as elaborated by Murchison. He pointed to the existence of three distinctive faunas among the Silurian rocks of the Bohemian basin. *Étages* E, F, and G together constituted the Third Fauna, and approximately paralleled Murchison's Upper Silurian and (perhaps) the lowest part of the Devonian. *Étage* D, or the Second Fauna, matched the original Llandeilo and Caradoc fossils in the *Silurian System*. Finally, and of special importance for the present story, there was *Étage* C which held the "First Fauna," or what Barrande referred to as the "Primordial Silurian." This ancient fauna, the oldest that he had found, was characterized by several trilobite genera, including twelve separate species of *Paradoxides*. With only one exception, the fossil genera found in this "Primordial" zone were distinct from those found in the overlying rocks of the Bohemian sequence, and the species were entirely different.[2]

From all appearances Barrande's adoption of the Silurian nomenclature in 1846 originally seems to have been made for ease of understanding. Like most palaeontologists, he wished to give his work the permanence offered by an established nomenclature. Murchison's classification appeared to offer precisely that, and the

[1] Barrande 1846, 1852-1911. For biographical details, see Horný 1980 and C. Lyell to Caroline Lyell, 28 Aug. 1856, in K. M. Lyell 1881, 2: 223-225. A full study of Barrande is greatly needed.

[2] Barrande 1852-1911: 57-99; also Murchison 1854: 340-349 for a brief summary in English.

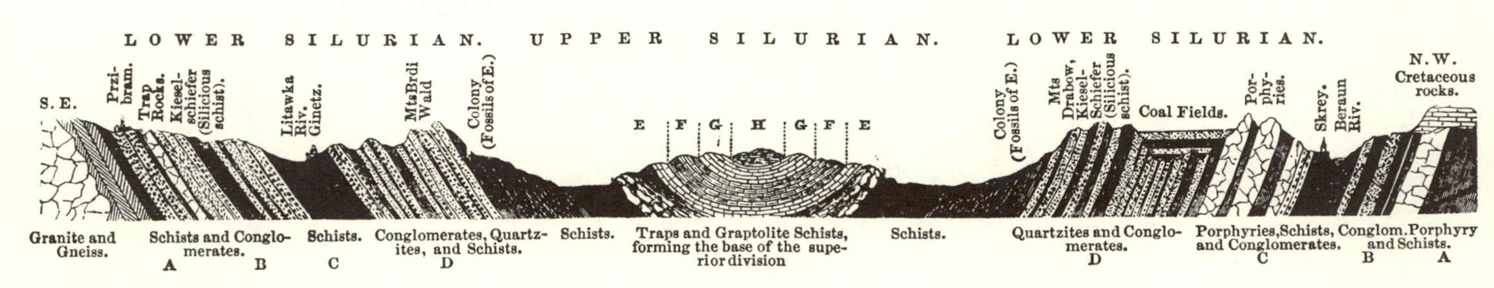

Fig. 9.1. Barrande's section across the Silurian basin of Bohemia.

Silurian chief had visited Barrande in 1843 to promote the extension of his system in person. He vigorously championed Barrande's work in Britain, and in exchange the Frenchman continued to be favorably disposed towards the incorporation of all the older fossiliferous rocks under a single title.[3] Although Barrande later claimed in a letter to Sedgwick that the nomenclature of the older rocks was an issue to be settled by British geologists, there can be little doubt that even as early as 1849 his own reputation had become securely tied to the names adopted in the title and first volume of his great work. As he told "Roderic Ier, illustre Roi des Silures" by letter:

> Vous savez bien combien mes sentimens sont monarchiques, et combien j'aime la véritable monarchie, celle du droit. J'ai donc sincèrement applaudi à la proclamation de votre Royauté scientifique, par l'évêque d'Oxford, qui a en ce jour là une excellente idée. . . . Votre droit est celui de la création, le plus légitime de tous, et vous avez fait votre Royaume, que personne ne songera à revendiquer. Pas de Rois détroné, pas de prétendant, voilà ce qui doit assurer le calme de votre possession et de votre conscience. Vous avez en autre l'avantage d'avoir fait la conquête pacifique, et par la seule voie de la raison, du pays Cambrien, qui seul pouvait songer à élever contre vous une ambitieuse rivalité. Vos colonies ont déjà envahie tout le nord de l'Europe et le nord de l'Amérique; elles sont établies dans la France, & j'ai l'honneur de les fonder en Bohême.

> You well know what a monarchist I am, and how I love the true monarchy—that of right. So I heartily approved the proclamation of your scientific kingship by the Bishop of Oxford, who had an excellent idea on that day. . . . Your perogative is that of creation, the most legitimate of all, and you have made your kingdom something that no man could think of claiming. No dethroned king, no pretender, these ought to assure you of peaceful possession and an easy conscience. You also have the advantage of having made a peaceful conquest, and solely by the force of reason, over the Cambrian country—which can do no more than dream of raising an ambitious rivalry against you. Your colonies have already encroached on all of northern Europe and North America; they have been established in France, and I have the honor of founding them in Bohemia.[4]

[3] Barrande to Sedgwick, 24 Mar. 1856, CUL: Add. ms 7652IIA1c. An explicit statement from the end of his life can be found in Barrande 1880. Geikie 1875, 2: 12 describes Murchison's visit.

[4] Barrande to Murchison, 7 Nov. 1849, GSL: M/B7/1.

Even at a distance, Barrande could not help becoming entangled in the debate over the older rocks.

As might be expected, the tracing of the Primordial in Britain was even more explicitly related to the controversy. The relationship between the three faunas of the Prague basin and the standard sequence in Wales emerged only gradually. After visiting Britain in 1851, Barrande tentatively correlated the Primordial fossils with those of the Lingula Flags, although neither the Woodwardian Museum nor the Survey had enough specimens to make the comparison really secure. As a result, De la Beche sent Salter back to North Wales in 1853 to reexamine the fossil succession in the Lingula Flags, in particular to see how many species or genera were held in common between this fauna and the overlying fossils of the Llandeilo formation. He returned with a large complement of specimens from the Tremadoc Slates and the Lingula Flags, and although the total number of species remained less than two dozen, no one doubted that Barrande's Primordial zone had been successfully found in the British Isles.[5]

Salter, who had worked extensively for both Sedgwick and Murchison, was almost desperate to find evidence of the Primordial fauna in Britain. He had assisted Sedgwick in the museum and in the field during the 1840s, yet had served his apprenticeship drawing fossils for Murchison's *Silurian System* and was busy in 1854 with palaeontological work for the first edition of *Siluria*. As he once told Murchison, "I am a hybrid animal, & no wonder, seeing as I am honoured by the acquaintance of the two leading geologists of England, and permitted to learn from both in turns."[6] Caught thus between his two patrons, Salter was eager to resolve the controversy by using the Primordial fauna as a way of vivifying the Cambrian without subtracting substantially from the fully extended Silurian. He proposed his preferred solution in May 1854 during the discussion of one of Sedgwick's May Hill papers at the Geological Society. In this scheme the upper limit of the Cambrian would be drawn at the top of the Tremadoc beds, just where Salter's latest survey had shown a major faunal break. According to Samuel P. Woodward of the British Museum, the spectacle of a figure of Salter's status "of-

[5] Salter 1866: 244-246; also Salter to Sedgwick, 29 Sept. 1853, 11 Nov. [1853], CUL: Add. ms 7652IIM6b, c.

[6] Salter to Murchison, 12 Jan. [1854], GSL: M/S3/1. Or as he told Ramsay, "I've made up my little mind into a ball, & shall let it roll backwards & forwards just as they kick it." Salter to Ramsay, 15 Sept. [1853], ICL(R).

fering to set things straight" in a matter of such importance occasioned "great amusement" at the Society, and his proposal met with a cool reception from both the principals—although not, as we shall see, from other interested parties.[7]

In continuing behind the scenes to advocate Barrande's "Primordial Silurian" fauna as a basis for the Cambrian, Salter also actively tried to increase its importance by tracing organic remains to ever lower levels of the geological column. Perhaps, as Lyell and Forbes had always claimed, the very oldest stratified rocks called "Cambrian" by the Survey had not in fact been formed prior to the existence of life on earth. A few trace fossils already known from strata of this horizon indicated that yet another ancient fauna might be found, one even older than Barrande's. Such a discovery would have theoretical consequences of the first order, adding a still earlier chapter to the history of life.[8] Of more immediate relevance to Salter's program of reconciliation, any fossils found in these "bottom rocks" would make a limited Cambrian (perhaps even one so limited as that of the Survey) potentially more acceptable to Sedgwick without impinging unduly upon Murchison's domain.

To forge such a compromise, Salter had to find as many fossils in the ancient strata as possible. His searches in the Longmynd offer another excellent example of discovery through controversy. Ever since the early 1830s the Longmynd strata had been believed to be destitute of organic remains, and the Survey had placed them below the Lingula Flags as the types of their unfossiliferous "Cambrian." In more recent years the Longmynd has once again been seen as a completely barren sequence. But Salter, so anxious to find fossils in his summer tour of 1855, found at least some indications of fossil remains: "The Cambrians are not barren of organic life," he wrote to Murchison after a concentrated search of three weeks. "I came down here determined to find some fossils in the old greywacke. . . ."[9] His most interesting discovery, which he reported to Sedgwick and Murchison on the same day, was the impression of part of a primitive trilobite, afterwards named *Palaeopyge Ramsayi*. Three illustrations of this ambiguous specimen—from the field, a

[7] Woodward to Sedgwick, 1 July 1854, CUL: Add. ms 7652IIX81. For Salter's own description of his proposal, see Salter to Sedgwick, 11 May [1854], CUL: Add. ms 7652IIX88. Sedgwick was not present when this paper (published as Sedgwick 1854b) was read on 3 May 1854.

[8] For the contemporary state of this problem, see Murchison 1854: 17-35; [Forbes] 1854b: 367-371.

[9] Salter to Murchison, 12 Sept. [1855], GSL: M/S3/5.

contemporary publication, and a modern photograph—are included here in Figure 9.2. Salter's difficulty in dealing with it is illustrated in his letter to Sedgwick:

> They are not I think rare, but the rocks are, as you call them so bedevilled that it is all but hopeless to look for them. I believe I have seen hundreds & can't be sure they are not ochreous stains. However there's no mistake about our friend on the last page [Fig. 9.2, top].
>
> Now look, my dear Professor at the cut of this gentleman. If he agrees with any trilobite at all, he's a primordial chap allied to *Olenus*. And if this be so—and the primordial zone be as it is generally believed truly distinct—what will you do with it. Will you have Lingula flags & all below or stand out for Llandeilo flags?[10]

As with the unconformities at Onny, Glyn Ceiriog, and Mathrafal discussed in the previous chapter, Salter's perception of the Longmynd fossils had been primed by strong theoretical expectations, in this case his wish to find a Cambrian fauna in rocks generally believed to be azoic. The role of high-level scientific theory in observation becomes especially evident in such borderline cases of trace fossils and "problematica." That these famous Longmynd specimens are discounted by most (although not all) modern palaeontologists only adds to the interest of this episode, for that dismissal is itself based in part on prior expectations of the same sort that motivated Salter's search. As one modern expert put the question: "it would be difficult to prove or disprove whether the remains are of organic origin. Perhaps their stratigraphic position implies that they are *Lusus naturae*."[11]

It is important to emphasize, however, that Salter's work with the Longmynd trace fossils presented Victorian palaeontologists with an unusually difficult set of decisions. On a great many issues a consensus had been firmly established long before. In the seventeenth century the discovery of weird remains like Barrande's Primordial trilobites would have added fuel to a raging controversy about the organic origins of all "fossil objects," but in the nineteenth all palaeontologists took it for granted that such strange and extinct forms had once been alive. The organic origins of *Palaeopyge Ramsayi*

[10] Salter to Sedgwick, 12 Sept. [1855], CUL: Add. ms 7652IIL66c. Also Salter 1856 and 1857.

[11] A.W.A. Rushton of the Institute of Geological Sciences, quoted in Sarjeant and Harvey 1979: 202. Dr. Rushton (personal communication) writes that interest in the potentially organic character of Salter's specimen has very recently been revived.

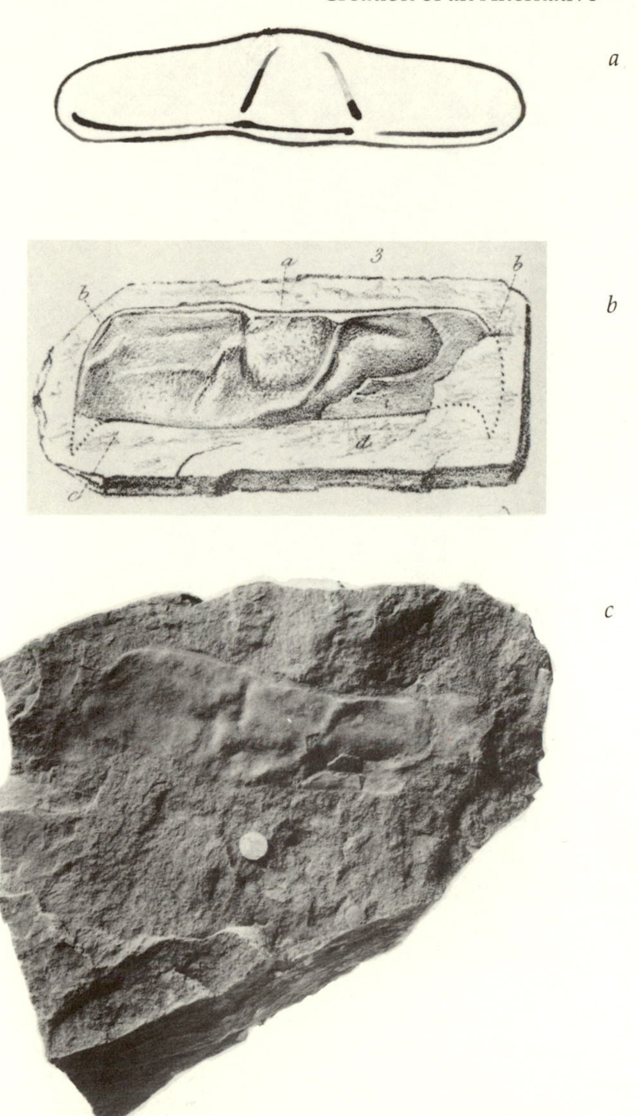

Fig. 9.2. *Palaeopyge Ramsayi*: (*a*) pen drawing sketched by Salter in letters from the field, when he thought it was the cephalon (head) of a trilobite allied to *Olenus*; (*b*) lithographed version that appeared in the Geological Society's *Quarterly Journal*, by which time Salter thought it was probably the pygidium (or tail-shield) of a genus related to *Dikelocephalus*; (*c*) a modern photograph.

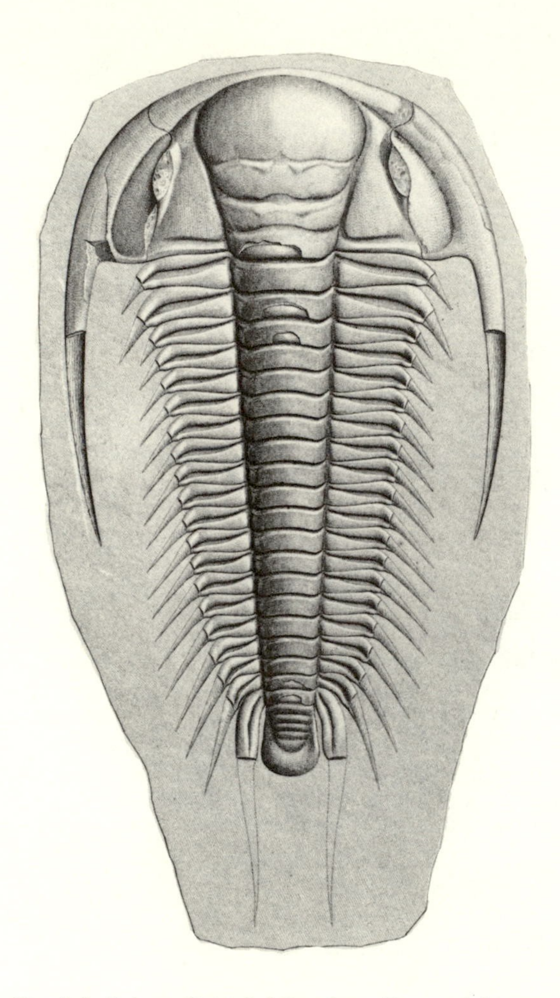

Fig. 9.3. Primordial trilobites from Barrande's great work on Bohemia: (this page) *Paradoxides Bohemicus* and (facing page) *Agnostus rex*.

may have been doubted by some Victorians, but the clearer examples of *Paradoxides Bohemicus* and *Agnostus rex* (Fig. 9.3) were not. From a relative handful of fossil genera, specialists could even agree that Barrande's Primordial fauna had existed in the British Isles, though none of his species had as yet been uncovered there. Consensus on an issue like this demanded agreement on a range of criteria that linked particular species into broader groups. The characteristic aspect, or "facies" (not the modern sense of the word), of

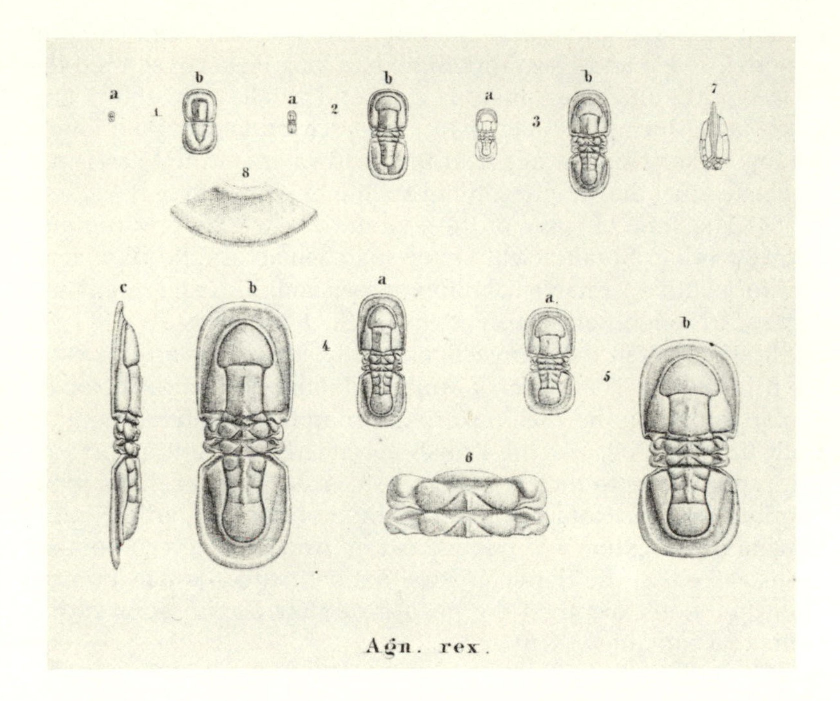

the Primordial trilobites was unmistakable. Referred to as "rudimentary" and "primitive" forms, they either exhibited an unusually large number of segments and a small pygidium (tail-shield), as in *Paradoxides*, or conversely very few segments and an enlarged pygidium, as in the blind *Agnostus*.[12]

By the mid-1850s these Primordial fossils were being uncovered in many countries, Barrande's work having stimulated searches not only in Britain but throughout the world. Nils Peter Angelin wrote in his *Palaeontologica Scandinavica* (1852–1854) of Swedish trilobites bearing characters like those found in Bohemia, and Theodor Kjerulf described fossils from a similar horizon in central and northern Norway and explicitly compared his sequence with the British Silurians.[13] In North America the Canadian Survey geologists William Logan and Elkanah Billings published detailed descriptions of the

[12] A clear explanation of these points is in Lyell 1865: 574-576; see also Barrande 1852-1911, *1*, which discusses the Bohemian trilobites in detail. Rudwick 1976a offers an illuminating analysis of trace fossils more generally. A parallel case of considerable interest is described in O'Brien 1970.

[13] The European work, a vast and unstudied literature, is briefly characterized by Zittel 1901: 445-448; e.g. Angelin 1852-1854.

ancient fossils of the Quebec Group. The Potsdam Sandstone of eastern Canada and New York State had long been considered the New World's oldest fossiliferous zone and paralleled with the Lingula Flags; during the late 1840s, research by David Dale Owen among rocks of similar age near the headwaters of the Mississippi suggested that they might contain an immense wealth of new fossil forms. The scale of some of these inquiries dwarfed any contemporary work in Britain itself. Owen marched across the Wisconsin Territory with a veritable battalion of assistants and an armed military escort to fend off Indian attacks.[14] Such inquiries dwarfed the publications of all those investigating the ancient invertebrates in the British Isles. The works of Morris, Phillips, and Salter look particularly puny on the shelf next to the sumptuous volumes issuing forth from Prague. But the British specimens held an importance disproportionate to their numbers. Wales, as home of the original Silurian fauna, inevitably remained the proving ground for questions of classification and nomenclature. And although contemporaries agreed on the characteristics that defined a fossil as Primordial, they hotly disputed the need for a new classification of the strata that contained them.

The Dissolution of a Consensus

Until 1853 almost all British geologists save Sedgwick had followed the Murchisonian nomenclature and placed all the older fossiliferous rocks in a single system. After that date, the discoveries at May Hill and in Bohemia facilitated the dissolution of the previous consensus. Although many geological papers continued to employ the names used by Murchison and Barrande, there was a growing feeling that the subject was due for revision. In particular, Salter's view that the Primordial fossils provided the basis for a limited version of the Cambrian became a common one in the mid-1850s. From this point onwards, one can trace the emergence of a variety of intermediate positions in the dispute, as shown in Figure 9.4. According to such schemes, the ancient fossiliferous rocks—Barrande's Primordial and the even older Longmynd beds—became the "Cambrian"; the strata from the May Hill Sandstone to the base of the Devonian became the "Upper Silurian," or "Silurian proper," and the intervening beds were called "Lower Silurian," "Cambro-Silurian,"

[14] Owen 1852, biographical information in Hendrickson 1943. Schneer 1978 offers essential background on the discovery of the Primordial in America; see also Merrill 1924. For the Canadian work, see Zaslow 1975, Logan 1861, Billings 1861-1865.

<table>
<tr><th></th><th>1</th><th>2</th><th colspan="2">3</th><th>4</th><th>5</th><th>6</th><th>7</th><th>8</th><th>9</th></tr>
<tr><td>SEDGWICK 1855</td><td colspan="3">Silurian</td><td colspan="2">Upper Cambrian</td><td colspan="4">Middle Cambrian</td><td>Lower Camb.</td></tr>
<tr><td>MURCHISON 1859</td><td colspan="3">Upper Silurian</td><td colspan="4">Lower Silurian</td><td colspan="2">(Primordial Silurian)</td><td>Camb.</td></tr>
<tr><td>GEOLOGICAL SURVEY 1866</td><td colspan="3">Upper Silurian</td><td colspan="6">Lower Silurian</td><td>Camb.</td></tr>
<tr><td>JUKES 1857</td><td colspan="3">Upper Silurian</td><td colspan="6">Cambro-Silurian</td><td>Camb.</td></tr>
<tr><td>PHILLIPS 1855</td><td colspan="3">Upper Silurian</td><td colspan="4">Lower Silurian</td><td colspan="3">Cambrian</td></tr>
<tr><td>LYELL 1865</td><td colspan="2">Upper Silurian</td><td colspan="2">Mid-dle Sil.</td><td colspan="3">Lower Silurian</td><td colspan="3">Cambrian</td></tr>
<tr><td>LYELL 1871</td><td colspan="4">Upper Silurian</td><td colspan="3">Lower Silurian</td><td colspan="3">Cambrian</td></tr>
<tr><td>HICKS 1874</td><td colspan="2">Upper Silurian</td><td colspan="2">Mid-dle Sil.</td><td colspan="3">Lower Silurian</td><td colspan="3">Cambrian</td></tr>
<tr><td>LAPWORTH 1879</td><td colspan="4">Silurian</td><td colspan="3">Ordovician</td><td colspan="3">Cambrian</td></tr>
<tr><td></td><td>1</td><td>2</td><td colspan="2">3</td><td>4</td><td>5</td><td>6</td><td>7</td><td>8</td><td>9</td></tr>
<tr><td>Principal Formations</td><td>Ludlow</td><td>Wenlock</td><td>Upper Llandovery = May Hill</td><td>Lower Llandovery</td><td>Bala = Caradoc Sandstone</td><td>Llandeilo</td><td>Arenig</td><td>Tremadoc</td><td>Lingula Flags</td><td>Longmynd</td></tr>
</table>

Fig. 9.4. Alternative classifications for the Lower Palaeozoic rocks of Britain, 1855-1879.

or "Siluro-Cambrian." Although these attempted compromises failed to placate the two principals in the controversy, they were accepted in various forms by a large number of intermediate parties.

Many younger men felt that neither of the two main opponents ever really appreciated the distinctiveness of the Primordial fossils. Just as he had done ever since the early 1840s, Murchison simply extended the base line of his Lower Silurian to take the new discoveries into account. By the terms of his argument for the unity of the Silurian system he could do little else, for as I have indicated, his definition of a "system" was by this time formidably comprehensive: to qualify as the basis for a distinctly named "Cambrian," the Primordial fossils would have had to present taxonomic differences above the *generic* level.[15] During the 1860s, when Salter and Henry Hicks found even older specimens than those of the Lingula Flags among the slaty rocks of St. David's in Pembrokeshire (including the spectacular trilobite *Paradoxides Davidis* over two feet in length), Murchison naturally included them in the Silurian as well. Salter was understandably annoyed. Increasingly troubled by violent shifts of mood, he had been forced out of his Survey post by Murchison and Huxley in 1863 and was temporarily in Sedgwick's employ cataloguing the Palaeozoic fossils of the Woodwardian Museum. He explained in a letter to the professor what had happened after the announcement of the new findings at Birmingham in 1865. "The paper was read at the Brit. Assocn.—and Murchison coolly explained to the audience that these were the basement beds of his lower Silurian!!! I suppose when he gets to heaven he'll know better."[16]

Pointing to this downward march of the lower boundary of the Primordial Silurian from 1842 on, Martin Rudwick has suggested that Murchison's wish to have his system encompass the origin of life lies at the heart of the entire controversy with Sedgwick. According to this view, rather than using Barrande's Primordial as the palaeontological basis for a revivified Cambrian as Salter had suggested, Murchison simply annexed the fauna to maintain his claims as discoverer of the only scientific proofs of a beginning to organic creation.[17] Especially after his Russian and Scandinavian tours Mur-

[15] See Chapter Six. The point is reiterated at length in Murchison 1854: 176-206.

[16] Salter to Sedgwick, 25 Oct. 1865, CUL: Add. ms 7652IIY39b; also a clipping at IIY44a reporting the discussion. The paper in question was Hicks and Salter 1866. See also Salter 1865, Hicks and Salter 1867 for additional reports on their findings in Pembrokeshire.

[17] Rudwick 1976a: 198-199, and 1976c.

chison certainly did insist that Siluria held the answer to this crucial question. But when his attitudes are considered over the longer term, a metaphysical issue such as life's origins seems extremely unlikely as the principal motive for the Silurian extensions. Already in 1848 an important discovery had challenged the simple picture drawn three years earlier in the *Geology of Russia*. Before leaving to direct the Indian Survey, Thomas Oldham had found two species of a primitive zoophyte in the ancient rocks of Bray Head in Ireland (Fig. 9.5), subsequently named *Oldhamia antiqua* and *Oldhamia radiata* by Forbes. The strata in question paralleled Murchison's "bottom rocks" in the Longmynd, and therefore Oldham's discovery pushed the origin of life out of the Silurian altogether. Murchison minimized the significance of this finding in his discussion of the issue in *Siluria*, but made no change in his lower boundary.[18] As a result, his system no longer held the origin of life. The second edition of *Siluria*, published in 1859, even adopted Survey usage and called these and other scantily vivified rocks "Cambrian," thereby demonstrating a reluctant willingness to cede this point to a name other than his own. One might contrast his position with the more rigid stance adopted by Barrande, who always included the unfossiliferous basement rocks of the Prague basin with the Silurian as insurance against future discoveries at the base of the geological column. As Lyell recognized, Barrande was even more of a "finality man" than Sedgwick or Murchison.[19]

Though Murchison left certain obscure but unquestionably organic forms out of his system, he utterly refused to do the same with anything approaching a distinct fauna—hence his subsuming of the St. David's fossil findings and his acceptance of Barrande's "Primordial Silurian" terminology. If *Oldhamia* ever proved for the ancient rocks to be what Forbes anticipated—a sure sign of others, like Robinson Crusoe's discovery of a footprint in the sand—Murchison could then consistently incorporate them into Siluria.[20] There can be little doubt he cared above all, not that life had been created in the Silurian, but that readily recognizable trilobites, cephalopods, and brachiopods, for example, commenced there. For Murchison, these fossils gained most of their significance from their

[18] Murchison 1854: 32; *Oldhamia* was first described in Forbes 1848. In the 1860s *Eozöon Canadense* put the origins of life even further below the lowest Silurian base; see O'Brien 1970 and Geikie 1875, 2: 322.

[19] Lyell to J. Fleming, 31 Aug. 1856, in K. M. Lyell 1881, 2: 225-226. For the change in *Siluria*'s second edition, see Murchison 1859.

[20] [Forbes] 1854b: 79-80.

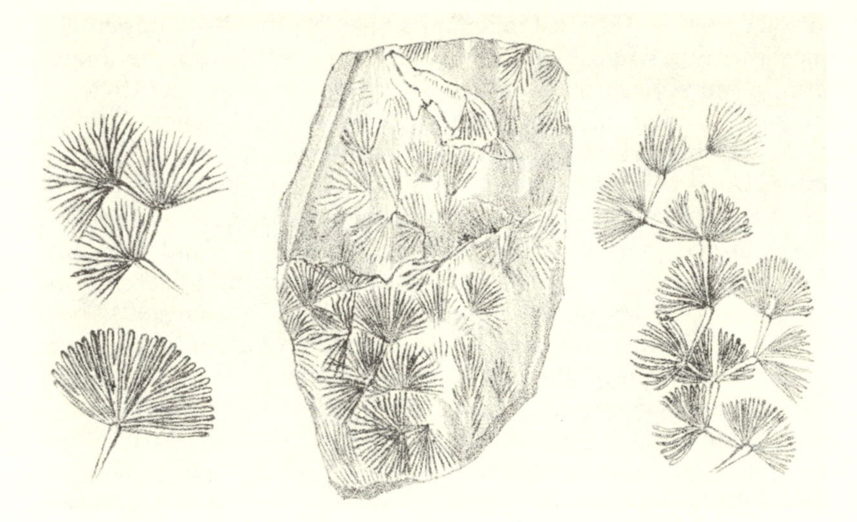

Fig. 9.5. *Oldhamia antiqua*, Forbes. The oldest known fossil in the late 1840s and early 1850s.

utility for his great enterprise of international correlation. The concept of a broadly defined fauna rested at the heart of his scientific achievement, his "Empire of Siluria" that could be traced throughout the world. As long as he kept that single Silurian fauna secure, "lesser" questions could be set to one side.[21] Thus the issue of life's origins, despite its profound significance for the Victorian debates about man's place in nature, played only a subsidiary part in the Cambrian-Silurian dispute.

Despite their differences, Sedgwick reacted to the Primordial discoveries much as Murchison had. In his view, these findings merely added to the fossil riches of his Cambrian, whose various divisions comprised by 1853 all of the stratified rocks from the sparsely fossiliferous beds at Harlech, Tremadoc, and the Longmynd, up to the base of the May Hill Sandstone. William Ash and David Homfray of Portmadoc in North Wales, as well as Salter and Hicks, made certain that the Woodwardian received its fair share of the precious organic remains from these rocks, but from this point onwards it is difficult to imagine enough specimens arriving at the museum to

[21] His withdrawal from the last vestiges of a claim that the Silurian held the oldest fossiliferous rocks is evidenced in the changing subtitle of *Siluria*: compare Murchison 1854, 1859, and 1867.

drive Sedgwick from the grand dimensions of his Cambrian.[22] By the 1850s he was even less concerned than Murchison about the inclusion of the origin of life within his classification, judging from the lack of any private or public statements on the issue. His main interest, amounting to an obsession, was to obtain recognition for a Cambrian ending with the May Hill break. Fired by a deeply felt sense of injustice, Sedgwick had completely abandoned the flexibility that had characterized his position in the 1840s. The change is evident as early as November 1852. In a letter to Phillips he recognized three "grand subdivisions" in the rocks of the Lower Palaeozoic, the lowest being based on Barrande's Primordial. In 1840 he would doubtless have demanded that all three be recognized by separate names; in 1852 he unhesitatingly subsumed the bottom two under "Cambrian." No wonder that men like Forbes, Lyell, and Phillips, who possessed a genuine interest in the wider bearings of the controversy for the ordering of nature, found Sedgwick's attitude irritating.[23]

In the usual course of events Sedgwick and Murchison, as the elder statesmen of Palaeozoic geology, would have been looked to for a generally acceptable classification of the older rocks. But the bitterness of their dispute made this impossible, and other figures stepped into the breach.

The most influential alternative was put forward by Charles Lyell. Although by no means a specialist on the ancient strata, he held immense prestige and was thereby uniquely well situated to overcome the deadlock over classification. His adoption in 1855 of a limited "Cambrianism," based on the ideas of Salter, is of particular interest in illustrating some of the factors that led many geologists to see the Primordial fauna as a new basis for a Cambrian. In this sense Lyell typifies a host of British scientists who began using Sedgwick's term (albeit in a sense utterly rejected by its author) after so many years of following the extended Silurian classification. Lyell's approach is of interest also because in another sense he is atypical, for the idiosyncratic uniformitarian views espoused in the *Principles of Geology* shaped his attitudes toward classification and nomenclature.

For almost twenty years Lyell had been one of the leading supporters of the Silurian. During his American tours of the 1840s he

[22] Sedgwick 1854b: 362. For these and other donors to the Woodwardian collections, see Sedgwick 1873: x.

[23] Sedgwick to Phillips, 27 Nov. 1852, ff. 222-225, 249-252, UMO.

identified the fossils of the Potsdam Sandstone in New York as Lower Silurian, and he frequently defended Murchison's classification in the Geological Society debates.[24] Yet Lyell disagreed with certain elements of Murchison's work. In particular, Murchison had always shared Sedgwick's support for the theory of progression and had gradually loaded the Silurian concept with theoretical ideas unacceptable to Lyell. Not only did Murchison claim to have found proofs of a "beginning," he also argued that the great extent and thickness of the Silurian strata indicated a world fundamentally different from the present, with a virtually stable climate and a relatively uniform fauna prevailing over the entire planet through long ages uninterrupted.[25] Such ideas were anathema to Lyellian notions of uniformity. He measured the passage of time in the older rocks by a chronometer of organic change almost clocklike in its regularity, and could find no modern analogy for the deposition of 45,000 feet of strata while a single fauna inhabited the earth.[26] From the earliest editions of the *Elements*, Lyell had preferred to speak of a Silurian "group" rather than use the more theoretical term "system." Although the Silurian could be accepted in this sense without implying agreement with its progressionist overtones, Lyell would have preferred a nomenclature that better coincided with his overall theoretical program.

Sedgwick's substitution of Cambrian for Lower Silurian scarcely offered a real alternative. Although Sedgwick reduced the anomalous length of the Silurian by dividing the older rocks into two groups, he placed even more emphasis than did Murchison on the long-continued persistence of the older fossil types. Moreover, as discussed earlier, the progressionist theory so strongly opposed by Lyell had guided much of Sedgwick's effort to subdivide the "Protozoic" and "Cambrian" classifications during the 1840s. For all their disagreements, Sedgwick and Murchison held almost identical views about progression, and in consequence Lyell found the theoretical associations of both their classifications equally unacceptable. He also rejected the Cambrian on the grounds that it lacked priority. He shared the common belief that geological clas-

[24] C. Lyell 1845, *1*: 157-158, 245-246; 1849, *2*: 263. Even after Lyell readopted a version of the Cambrian, Murchison continued to refer back to these earlier works as though he still supported the full extension of the Silurian. See Murchison 1866: 20-21.

[25] Murchison 1854: 459-483, and J. Browne 1983: 101. See also Bowler 1976: 95-98, and Page 1976.

[26] C. Lyell 1855: 452-453.

sifications had to be founded on fossil types and fossil types alone, and in consequence felt that the right of naming groups of strata devolved upon those who had first defined them in terms of a distinctive fauna. According to this criterion, Sedgwick's structural, lithological, and geographical descriptions of the Cambrian had no priority. "Sedgwick's attempt to take the Lower Silurian into his Cambrian," Lyell wrote to Charles Bunbury in 1855, "is even worse than Murchison claiming all that is older than the Devonian as appertaining to his Silurian." In the successive editions of his *Manual of Elementary Geology*, Lyell argued that only a separate group of fossils below the Lower Silurian would justify the revival of the Cambrian.[27]

In Lyell's eyes, Barrande had provided precisely such a distinct fauna. For help in revising the relevant sections in the fifth (1855) edition of his *Manual*, Lyell obtained the services of Salter, and in doing so he accepted Salter's attempted resolution of the controversy between Sedgwick and Murchison. In part he adopted a limited Cambrian for the same reasons of compromise that appealed to Salter. In addition a fauna lower than the Silurian had important theoretical consequences for Lyell. It suggested that the first beginning to well-characterized forms of life had not been found after all and showed that the long-continued sameness of the oldest known fauna claimed by both Murchison and Sedgwick was largely illusory. Lyell felt that Barrande, as chief discoverer of this new period in earth history, had earned the right to give it a name, such as "Bohemian." However, since Barrande insisted on calling these strata by the loaded term "Primordial," Lyell made use of "Cambrian" instead—not so much because of its advocacy by Sedgwick, but rather as an alternative to the progressionist implications of a further downward extension of the Silurian.[28]

This reinterpretation of the "Primordial Silurian" of Barrande and Murchison, as espoused in subsequent editions of the *Manual*, carried great weight in the geological community. Karl von Zittel, professor of geology and palaeontology at Munich, noted at the end of the century that Lyell's adoption of "Cambrian" played a greater

[27] Lyell to Bunbury, 6 Oct. 1855, in K. M. Lyell 1881, *2*: 205-206. For the *Elements*, see Lyell 1841, *2*: 176, and 1851: 361. (The fourth edition of 1852 was unchanged in this and most other respects.)

[28] Lyell 1855: 451-457; Lyell to Darwin, 23 Apr. 1855, ICL(H). Lyell's dependence on Salter is evident from the series of letters from Salter to Lyell, 1853-1864, EUL: Lyell 1/5174-5223; and Murchison, Journal "Wanderings in Russia," VII, GSL: M/J14, opposite p. 768.

role in the eventual inclusion of the term in the received classification than did Sedgwick's polemical efforts.[29] Thus Sedgwick, the arch-progressionist, saw his Cambrian name gradually introduced into geological usage in part because of the support it provided for an *anti*-progressionist conception of earth history. The irony was heightened after Lyell gave his qualified acceptance to Darwin's *Origin of Species*, a work that Sedgwick loathed. In the 1865 edition of the *Elements of Geology*, the discovery of a Cambrian fauna was used by Lyell as support for evolution, demonstrating that the fossil record would be pushed further and further back in time.[30] For his part, Darwin mentioned the Bohemian and Longmyndian discoveries in the first edition of the *Origin* after corresponding with Ramsay. Although initially accepting the Survey's views on nomenclature, he gradually came around to a Lyellian position. As Darwin told Lyell in 1860, "It is indeed most important that the Cambrian or Barrandes primordial should be getting so much separated (in spite of Murchison) from Lower Silurian: I did not at all know this." Only in 1869, however, in the *Origin*'s fifth edition, did he fully incorporate the Lyellian version of the Cambrian into his published text.[31] Both Darwin and Lyell considered stratigraphical classification a matter of convenience, but they evidently recognized that it could also serve theoretical interests.

Although few geologists shared Lyell's strict notions of uniformity as a basis for accepting a limited Cambrian, many during the 1850s took up positions in the debate similar to the one advocated in the *Manual*. The process of conversion to the limited use of Cambrian is illustrated with particular clarity in the case of Lonsdale, retired curator for the Geological Society and a specialist on fossil corals. Ever since the resolution of the Devonian controversy, Lonsdale had been one of Murchison's closest confidants and a great supporter of the expanded Silurian classification, as shown in a letter of December 1851:

> You may well say, that you "never felt more secure of any position" than that of the existence and in its full integrity of the Silurian System. It is yours historically and palaeontologically; and if the extension of Silurian fossils downwards prove any thing, it is, that other deposits should be added to your lower strata, and not that the latter should be abstracted from the Silurian System.

[29] Zittel 1901: 444. [30] Lyell 1865: 585-586.

[31] Darwin 1959: 512-520; for the quotation, see Darwin to Lyell, [8 May 1860], APS(D/L): B/D25.L; see also Darwin to Ramsay, 1 July [1860], [1860], ICL(R).

Four years later, when reading the proofs for the second edition of *Siluria*, Lonsdale no longer spoke with such certainty.

> The all important point is the position of the "Lingula Flags" and equivalent deposits; and it appears to me, that the fossil evidence is in favour of the commencement of a distinct fauna. As it at present stands, it is opposed to the incontrovertible argument, whereby the unity of the Silurian System exists; and the nature of the Fauna is precisely that which marks a boundary—some generic continuations, but specific differences throughout.

Clearly Lonsdale did not accept the Murchisonian argument that a "system" like the Silurian encompassed whole families and classes of animals, and believed instead that smaller differences could be used for drawing boundaries. After receiving a letter from Lyell in 1859, Lonsdale was completely converted:

> The extension of the Cambrian system will even be acceptable. It was not till my attention was drawn to the 5th Edit[ion] of the Manual, that I rightly understood the evidence respecting a Cambrian fauna; but after reading what is there stated, no doubt remained upon the subject. Mr. Salter's account of the Longmynd fossils obtained by himself was not seen till very recently. His Trilobite is undeniable, and the other strange impressions cannot be inferred to inorganic moulds. Whatever may be their nature, the establishment of a single vegetable or animal, proves the existence of others.[32]

Although Lonsdale was by his own admission almost a geological recluse since his departure from the metropolis, and men like Lyell or Darwin scarcely specialized in the older rocks, views similar to theirs began to be adopted by those on the leading edge of Palaeozoic knowledge. The most influential among this group was Phillips, who in 1853 had been appointed deputy reader of geology at the University of Oxford. Working independently of Lyell and Salter, Phillips drew the base line of the Lower Silurian just above the Lingula Flags in editions of his popular textbooks published from 1855 onwards. He considered the Primordial fauna sufficiently distinct for it to sanction a Cambrian "group," but not a "system" in the full-fledged sense of a unit like his "Palaeozoic system." This conclusion resulted from a simple statistical comparison of the numbers of fossil species and genera held in common by adjacent

[32] Lonsdale to Murchison, 29 Dec. 1851, 9 May 1855, GSL: M/L13/18, 21; Lonsdale to Lyell, 19 Sept. 1859, EUL: Lyell 1/3601.

stratigraphical units, the same technique that had led him to name the Palaeozoic so many years before.[33] Such a method was becoming increasingly common in geology—McCoy and Salter both employed it—but Phillips felt that specialists on both sides of the controversy had compiled their fossil lists with the aim of supplying ammunition to conflicting parties rather than of finding the truth. As a result the disputants sometimes seemed to speak a babble of different palaeontological tongues. "It seems to me also that you rely too securely on M'Coy's analysis, and particular lists," he warned Sedgwick in March 1855; "for as long as Salter & Morris speak another language, you can not expect *them* to be set aside, and Murchison relies on them. For myself, I doubt them all except in cases where we can be sure they have the requisite data, & plenty of paper, which is not my case, this being the last scrap."[34] Here again one sees Phillips's characteristic caution towards premature classification. In his view the problem required not only more paper but a much fuller census of the population of the Primordial fossils, coordinated by an "impartial person" (himself) who would not lump or split particular species merely to meet the needs of some controverted classification of the strata.

A reader of the published maps and memoirs of the Geological Survey could easily assume that its members remained aloof from discussions about the potential basis for a Cambrian with a fauna of its own. Any appearance of controversy in an official scientific publication was to be avoided at all costs. But the correspondence between the Survey men tells a very different story.

We have seen that De la Beche had departed from standard Murchisonian usage by coloring almost unfossiliferous tracts like the Longmynd as "Cambrian" (Fig. 7.3). When Murchison assumed control of the Survey in 1855 several members of the staff suspected that he might use his position to quash even this tiny remnant of his opponent's claims. At the end of the year their worst fears seemed in danger of being realized. "Sir Roderick," Jukes told Ramsay, "in what looks a very wily pre-arranged way has been laying a foundation for obliterating 'Cambrian.' "[35] The scheme, as so often in the past, involved the sensitive question of the coloring of maps. Murchison's proposal retained the name "Cambrian" in the key, but subsumed the strata under the same color as Lower Silurian—thereby implying that the Cambrian was valid only locally. This

[33] Phillips 1855: 101-134.

[34] Phillips to Sedgwick, 11 Mar. 1855, CUL: Add. ms 7652IIL28.

[35] Jukes to Ramsay, 26 Dec. 1855, ICL(R).

plan shut the very door that Salter's search for fossils in the Longmynd had begun to open. Jukes was furious.

> If he still persist in carrying out his plans what are we to do?—Shall we write official protests to be entered on the books—I think we ought:—we may both say that we are so strongly opposed to the proposed change that we should prefer a record of its being against our wish & express desire being put upon our books—If he comes the Autocrat, receives the protests & goes ahead, why we can only obey, but we need make no secret of our having handed in these protests, & he must bear the brunt.—It can all be done good humouredly—that is his superiority over Sir Henry who got with a fidget & a fume the moment you opposed him.[36]

A few days after protests from both Jukes and Ramsay, Murchison quietly backed down, much as he would shortly do over the "Upper Caradoc" question. A mere trilobite fragment and some worm burrows were just not worth the trouble. "I suppose he hardly calculated on our both setting up our backs simultaneously," Jukes wrote in triumphant relief.[37]

The incident had a lasting importance, however, in driving the first significant wedge between Murchison and his two chief lieutenants on questions of classification and nomenclature. Previously Jukes had always favored De la Beche's version of the Murchisonian Silurian, principally on grounds of priority and utility. He did so, as he explained in a presidential address to the Geological Society of Dublin in 1854, "in spite of my natural feeling, which would lead me to back the opinions of Professor Sedgwick against all the rest of the geological world."[38] After the discoveries of the Primordial fossils and the splitting of the Caradoc, Jukes could at last let his heart and his science speak with a single voice. He outlined his heresy for the first time immediately after learning of Murchison's plan to color Cambrian and Lower Silurian together. Jukes's scheme retained the Cambrian as defined by the Survey; unlike Lyell and Phillips he placed an upper limit *below* the Lingula Flags. He suspected, however, that additional discoveries of Primordial fossils might eventually move the boundary to a higher level. But Jukes was primarily interested in physical and structural phenomena, an

[36] Jukes to Ramsay, 1 Jan. 1856, ICL(R); also Ramsay to Aveline, 31 Jan. 1855, BGS: GSM 1/420(A), f. 308.

[37] Jukes to Ramsay, 2 Jan. 1856, ICL(R); further comments in Ramsay to Aveline, 4 Jan. 1856, BGS: GSM 1/420(A), f. 309.

[38] Jukes 1854: 90. Also Jukes to Sedgwick, 10 Mar. 1853, CUL: Add. ms 7652IIX42.

emphasis derived from his training at Cambridge under Sedgwick. Like the man he addressed as his geological "father," Jukes was particularly impressed by the persistent unconformity at the base of the May Hill Sandstone. This discovery, he felt, had given Sedgwick equal rights with Murchison to the disputed strata that the latter called Lower Silurian and the former Upper Cambrian. In a compromise clearly reflecting his position as a Survey man trained at Cambridge, Jukes called this intermediate body of strata "Cambro-Silurian." From 1855 his publications as Irish Survey director necessarily continued to toe the official line, but his articles, comments at scientific gatherings, and influential textbooks gave this viewpoint considerable currency.[39]

None of the other Survey men went so far as to publicly differ from their Director General. But privately Ramsay had also begun to harbor doubts about the wisdom of classing all the older fossiliferous rocks under a single heading. As he confessed in a letter to Aveline at the end of 1855—"though I say nothing about this to any one but you"—he suspected that the palaeontological break between the two halves of the Silurian overshadowed that between the Silurian and the Devonian.[40] Many years later, Ramsay came to accept a view similar to Lyell's, with the Cambrian ending at the top of the Tremadoc beds. "*Officially* I still call the Lingula Flags Lower Silurian," he told Geikie in 1871, "because of the Director-General's classification, but theoretically I consider the Lingula Flags more closely allied to the Cambrian." For obvious reasons Ramsay kept such opinions to himself, and even after assuming the directorship after Murchison's death he retained the full Silurian nomenclature.[41]

But Ramsay, while continuing to support the Silurian, at a more fundamental level shared Lyell and Darwin's doubts that geological classifications would eventually prove more than convenient labels. In this he typifies an increasingly large number of geologists during the second half of the nineteenth century. Like many others, he preferred to interpret the strata as the transcript of a changing and diversified succession of environments. From such a viewpoint the definition of classificatory boundaries was often dismissed as an arbitrary exercise. As William Pengelly wrote to his wife from the 1856 British Association meeting, "The attempt is in my judgment ab-

[39] E.g. Jukes 1857: 404-426.

[40] Ramsay to Aveline, 31 Dec. 1855, BGS: GSM 1/420(A), f. 308.

[41] Ramsay to Geikie, 24 Mar. 1871, in Geikie 1895: 308.

surd, such lines have no existence in nature."[42] In Britain, aspects of this approach had been pioneered in the 1830s and applied in earnest to the older rocks by De la Beche, Phillips, and other geologists (many of them in the Survey) during the 1840s and 1850s. Their environmentalist program of research emphasized the reconstruction of palaeogeographies and the habitats of ancient populations of organisms; the boundaries of ancient continents and oceans seemed of more importance than those delimiting "systems" or "groups" of strata. Ramsay, preeminent heir to this tradition, could even predict in 1878 that the great Cambrian-Silurian controversy would "by-and-by be forgotten along with other minor debates, that in their day were of equal or more importance."[43] The triumph of evolutionary theory after 1859 only encouraged increasingly widespread suspicions that the fossil record was extremely imperfect. For many younger scientists, violent debate over a natural classification was beginning to lose its substantive interest.

The Invention of the Ordovician

Murchison died in October 1871, Sedgwick fifteen months later in 1873, and Phillips and Lyell within the following two years; Jukes and Salter, preceding their elders, both died in 1869.[44] But far from ending the dispute, the disappearance of so many major figures from the scene set the stage for a new round of controversy. I shall not follow the detailed story of the debate into this later period. The relative scarcity of manuscripts, the increasingly international character of geology, the volume of publication, and the large numbers engaged in the study of the Lower Palaeozoic would dictate a different method of treatment from that used here. The cosy if competitive world of the gentlemen geologists was fast disappearing, with an ever-increasing horde of paid professionals taking its place. This new generation looked back to the early and middle years of the century as the golden age of the science and felt dwarfed by the

[42] W. Pengelly to H. Pengelly, 7 Aug. 1856, in Pengelly 1897: 518-559. Darwin 1964: 279-311 comments "On the Imperfection of the Geological Record." As in so many other areas of natural history, Darwin's work grew out of developments dating from the 1840s and 1850s; see esp. Ospovat 1981, as well as J. Browne 1983, Desmond 1982, Rehbock 1983, Rudwick 1976a, and Secord 1986. For Ramsay's views on classification, see Ramsay 1863 and 1866: 2-3.

[43] Ramsay 1878: 58.

[44] The passing of the older generation was widely remarked upon by contemporaries, as in Bigsby to Hall, 6 July 1871, printed in Clarke 1923: 432-433. See also M. Greene 1982: 192-193.

Fig. 9.6. The old and new generations: Murchison and his protégé Archibald Geikie on tour in Scotland in 1860, as seen by the French author and amateur artist Prosper Mérimée.

reputations of their predecessors. "They were philosophers," one wrote, "we are only scientific specialists."[45] As a result Sedgwick and Murchison cast long shadows over the debate even after their deaths. Just as retrospectives of their own work (and of geology as a whole) had often served as weapons in controversy, so now did their successors employ posthumous reputations as elements in a new sequence of research and dispute. The final decades of the debate, although carried out in circumstances so entirely altered, show how a priority controversy can be carried on long after the deaths of the original parties.

The new phase of the controversy had actually begun with the introduction of the Primordial fauna. Both Sedgwick and Murchison had found the compromises of Lyell, Salter, Phillips, Jukes, and other geologists totally unacceptable—particularly in the case of Lyell, who had never specialized in the older rocks. "Lyell knows nothing of them," Murchison wrote in 1866, "and had no more right to interfere with my views & labours than I should have had to meddle with his Eo's Mio's & Plio's."[46] With every reiteration of the points at issue, the positions of the two principals had become more irreconcilable. Sedgwick's last statements on the controversy contained by his own admission "nothing but what is found in my old Preface," a work published with McCoy eighteen years earlier. Similarly, papers left by Murchison for posthumous publication merely repeated arguments that he had been using since 1842.[47]

Murchison's death sparked a major new offensive from the growing band of those sympathetic to Sedgwick's Cambrian. The central document in the campaign came from the pen of Sterry Hunt, a pugnacious "chemical geologist" from Canada. Hunt claimed to have begun historical research on the controversy while writing an obituary notice of Murchison, a task "that brought to light facts which both surprised and pained me." But far from being surprised at his discoveries, he had actually started work convinced of his subject's turpitude.[48] Hunt based his lengthy anti-Murchisonian history almost entirely on Sedgwick's retrospectives and for good

[45] [Dawkins] 1875: 200-203. The same nostalgia is expressed in Woodward 1908 and 1911, and in Geikie 1924.

[46] Murchison to T. R. Jones, 16 Aug. 1866, BL: Add. ms 45926, f. 172. See also the letters quoted in Page 1976: 158, 162.

[47] Sedgwick to McCoy, 5 Oct. 1872, ML: frames 716-719, comparing Sedgwick 1855 and 1873; Murchison, "Document to be referred to by executors," in Craig 1971: 496-498, and "Silurian Appendix, never printed, R. I. M. 1858," GSL: M.

[48] Hunt to Lyell, 26 Mar. 1871, EUL: 1/2741-2743; Hunt 1872a, republished with additions as Hunt 1878 (see p. 349).

measure added a few twists of his own to the story. The aging Sedgwick not only approved these, but helped to arrange for partial republication of the piece in the recently founded journal *Nature*. Supporters in the Woodward family of the British Museum also placed it in the *Geological Magazine*, and the Belgian geologist G. J. Gustave Dewalque planned to translate it into French.[49] Sedgwick, nearing ninety and almost entirely deaf, made his own valiant effort "to speak out in the cause of historic truth and reason" by dictating a preface to a new catalogue of the Palaeozoic fossils at Cambridge. "And it was high time," he said, "that a shameful incubus should be shaken off from the breathing organs of the older Palaeozoic rocks; and that they should express themselves once again in the language of truth and freedom."[50] Finished a few months before his death, the preface was his last publication.

Sedgwick had barely breathed his last before aspirants to his chair were rallying forces for testimonials and votes. As Henry W. Bristow wrote to Osmond Fisher the day after the old geologist's demise:

> I do not know what is the usual way of proceeding in such cases; but it seems to me to savour of indecent haste & to show a want of proper feeling to take active measures to fill up a vacancy so soon after death—before the dead man has been carried to the grave: but I suppose it only amounts to a practical carrying out of the cry "The King is dead! Long live the King."

Although testimonials from expert specialists now played a part in the jockeying for position, the Woodwardian election had in one respect changed little from 1818: voting divided along college lines, with Thomas McKenny Hughes, like Sedgwick a Trinity man, the victor by a narrow margin.[51]

With both Murchison and Sedgwick in the grave, a stream of obituaries, biographical sketches, and religious tracts began the process of literary embalming. Just as the two men had exploited images of William Smith and Abraham Werner, so now were their own reputations crystallized for the edification of future generations. The

[49] Hunt 1878: 349; 1872b; 1873. For Sedgwick's adoption of Hunt's new accusations, and his efforts to publish the piece in England, see Sedgwick to Hunt, 13 Apr. 1872, 23 Apr. 1872, 25 July 1872, 14 Oct. 1872, CUL: Add. mss 7652IIIB1-4 (copies).

[50] Sedgwick 1873: xiii-xiv.

[51] [Anon.] 1873; the quotation is from Bristow to Fisher, 28 Jan. 1873, CUL: Add. ms 7652VR3; see also H. N. Middleton to Fisher, 23 Feb. 1873, CUL: Add. ms 7652VR19, and Sheets-Pyenson 1982: 194.

procedure was a standard one, having developed in the eighteenth century through the tradition of the French *éloge*.[52] In many respects the obituarists only completed a process that had begun many years before under the direct supervision of Murchison and Sedgwick. Both had attempted autobiographies: Murchison dictated some twenty-six volumes of reminiscences based on his field notebooks and appointed Archibald Geikie as his literary executor when he realized "that their details could hardly possess sufficient interest for general readers." On the other hand, Sedgwick's autobiography—like so many things he began—was left unfinished. Written by Salter from dictation, it is largely a blank book that reveals more about the state of an eighty-year-old man's memory than about the history of his life.[53] But both combatants recognized the need for public men of science to have a recognizable image to pass down to posterity.

After the deaths of Sedgwick and Murchison the old contrasts between them were inevitably drawn more sharply. Colorful stories like Wilberforce's proclamation of Murchison as "King of Siluria" in 1849 became legend. Some even wondered what type of crown had been used and whether there had been any other regalia at the ceremony. In a speech at a memorial service, Sedgwick became not merely "first of men," but also "a block of that primitive granite of English right and liberty upheaved at Runnymede."[54] Moreover, the research papers of Sedgwick and Murchison were no longer read or referred to; descriptions of their work derived entirely from retrospectives. As John Challinor has pointed out, this process worked largely to Murchison's disadvantage. Since Murchison had never issued a point-by-point refutation of Sedgwick's history, the latter's version became accepted as standard, particularly after

[52] The French tradition is described in Outram 1978. Obituaries appeared in almost all the important periodicals of the day. Early accounts of Sedgwick include Carlisle 1880, Pattison n.d., Phillips 1873; and for Murchison, [anon.] 1871 and Geikie 1871; see also the listings in Sarjeant 1980: 1768-1769, 2086-2087. Many significant notices appeared in 1875 and 1890 as essay reviews of the respective "lives and letters" of the two men. The uses of scientific biography are well explored in Outram 1976 and J. Browne 1978 and 1981; Geikie's views as a historian of geology are placed in context in Oldroyd 1980.

[53] Murchison's journals are at GSL; see Geikie 1875, *1*: vi. For Sedgwick's autobiography, see "Life of Adam Sedgwick . . . written at the request of his friends by himself in 1865," CUL: Add. ms 7652IIIH2. For further details, see Speakman 1982: 123-124.

[54] For Murchison: [anon.] 1891; for Sedgwick: W. Selwyn, in Clark and Hughes 1890, *2*: 483. A useful orientation to similar episodes in the "invention of tradition" is available in the essays gathered in Hobsbawm and Ranger 1983.

Hunt's well-publicized and seemingly exhaustive account.[55] Geikie attempted to counteract this tendency in his 1875 biography of Murchison, partly by bending over backwards to treat both sides fairly. (John Willis Clark, a nongeologist who wrote most of Sedgwick's life, even agreed with Geikie's interpretation of the controversy.)[56] But as far as the really committed were concerned the effort was in vain; most supporters of the full Cambrian remained as implacable as their master had been. As a result, histories of the dispute froze into extreme versions similar to those espoused by the principals in their final years.[57]

Especially notable is the way in which classifications began to be explicitly associated with particular institutions. Sedgwick's successors at Cambridge apotheosized the "grand old man" into a mythical father figure as part of their campaign to found a "Cambridge school of geology" like those achieving such contemporary renown in physics and physiology.[58] Sedgwick, of course, had always drawn his support from Cambridge—witness the examples of Hopkins, McCoy, and Jukes—but had never wished to found a research school. Hughes, having secured the Woodwardian chair in 1873, set about turning his predecessor's reputation towards precisely that end. He co-authored the *Life and Letters* with Clark, scouted for funds to build a memorial museum, and encouraged the Geology Department's extracurricular "Sedgwick Club." Support for an extended version of the Cambrian became an important aspect of this plan, although the term was carefully stripped of its moral and philosophical meanings.[59] Conversely, Murchison's successors at the Survey maintained the Silurian classification long after various intermediate solutions were being adopted elsewhere. Both Ramsay

[55] Challinor 1969: 7; for examples, see D. C. Davies 1872, and Charles D. Walcott to Hughes, 7 Feb. 1882, CUL: Add. ms 7652VL122, which shows that Walcott accepted the Cambrian after reading Sedgwick 1855 and Hunt 1872a.

[56] Clark to Geikie, 2 Feb. 1890, responding to Geikie to Clark, 29 Jan. 1890; both GSL. For Geikie's wish to be fair, see Geikie to T. R. Jones, 15 Dec. 1874, BL: Add. ms 42580, f. 46, and Geikie 1924: 163-164. Challinor 1969: 7, mentions the success of the Sedgwickians' efforts.

[57] This trend has continued in muted form even in the most insightful commentaries in the modern secondary literature; for example, compare J. C. Thackray 1976 (written from a Murchisonian perspective) with Rudwick 1976c (from an essentially Sedgwickian viewpoint).

[58] The Cambridge school of geology is described in Porter 1982a; Geison 1978 provides information on the others and discusses the growth of research schools in the English academic environment during this period more generally.

[59] For Hughes, [anon.] 1906; the most prominent among his many defenses of the Cambrian is in Clark and Hughes 1890, 2: 502-563.

Fig. 9.7. McKenny Hughes and the Sedgwick Club on a field excursion in the Malvern Hills.

and Geikie kept the extended Silurian on Survey maps throughout their tenures of the directorship. "For my own part," wrote Geikie in 1889, "I have no sympathy with the Sedgwickians who from some (mistaken) notion that Murchison was unfair to their demigod would upset the long established nomenclature, and introduce new names which in most cases are mere synonyms for names already in use."[60]

That the later history of the debate should be expressed so clearly in institutional terms is typical of the changing face of British geology during this period. To some extent, the idiosyncracies of a new generation of Sedgwicks and Murchisons were being ironed out by the pressures of making a career within geology. The science became more fully professionalized, increasingly organized around institutions (particularly the universities and the Survey) rather than a few outstanding individuals. As part of these developments, the Geological Society—symbol of the old style of scientific sparring between independent gentlemen—lost its central role as a debating arena. In this new environment, there were frequent clashes between academic and government geologists. Relations between the two groups were obviously complex; for example, the Survey recruited extensively from Cambridge, and Hughes himself had been both a student of Sedgwick and a member of the Survey. But when conflict did arise, Cambria and Siluria were weapons well-tempered and close at hand.[61]

By the 1870s, then, the institutional settings, cast of characters, and issues in the debate had all been substantially transformed. But in one respect things remained unchanged. Dispute and discovery proceeded hand in hand much as they had from the very beginning. Indeed, many new developments took place in the study of the older rocks during this period. The Primordial fauna was greatly enriched, both in the British Isles and abroad. Much of the Survey's seemingly unassailable mapping of the older rocks was chipped away piece by piece, not only by individuals, but also through renewed investigations undertaken in response by the Survey itself. Six-inch-to-the-mile mapping replaced one-inch, and for all but the

[60] Geikie to J. D. Dana, 14 Apr. 1889, YUL: Dana papers, carton 1, folder 16. In the first edition of his *Textbook of Geology* (1882: 647-660), Geikie had gone some way towards a compromise, but after this was rejected he returned to the complete Murchisonian classification.

[61] The history of British geology during this period has been little studied, but see Desmond 1982, Porter 1978, and Rudler 1887. Sheets-Pyenson 1982 gives the best brief survey.

most ancient strata graptolites replaced brachiopods and trilobites as the preferred method of correlation. (Fig. 9.8). One of the leading figures in these developments was Charles Lapworth, who began his career as a schoolmaster in the Scottish Borders. Drawn to the region by its associations with Sir Walter Scott, he soon took up geology as an outdoor remedy for poor health. By the mid-1870s he was doing exacting and specialized work of great importance, pioneering the use of graptolite zones as a means for correlation in the older rocks. This technique, which involved mapping on a minute scale unheard of even in the Survey, proved of great utility in unfolding the intricacies of the older strata. At Dobb's Linn in the Moffat district of the Southern Uplands, for example, he showed that a large part of the entire Lower Palaeozoic sequence was condensed into a single exposure about 250 feet thick. Sedgwick, Murchison, and their contemporaries had often practiced geology by the mile; Lapworth did it by the inch.[62]

The famous paper on Moffat, published in 1878, made no pronouncements on the vexed questions of nomenclature and classification. But there can be no doubt that Lapworth had wished to establish his right to do precisely that, for within a year he proposed a new term—"Ordovician"—for the rocks between the May Hill Sandstone and the Lingula Flags (Fig. 9.4). Although Lapworth presented himself as a cosmopolitan man of knowledge, caring "nothing for schools, but everything for the facts," his proposal clearly had close affinities with those of Lyell, Salter, Phillips, and Hicks. Like them, Lapworth agreed that the Primordial fauna provided a palaeontological basis for the Cambrian. But at the same time he also saw that Hughes and his followers at Cambridge would never agree to call any rocks below the May Hill Sandstone by the name of Silurian. Thus rather than calling the disputed strata of Barrande's Second Fauna "Lower Silurian" as Lyell had done, or by some hybrid term like "Cambro-Silurian" or "Siluro-Cambrian," Lapworth argued that they deserved a name of their own. "Ordovician" offered a brilliant compromise for this purpose. Like Murchison's Silures, the Ordovics had been a fierce Welsh tribe during the time of the Romans; they had lived in the heart of the Bala district first examined by Sedgwick. Lapworth also emphasized that a threefold classification of Cambrian, Ordovician, and Silurian would aid in the process of international correlation by marking

[62] Biographical information on Lapworth is available in Watts 1939; for Moffat, see Lapworth 1878. Elles and Wood (1901-1918: i-clxxi) provide a history of graptolite research.

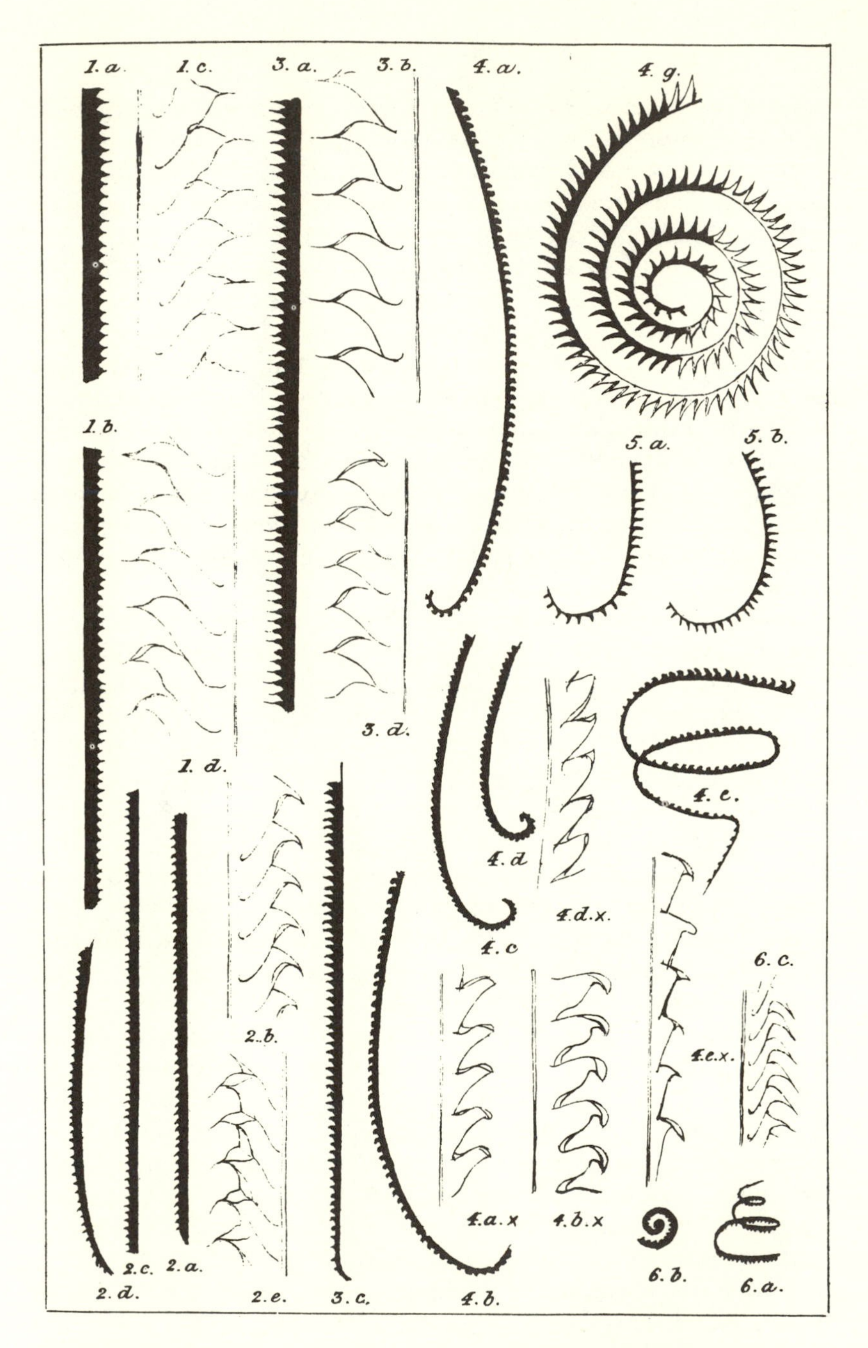

Fig. 9.8 Scottish graptolites described by Charles Lapworth in 1876. Close study of the different forms made possible the correlation of many of the older strata with unprecedented detail and precision.

each of the three major faunas of the Lower Palaeozoic with a distinctive name.[63]

The seeds of Lapworth's solution, of course, lay in the discoveries made in the older rocks during the early 1850s. Even by then, the positions of the two principals had already diverged so completely that the emergence of some kind of alternative had become almost inevitable. Like Lyell, many men of science were caught in the middle and required an escape route from the two extremes. Years later when Geikie, Hughes, and other followers of Murchison and Sedgwick were still fighting for the positions of their respective heroes, Lapworth recognized the need for a mediating classification, and geologists eventually accepted his suggestion (from many others that were made) partly because it accomplished the delicate task so well. In particular, he realized that any successful classification would have to take the intertwined social and technical roots of the debate into account. The two considerations are strategically balanced in the final sentence of Lapworth's famous paper:

> Every geologist will at last be driven to the same conclusion that Nature has distributed our Lower Palaeozoic Rocks in three subequal systems, and that history, circumstance, and geologic convenience, have so arranged matters that the title here proposed for the central system is the only one possible.[64]

Thus when Lapworth urged the adoption of the Ordovician, he did so "in the name of science" and on the grounds that a threefold division "imitates Nature herself"; but he also explicitly recognized that his proposal offered a politic compromise between individuals, interests, and ideals locked in fundamental opposition. At the same time, both Murchison and Sedgwick would have been aghast at the idea that the Ordovician compromise satisfactorily recognized their achievements. Under such a scheme Silurian became the "mere band" of Murchison's worst nightmares, while the Cambrian was but a pitiful fraction of what Sedgwick had always demanded. Their response would have been that of their successors; as Hughes asked, "One shell is given to Sedgwick, the other to Murchison, but who gets the oyster?"[65]

The question was not unreasonable. Why should Lapworth, born eleven years after Murchison and Sedgwick started work in Wales,

[63] Lapworth 1879.

[64] Lapworth 1879: 15.

[65] Clark and Hughes 1890, 2: 555. "Nature" and "science" are invoked in Lapworth 1879: 8, 15.

be rewarded with a name on the geological column—especially when so many earlier proposals had failed? The answer becomes evident if one looks at those who were actively engaged in research on the older rocks in the final decades of the century. In many respects Lapworth was the leading member of a new group of geologists who gained reknown in the 1860s, 1870s, and 1880s for pioneering work on the Palaeozoic strata. Tied neither to the Survey nor to Cambridge, many of them taught at the new "redbrick" universities, and as authors of textbooks and trainers of large number of students, they exercised a growing power in the scientific community. Lapworth himself, although still serving as a schoolmaster in 1879, had already decided upon "a scientific life pure & simple" at the time he proposed the Ordovician, and two years later he successfully obtained the new geology chair at Mason College of Science in Birmingham.[66] Put in its simplest terms, just as Lapworth's "Cambrian" recognized the Sedgwickian diehards among the Cambridge men, and his "Silurian" the Murchisonians in the Survey, so "Ordovician" (and not so incidentally the scheme as a whole) marked the achievements of this third group only just coming into their own. One of them, William W. Watts, recorded the standard riposte to Hughes's molluscan analogy: Sedgwick and Murchison deserved mere shells, it was said, while Lapworth got the oyster because he had discovered a pearl inside.[67] For the rising generation, the invention of the new system was closely bound up with fruitful scientific achievements.

Even so, Lapworth's solution achieved acceptance only very gradually. Particularly important, as Michael Bassett has suggested, was the conversion of John Marr of Cambridge, and (more slowly) many others in the Sedgwickian party. With the recognition of the Ordovician by the Survey after Geikie's retirement from the directorship in 1901, the debate over Cambrian and Silurian in Britain was essentially over. However, on the international level, its acceptance was by no means a foregone conclusion. In fact, the mediating term did not gain full international approval until 1960, when the Commission on Stratigraphy of the International Geological Congress finally gave Ordovician a place in the column between Cambrian and Silurian.[68]

[66] Lapworth to Hughes, 26 July 1878, CUL: Add. ms 7652VL64; Watts 1939: 237.

[67] Watts 1939: 251; also Callaway 1879: 142, for a similar recognition of Lapworth.

[68] An excellent brief summary of later work on the Ordovician (and its gradual acceptance) is provided in M. Bassett 1979. Marr's early views are in his 1883 "Sedgwick Prize Essay," which also includes a lengthy bibliography. The reports of the in-

For all the importance of these developments, one notes a gradual decline of interest in the dispute during the latter decades of the nineteenth century. This was to some degree inevitable after the deaths of Sedgwick and Murchison, but in a more important sense the waning of the controversy mirrors a slow stagnation of the wider traditions of stratigraphical geology. The first half of the century offers an excellent example of the exploitation and increasingly specialized refinement of a new scientific technique. In 1831 the methods of classificatory stratigraphy offered results of considerable interest and importance. By the middle years of the Victorian era, especially after the detailed mapping of the Survey, this was often no longer the case. Among scientists at large, stratigraphy was thought to have achieved many of its original goals and the order of the British rocks was known with a tolerable degree of certainty. After the work of men like Lapworth, this was believed true even for the most ancient fossiliferous rocks. Moreover, those involved in the final phase of the debate—Hughes, Lapworth, Hicks, Marr, Watts, Charles Callaway—were highly competent stratigraphers, but for most of them the task of ordering the rocks lacked the broad cultural significance it had possessed for an earlier generation. Although stratigraphical memoirs continued to appear at an ever-increasing rate, many scientists of skill and ability were turning their attention to other subjects: tectonics, petrology, geomorphology, vulcanology, geophysics.[69] The classificatory focus characteristic of the first half of the nineteenth century had lost much of its original intellectual, moral, and social importance.

---

ternational congresses which began to be held at this time usefully survey contemporary opinion and were critical in establishing a consensus on classification; see Ellenberger 1978.

[69] For these new foci of research, see M. Greene 1982, B. M. Hamilton 1982, Porter 1978: 829-830, O'Connor and Meadows 1976, Rudler 1887, and Zittel 1901.

Conclusion

VICTORIAN SCIENCE was geared up for the production of facts. Facts—from a single classified trilobite to the entire geological column—could be displayed in museums and at lectures; they could license speculation and advance scientific careers. For the public at large, facts blended the commonplace and the marvelous into one. But the wonders of science did not lie on the surface of things; they needed to be dug up, dredged for, searched out, by "laborious and patient investigation." As Gillian Beer has recently suggested, Victorian facts were perceived not merely as inert entities, but as strenuous performances: a fact was something *achieved*.[1] Naming was an essential feature in this active process of achievement. To name something was to place it, to possess it, to confer upon it the status of a natural object.

Because of the central importance of naming, Victorian naturalists fully recognized that the Cambrian-Silurian controversy was about words. At stake—here and in all debates about classification—was one of the most fundamental questions in science, the choice of a language appropriate to the perceivable phenomena of the natural world. In recent years, scholars from a variety of disciplines have pointed to the rich array of meanings carried by the language of science. Even in developed scientific specialties, words possess a remarkable range of metaphorical associations that resonate with the surrounding culture. They are evocative "natural symbols," laden with meanings drawn from the immediate specialist milieu and from the wider society.[2] As I have emphasized throughout this book, the words "Cambrian" and "Silurian" each accumulated meanings far beyond their immediate purpose as place holders in the geological column. These linguistic associations were not transcendental or *a priori* ones, inherent in the names themselves for all time; rather, they depended on the specific historical circumstances in which the dispute developed. In the context of controversy, the use of one nomenclature rather than another—

[1] Beer 1983: 81-82.

[2] Douglas 1973.

even for reasons of convenience—necessarily implied an acceptance of some of these meanings.

The preceding chapters have argued that Cambria and Siluria were conceived of and valued as labels for scientific properties, signals of individual accomplishment and success in the quest for originality in science. Sedgwick eventually fought for Cambria in part as a compensation for his failure to complete a major geological book, while Murchison defended his Silurian territories partly out of a wish to preserve his scientific accomplishments intact. At the same time, the competing classifications achieved their value as property by virtue of the associations that had gathered around the words "Cambrian" and "Silurian" during the controversy. Some of the most significant of these meanings involved questions of method fundamental to geological practice. Were fossils or physical characteristics to be used as the determinants of stratigraphical classifications? Was geology, which was generally agreed to be part of natural history, also to have ties with the physical sciences? Which classification best reflected the progressive history of life on earth? Was the Palaeozoic the only system in the older rocks or were there several, the lowest being the Silurian? Were classifications to be framed first with reference to local circumstances of original type sections and their geographical limits, or as worldwide standards taking data from around the globe into account? In each case, scientific nomenclature encoded a host of associated ideas.

Given the existence of a well-developed discipline of geology, it is to be expected that the contingent meanings of Siluria and Cambria related most immediately to issues of importance within the boundaries of specialized scientific discourse. Geology had been securely established by the early decades of the nineteenth century, and the communal goals of the stratigraphical enterprise gave a common measure to scientific debate about the earth. Without such a consensus, fruitful disagreement on questions like those so prominent in the Cambrian-Silurian dispute could never have arisen. The word "system," always a crucial term in geology, changed its meaning in ways that illustrate precisely how this consensus had been established. In the eighteenth century a "system" had been a cosmological theory, a comprehensive account of the past and future of the planet; by the nineteenth century it had become a taxonomic term for a set of rock formations. The word remained a focus of controversy in both centuries, but by the time of Sedgwick and Murchison the arena for potential conflict had been vastly reduced. Where Enlightenment cosmogonists had argued to a stalemate about the entire world, Victorian geologists fruitfully focused attention on some crucial strata in a single part of Britain. To inflate the Sedgwick-Murchison debate into a conflict between incommensurable world views would be to ignore the very considerable range of issues upon which all members of the geological community were agreed.

Yet the boundaries between the study of the earth and other areas of thought remained far from rigidly drawn. All the participants in the dispute had devoted much of their lives to geology, but they were simultaneously engaged in a wide variety of other activities. Sedgwick was an ordained priest in the Anglican church, and Murchison began as a military man and became an active promoter of colonial development. Sharpe was active in business, Phillips a provincial orphan who made good, and De la Beche the promoter of a new kind of scientific career. Such differing commitments and the idiosyncratic nature of recruitment into science during this period meant that concerns from the general culture of Victorian England could inform even the most esoteric disputes. Among the sciences, geology had a vocabulary that remained particularly open to the outside world. At one level, alternative nomenclatures proclaimed alternative methodologies for geological practice. But the protagonists themselves held that such contrasting technical commitments could be (indeed *should* be) supported and sustained by parallel social, religious, and philosophical ones. Certainly it was clear by the 1850s that Siluria and Cambria had come to represent differing visions of the nature and function of science. In connection with his career as a leading metropolitan scientist, Murchison had defined the value of his system (and the associated method of correlation by fossils) in political and social terms. Siluria acquired overtones of agressive imperial expansionism, with the maintenance of order and consensus in scientific matters given a high value, even if achieved through a certain measure of coercive control. For Murchison, geology was closely allied to geography, and his interest in extending his Silurian territory was part of a broader effort to encourage the growth of the British empire. For Sedgwick, on the other hand, geology permitted the study of a divinely created nature through the principles of induction which he taught his students at Cambridge. In his view, dependence on the secure evidence of sections was a methodological imperative. Physical evidence grounded the procedures that validated geology as a science—and science could be valued precisely because it represented the process of rational thought that "raised us above the dregs of matter."[3] His vehement defense of his Cambrian territory has much of the air of a moral crusade, a battle for truth and justice all the more necessary because fought against a strongly supported opponent. Only if the "facts" of geology were securely constructed could they be used as weapons in the battle against utilitarianism, materialism, and transmutation.

Obviously Murchison's and Sedgwick's views of the ends of science were not inherently contradictory. Indeed, at its most fundamental level the controversy between them involved alternative emphases within the wider program of enlightened conservatism being actively forwarded by the leaders of early Victorian science. This program, which has been analyzed in depth by recent historians, appealed to science as a means of maintaining social stability in an age of industrial revolution and widespread discontent among the working class.[4] Thus, like most of the gentlemanly Fellows of the Geological Society, Murchison and Sedgwick wholeheartedly supported the British Association, the Mechanics Institutes, and the Society for the Diffusion of Useful Knowledge. They both applauded David Livingstone's missionary expeditions in Africa (although with revealing differences in emphasis) and combatted the incipient materialism of the early evolutionists. Most significantly, they collaborated on a variety of technical papers in geology, contributing to the wider goal of taming the "chaos" of the strata. Among many other functions, all these endeavors were explicitly designed to celebrate and sustain the established social, political, and religious order. On the importance of this task, the two men could speak with a single voice. But when differences in their approaches led to conflict over the boundary between Siluria and Cambria, the two men fought for almost thirty years and died as bitter enemies.

It would be unfortunate if the final estrangement obscured the remarkable fruitfulness of their dispute. All too often longstanding controversies about property, priority, and nomenclature are condemned as inimical to the progress of science. This view is a relic of a historical tradition that places the social and technical aspects of scientific activity in sharp opposition, with the social world having nothing to do with the making of knowledge except in a negative sense. It will now be abundantly clear just how misleading this kind

[3] From a report of a toast by Sedgwick at an anniversary meeting of the Geological Society, in Lyell to Horner, 24 Feb. 1838, K. M. Lyell 1881, 2: 39.

[4] Morrell and Thackray 1981, esp. pp. 29-34; Desmond 1982; Neve 1983; and Cannon 1978. For a review essay on some of this literature, see Miller 1984.

of Manichean duality really is. Relegating social processes like controversy to the "pathology" of science only demonstrates a basic misconception of the nature of scientific activity.

In fact, the major accomplishments of Lower Palaeozoic geology in Britain prove to have been *part* of a debate, not something separate from it. From the beginning, controversy had defined the stakes in the scramble for scientific success. In the 1830s differences in the approaches of Sedgwick and Murchison had served as a source of creative innovation, producing significant concepts like the collaborative boundary. When outright disagreement emerged in the 1840s, new fieldwork was undertaken and new information about fossils and strata was brought to light by Sedgwick, Sharpe, Bowman, Murchison, and the Survey. The splitting of the Caradoc formation initiated by McCoy and the unexpected fossil riches from the older rocks found by Barrande, Salter, and others were likewise direct outcomes of the dispute in the 1850s. All parties to the controversy accepted these findings on an observational level, according to the communal rules and tacit understandings of the stratigraphical enterprise—even while their relevance to wider questions of classification remained open to further negotiation. And of course, empirical novelties could never in and of themselves force a change in an individual's position. The stubborn persistence of Sedgwick and Murchison in their final years demonstrates that no scheme of classification was ever quite beyond repair. But when skillfully deployed by those less committed than the two principals, observational findings did prove of great importance. Most significantly, they provided an essential technical rationale for intermediate positions. In the 1850s and 1860s the widespread acceptance once enjoyed by Murchison's Silurian classification faded, as men like Lyell, Darwin, Lonsdale, Jukes, Phillips, and Salter transferred their allegiance to a threefold ordering of the Lower Palaeozoic. Sedgwick's vigorous attempts to keep alive the idea of a separate Cambrian, the battle of May Hill, the Primordial fossil findings, Murchison's autocratic heavy-handedness, the rise of a new generation of geologists—all had played a part in establishing the contours of a new consensus. Conflict fostered new research and discovery with remarkable consistency, and not only by Sedgwick and Murchison, but also by a host of other investigators.

Although the widespread lack of appreciation for the role of controversy in the classificatory sciences is regrettable, the reasons for it are abundantly clear. Built into the entire taxonomic enterprise is an emphasis on permanence, impersonality, and neutrality. In works printed for long life and standard reference, explicit polemicizing is seen as a guarantee of temporary status. As a result, every Victorian naturalist adjusted his descriptive practice to a series of conventions that smoothed out the rough edges of disagreement. In its chief forums, dispute was ephemeral, typically confined to discussions at scientific meetings or private letters. Modern readers familiar only with the bland neutrality of finished monographs are thus bound to be misled. Nothing looks more straightforward than a geological map (especially as it employs a powerfully convincing visual medium of expression), yet nothing could arouse more passionate controversy. Only through careful historical inquiry will the full range of intentions behind such descriptive statements be recovered. Not only must individual maps and memoirs be examined with care, but also this study must be accompanied by the closest possible understanding of their immediate context. In this way, scientific classifications can be recognized, not as unmediated accounts of observation, but as specific choices of language from a variety of potential alternatives.

These challenges, though occasionally tackled by historians of botany, zoology, chemistry, and palaeontology, have usually been sidestepped by those dealing with the sciences of the earth. Ignoring classification, they have presented geology as an essentially historical enterprise concerned with process and time. The standard accounts are organized around a succession of debates about major theories: neptunism, plutonism, catastrophism, diluvialism, progressionism, and uniformitarianism. Even revisionist historians of the past fifteen years, while offering a needed corrective to the overemphasis on James Hutton and Charles Lyell as the "founders" of modern geology, have continued to conduct the discussion in terms of abstract theoretical categories such as "directionalism" and "historical geology." This approach is exemplified by Mott T. Greene's excellent study of tectonics and mountain building, published as *Geology in the Nineteenth Century*, which focuses almost exclusively on the evolution of comprehensive theoretical syntheses. It is even more apparent in Anthony Hallam's recent *Great Geological Controversies*, which pictures the history of geology as a series of conflicts between underlying "scientific research programmes" of great generality. Hallam explicitly relates his study to the work of Imre Lakatos, a philosopher of science who has argued for the importance of competition between such programs in the growth of knowledge.[5] It will be evident that my own views share certain elements of this position, especially an emphasis on the productive roles of criticism, conflict, and tenacity in science. But like Lakatos, Hallam defines his competing research programs so broadly and so abstractly that they are impossible to relate to the practical traditions of research pursued in particular social locations.[6] An understanding of high-level theory is an essential part of the history of science, but it gives only limited access to the vast majority of the work that has produced new knowledge about nature during the past two centuries. In a major dispute like the Cambrian-Silurian controversy, questions involving the history of the earth arise, but only tangentially. Suggestions that the debate was primarily about the origin of life or the extension of the time scale cannot be sustained.

A basic reorientation of our view of the history of the natural sciences is in order. Rather than focusing exclusively on a handful of classic theoretical debates, the historical spotlight should turn to the concrete traditions of research that have shaped the maps, monographs, and textbooks which crowd the shelves of our libraries. Such a revised perspective would direct attention to the changing structure of the scientific community and the varieties of practice developed in particular local contexts. It would inquire into the ways in which research networks, new career patterns, and institutional schools have shaped new knowledge about nature. Histories of fieldwork, experiment, mapping, communication, and publication would assume center stage. Especially welcome would be greater sensitivity to scientific language, particularly its visual aspects and metaphorical contexts. And finally, close attention to what most practitioners actually do would enrich our understanding of the cultural and social relations of science. It is widely recognized that the power of scientific knowledge in society derives from its apparent success in giving the results of a human activity the status of objective natural facts. But the ways in which this has occurred are just beginning to receive attention. Not only in the history of geology, but throughout the history of science, fundamental issues await investigation. An approach through the study of practice opens up possibilities as yet barely explored.

[5] Greene 1982; Hallam 1983, esp. pp. 162-168; Lakatos 1970.

[6] More useful in this respect is the work of Jerome Ravetz (1971) and Thomas Kuhn (1970) on the role of concrete exemplars in science.

Abbreviations

BJHS *British Journal for the History of Science*
DNB *Dictionary of National Biography*
DSB *Dictionary of Scientific Biography*
JHB *Journal of the History of Biology*
JSBNH *Journal of the Society for the Bibliography of Natural History*
MGS *Memoirs of the Geological Survey of Great Britain*
PGS *Proceedings of the Geological Society of London*
QJGS *Quarterly Journal of the Geological Society of London*
TGS *Transactions of the Geological Society of London*

Proceedings of the annual meetings of the British Association for the Advancement of Science are cited as *British Association Reports,* preceded by the year of the meeting in question; for example, *1840 British Association Report* refers to *Report of the Tenth Meeting of the British Association for the Advancement of Science; Held at Glasgow in September 1840* (London: John Murray, 1841).

Bibliography

[Anon.] 1811. "Preface." *TGS*, 1st ser., *1*: v-ix.

[Anon.] 1839. "Review of New Books." *Literary Gazette*, no. 1153: 113-114.

[Anon.] 1841. "The Geology of Society." *Punch*, *1*: 157, 178.

[Anon.] 1847. Review of R. I. Murchison et al., *The Geology of Russia in Europe*. *British Quarterly Review*, *6*: 290-314.

[Anon.] 1854. "Murchison's *Siluria*." *Spectator*, *27*: 815-816.

[Anon.] 1855. "*Siluria*; or Glimpses of the Ancient World." *Hogg's Instructor*, *4*: 57-67.

[Anon.] 1870. "Eminent Living Geologists. John Phillips." *Geological Magazine*, *7*: 301-306.

[Anon.] 1871. "The Late Sir Roderick Murchison." *Times*, 23 Oct. 1871: 12, cols. 4-6.

[Anon.] 1873. "The New Woodwardian Professor of Geology for Cambridge." *Geological Magazine*, *10*: 144.

[Anon.] 1891. "Sir Roderick Murchison as King of Siluria." *Midland Naturalist*, *14*: 268.

[Anon.] 1906. "Eminent Living Geologists: Thomas McKenny Hughes." *Geological Magazine*, n. s., 5th decade, *3*: 1-13.

Albritton, Claude C. 1980. *The Abyss of Time: Changing Conceptions of the Earth's Antiquity after the 16th Century*. San Francisco: Freeman, Cooper, and Co.

Albury, W. R., and D. R. Oldroyd. 1977. "From Renaissance Mineral Studies to Historical Geology, in the Light of Michel Foucault's *The Order of Things*." *BJHS*, *10*: 187-215.

Allen, David Elliston. 1976. *The Naturalist in Britain: A Social History*. London: Allen Lane.

———. 1979. "The Lost Limb: Geology and Natural History." In L. Jordanova and R. Porter (eds.), *Images of the Earth*, pp. 200-212. Chalfont St. Giles: British Society for the History of Science.

———. 1983. "Life Sciences: Natural History." In P. Corsi and P. Weindling (eds.), *Information Sources in the History of Science and Medicine*, pp. 349-360. London: Butterworth Scientific.

———. 1985. "The Early Professionals in British Natural History." In A. Wheeler and J. Price (eds.), *From Linnaeus to Darwin*, pp. 1-12. London: Society for the History of Natural History.

Altick, Richard D. 1978. *The Shows of London*. Cambridge, Mass.: Harvard University Press.

Angelin, Nils Peter. 1852-1854. *Palaeontologia Scandinavica.* Leipzig: T. O. Weigel.

Ansted, David T. 1838. "On a New Genus of Fossil Multilocular Shells, Found in the Slate-Rocks of Cornwall." *Transactions of the Cambridge Philosophical Society, 6*: 615-622.

Appel, Toby A. 1976. "The Cuvier-Geoffroy Debate and the Structure of 19th Century French Zoology." Ph.D. thesis, Princeton University.

Babbage, Charles. 1830. *Reflections on the Decline of Science in England, and on Some of its Causes.* London: B. Fellowes.

Bailey, Edward B. 1952. *Geological Survey of Great Britain.* London: Murby.

Balan, Bernard. 1979. *L'Ordre et le Temps: L'Anatomie Comparée et L'Histoire des Vivants au XIXe Siècle.* Paris: Librairie Philosophique.

Barnes, Barry. 1974. *Scientific Knowledge and Sociological Theory.* London: Routledge and Kegan Paul.

———. 1982. *T. S. Kuhn and Social Science.* London: Macmillan.

Barnes, Barry, and Steven Shapin (eds.). 1979. *Natural Order: Historical Studies of Scientific Culture.* Beverly Hills: Sage.

Barrande, Joachim. 1846. *Notice Préliminaire sur le Systême Silurien et les Trilobites de la Bohême.* Leipzig: the author.

———. 1852-1911. *Systême Silurien du Centre de la Bohême.* 8 vols. in 29. Prague: the author.

———. 1880. "Du maintien de la nomenclature établie par M. Murchison." *Congrès international de géologie, tenu à Paris, du 29 au 31 Août et du 2 au 4 Septembre 1878*, pp. 101-106.

Barrett, Paul H. 1974. "The Sedgwick-Darwin Geologic Tour of North Wales." *Proceedings of the American Philosophical Society, 118*: 146-164.

Barrett, Paul H. (ed.). 1977. *The Collected Papers of Charles Darwin.* 2 vols. Chicago: University of Chicago Press.

Bartholomew, Michael. 1976. "The Non-progress of Non-progressionism: Two Responses to Lyell's Doctrine." *BJHS, 9*: 166-174.

———. 1979. "The Singularity of Lyell." *History of Science, 17*: 276-293.

Bassett, Douglas A. 1963. *Bibliography and Index of Geology and Allied Sciences for Wales and the Welsh Borders, 1536-1896.* Cardiff: National Museum of Wales.

———. 1967. *A Source-Book of Geological, Geomorphological and Soil Maps for Wales and the Welsh Borders.* Cardiff: National Museum of Wales.

———. 1969a. "Wales and the Geological Map." *Amgueddfa: Bulletin of the National Museum of Wales*, no. 3: 10-25.

———. 1969b. "Some of the Major Structures of Early Palaeozoic Age in Wales and the Welsh Borderland: An Historical Essay." In A. Wood (ed.), *The Pre-Cambrian and Lower Palaeozoic Rocks of Wales*, pp. 67-116. Cardiff: University of Wales Press.

Bassett, Michael G. 1979. "100 Years of Ordovician Geology." *Episodes: Geological Newsletter, International Union of Geological Sciences*, no. 2: 18-21.

Baxandall, Michael. 1972. *Painting and Experience in Fifteenth Century Italy.* Oxford: Oxford University Press.

———. 1980. *The Limewood Sculptors of Renaissance Germany*. New Haven: Yale University Press.

Becher, Harvey W. 1980. "William Whewell and Cambridge Mathematics." *Historical Studies in the Physical Sciences, 11*: 1-48.

[Beckford, Peter.] 1781. *Thoughts on Hunting: In a Series of Familiar Letters to a Friend*. London: P. Elmsly.

Beer, Gillian. 1983. *Darwin's Plots: Evolutionary Narrative in Darwin, George Eliot and Nineteenth-Century Fiction*. London: Routledge and Kegan Paul.

Berman, Morris. 1975. " 'Hegemony' and the Amateur Tradition in British Science." *Journal of Social History, 8*: 30-50.

Berry, William B. N. 1968. *Growth of a Prehistoric Time Scale Based on Organic Evolution*. San Francisco: W. H. Freeman.

Bigsby, John J. 1868. *Thesaurus Siluricus: The Flora and Fauna of the Silurian Period*. London: John van Voorst.

Billings, Elkanah. 1861-1865. *Palaeozoic Fossils: Vol. 1. Containing Descriptions and Figures of New or Little Known Species of Organic Remains from the Silurian Rocks*. Montreal: Dawson Brothers.

Bloor, David. 1976. *Knowledge and Social Imagery*. London: Routledge and Kegan Paul.

———. 1982. "Durkheim and Mauss Revisited: Classification and the Sociology of Knowledge." *Studies in History and Philosophy of Science, 13*: 267-297.

Boué, Ami. 1840. *La Turquie d' Europe*. 4 vols. Paris: Arthus Bertrand.

Bowler, Peter J. 1976. *Fossils and Progress: Paleontology and the Idea of Progressive Evolution in the Nineteenth Century*. New York: Science History Publications.

———. 1984. *Evolution: The History of an Idea*. Berkeley: University of California Press.

Bowman, John E. 1841a. "On the Great Development of the Upper Silurian Formations in the Vale of Llangollen, North Wales, and on a Plateau of Igneous Rocks on the East Flank of the Berwyn Range." *1840 British Association Report*, pt. ii, pp. 100-102.

———. 1841b. "Notice of Upper Silurian Rocks in the Vale of Llangollen, North Wales; and of a Contiguous Eruption of Trap and Compact Felspar." *Transactions of the Manchester Geological Society, 1*: 194-211.

Brannigan, Augustine. 1981. *The Social Basis of Scientific Discoveries*. Cambridge: Cambridge University Press.

Brett, Raymond L. (ed.). 1979. *Barclay Fox's Journal*. London: Bell and Hyman.

[Brewster, David.] 1846. Review of R. I. Murchison et al., *The Geology of Russia in Europe. North British Review, 5*: 178-221.

Briggs, Asa. 1959. *The Age of Improvement, 1783-1867*. London: Longman.

Britton, J., and A. W. N. Pugin. 1826-1828. *Illustrations of the Public Buildings of London: With Historical and Descriptive Accounts of each Edifice*. 2 vols. London: J. Taylor.

Brooke, John Hedley. 1979. "The Natural Theology of the Geologists: Some

Theological Strata." In L. Jordanova and R. Porter (eds.), *Images of the Earth*, pp. 39-64. Chalfont St. Giles: British Society for the History of Science.

Browne, C. A. (ed.). 1871. *Letters and Extracts from the Addresses and Occasional Writings of J. Beete Jukes*. London: Chapman and Hall.

Browne, Janet. 1978. "The Charles Darwin—Joseph Hooker Correspondence: An Analysis of Manuscript Sources and their Use in Biography." *JSBNH*, *8*: 351-366.

———. 1981. "The Making of the *Memoir* of Edward Forbes, F.R.S." *Archives of Natural History*, *10*: 205-219.

———. 1983. *The Secular Ark: Studies in the History of Biogeography*. New Haven: Yale University Press.

Buch, Leopold von. 1836. "Extrait d'une lettre de M. Léopold de Buch à M. Élie de Beaumont." *Bulletin de la Société Géologique de France*, *7*: 155-158.

[Buckland, Frank.] 1869. "Souvenirs of the Life of Lady Murchison." *Victoria Magazine*, *12*: 461-463.

Buckland, William. 1837. *Geology and Mineralogy Considered with Reference to Natural Theology*. 2d ed., 2 vols. London: William Pickering.

Buckland, William, and William D. Conybeare. 1824. "Observations on the South-western Coal District of England." *TGS*, 2d ser., *1*: 210-316.

Bunbury, Frances (ed.). 1890-1893. *Memorials of Sir C. J. F. Bunbury, Bart*. 9 vols. Mildenhall: privately printed; copy in Cambridge University Library.

Burchfield, Joe D. 1974. *Lord Kelvin and the Age of the Earth*. London: Macmillan.

Burd, Van Akin (ed.). 1973. *The Ruskin Family Letters: The Correspondence of John James Ruskin, his Wife, and their Son, John, 1801-1843*. 2 vols. Ithaca: Cornell University Press.

Burrow, John W. 1981. *A Liberal Descent: Victorian Historians and the English Past*. Cambridge: Cambridge University Press.

Bynum, William F. 1984. "Charles Lyell's *Antiquity of Man* and its Critics." *JHB*, *17*: 153-187.

Callaway, Charles. 1879. "The Tripartite Division of the Silurian and Cambrian Formations." *Geological Magazine*, 2d decade, *6*: 142-143.

[Campbell, Ina] Dowager Duchess of Argyll. 1906. *George Douglas, Eighth Duke of Argyll, K.G., K.T.: Autobiography and Memoirs*. 2 vols. London: John Murray.

Cannon, Susan Faye. 1978. *Science and Culture: The Early Victorian Period*. New York: Science History Publications.

Cannon, Walter F. 1960a. "The Uniformitarian-Catastrophist Debate." *Isis*, *51*: 38-55.

———. 1960b. "The Problem of Miracles in the 1830's." *Victorian Studies*, *4*: 5-32.

Cannon, W. Faye. 1976. "Charles Lyell, Radical Actualism, and Theory." *BJHS*, *9*: 104-120.

Carlisle, Harvey. 1880. "Adam Sedgwick." *Macmillan's Magazine, 41*: 476-484.

Carver, J. Scott. 1980. "A Reconsideration of Eighteenth-Century Russia's Contributions to European Science." *Canadian-American Slavic Studies, 14*: 389-405.

Challinor, John. 1969. "Geological Research in Cardiganshire, 1842-1967." *Welsh Geological Quarterly, 4*: 3-37.

———. 1971. *The History of British Geology: A Bibliographic Study*. Newton Abbot: David and Charles.

———. 1978. *A Dictionary of Geology*. 5th ed. Cardiff: University of Wales Press.

———. 1979. "History of the Study of the 'Pre-Cambrian' in Wales." In W. O. Kupsch and W.A.S. Sarjeant (eds.), *History of Concepts in Precambrian Geology. Geological Association of Canada Special Paper, 19*, pp. 33-49.

[Chambers, Robert.] 1969. *Vestiges of the Natural History of Creation*. (London: John Churchill, 1844.) Reprint. Leicester: Leicester University Press.

Chancellor, Valarie (ed.). 1969. *Master and Artisan in Victorian England: The Diary of William Andrews and the Autobiography of Joseph Gutteridge*. London: Evelyn, Adams, and Mackay.

Clark, John Willis, and Thomas McKenny Hughes. 1890. *The Life and Letters of the Reverend Adam Sedgwick*. 2 vols. Cambridge: Cambridge University Press.

Clarke, Edward Daniel. 1818. *A Syllabus of Lectures on Mineralogy*. 2d ed. London: T. Cadell.

Clarke, John M. 1923. *James Hall of Albany: Geologist and Palaeontologist, 1811-1898*. Albany: privately printed.

Close, Charles F. 1926. *Early Years of the Ordnance Survey*. Chatham: Institution of Royal Engineers.

Collins, Harry M. 1981a. "The Place of the 'Core-set' in Modern Science: Social Contingency with Methodological Propriety in Science." *History of Science, 19*: 6-19.

Collins, Harry M. (ed.). 1981b. "Knowledge and Controversy: Studies in Modern Natural Science." *Social Studies of Science, 11*: no. 1 (special issue).

Conybeare, William D. 1832. "Inquiry How Far the Theory of M. Elie de Beaumont Concerning the Parallelism of Lines of Elevation of the same Geological Era, is Agreeable to the Phaenomena as exhibited in Great Britain." *Philosophical Magazine*, 3d ser., *1*: 118-126.

———. 1833a. "Report on the Progress, Actual State, and Ulterior Prospects of Geological Science." *1832 British Association Report*, pp. 365-414.

———. 1833b. "On the Application to Great Britain and Ireland of that Part of the Theory of M. Elie de Beaumont, which Asserts that the Lines of Disturbance of the Strata, Assignable to the Same Age, are Parallel." *1832 British Association Report*, pp. 581-583.

[Conybeare, William D.] 1846. Review of R. I. Murchison et al., *The Geology of Russia in Europe. Quarterly Review, 77*: 348-380.

Conybeare, William D., and William Phillips. 1822. *Outlines of the Geology of England and Wales, with an Introductory Compendium of the General Principles of that Science, and Comparative Views of the Structure of Foreign Countries.* London: William Phillips.

Craig, Gordon Y. 1971. "Letters Concerning the Cambrian-Silurian Controversy of 1852." *Journal of the Geological Society of London, 127*: 483-500.

Crosse, Mrs. Andrew [Cornelia]. 1892. *Red-letter Days of My Life.* 2 vols. London: R. Bentley and Son.

Cumming, David. 1985. "John MacCulloch, High Priest, Blackguard and Thief, Reassessed." In A. Wheeler and J. Price (eds.), From *Linnaeus to Darwin*, pp. 77-88. London: Society for the History of Natural History.

Dana, James D. 1890. "Sedgwick and Murchison: Cambrian and Silurian." *American Journal of Science*, 3d ser., *39*: 167-180.

Darwin, Charles. 1839-1843. *The Zoology of the Voyage of H.M.S.* Beagle, *under the Command of Captain Fitzroy, R.N., during the Years 1832 to 1836.* 5 parts. London: Smith, Elder.

———. 1959. *The Origin of Species: A Variorum Text.* Edited by Morse Peckham. Philadephia: University of Pennsylvania Press.

———. 1964. *On the Origin of Species by Means of Natural Selection.* (London: John Murray, 1859.) Reprint. Cambridge, Mass.: Harvard University Press.

———. 1977a. "On the Geology of the Falkland Islands." In P. H. Barrett (ed.), *The Collected Papers of Charles Darwin*, vol. 1, pp. 203-212. Chicago: University of Chicago Press.

———. 1977b. "Geology." In P. H. Barrett (ed.), *The Collected Papers of Charles Darwin*, vol. 1., pp. 226-250. Chicago: University of Chicago Press.

Darwin, Francis (ed.). 1887. *The Life and Letters of Charles Darwin, Including an Autobiographical Chapter.* 3 vols. London: John Murray.

Davidson, Thomas. 1866-1871. *A Monograph of the British Fossil Brachiopoda, Part VII: The Silurian Brachiopoda.* London: Palaeontographical Society.

Davies, D. C. 1872. "The Names Cambrian and Silurian in Geology." *Nature, 6*: 222.

Davies, Gordon L. 1969. *The Earth in Decay: A History of British Geomorphology.* New York: Elsevier.

Davis, John E. 1846. "On the Geology of the Neighbourhood of Tremadoc, Caernarvonshire." *QJGS, 2*: 70-75.

[Dawkins, W. Boyd.] 1875. Review of A. Geikie, *Life of Sir Roderick Impey Murchison. Edinburgh Review, 142*: 173-203.

Dean, Dennis R. 1968. "Geology and English Literature: Crosscurrents, 1770-1830." Ph.D. thesis, University of Wisconsin.

———. 1981. " 'Through Science to Despair': Geology and the Victorians." In J. Paradis and T. Postlewait (eds.), *Victorian Science and Victorian Values*, pp. 111-136. New York: New York Academy of Science.

Dean, John. 1979. "Controversy over Classification: A Case Study from the

History of Botany." In B. Barnes and S. Shapin (eds.), *Natural Order*, pp. 211-230. Beverly Hills: Sage.

De Beer, Gavin (ed.). 1974. *Charles Darwin, Thomas Henry Huxley: Autobiographies*. London: Oxford University Press.

De la Beche, Henry T. 1831. *A Geological Manual*. London: Treuttel and Würtz.

———. 1832. *Handbuch der Geognosie*. Translated by Heinrich von Dechen. Berlin: Duncker und Humblot.

———. 1833. *A Geological Manual*. 3d ed. London: Charles Knight.

———. 1839. *Report on the Geology of Cornwall, Devon, and West Somerset*. London: Longman.

———. 1846. "On the Formation of the Rocks of South Wales and South Western England." *MGS, 1*: 1-296.

Desmond, Adrian. 1979. "Designing the Dinosaur: Richard Owen's Response to Robert Edmond Grant." *Isis, 70*: 224-234.

———. 1982. *Archetypes and Ancestors: Palaeontology in Victorian London, 1850-1875*. London: Blond and Briggs.

———. 1984a. "Interpreting the Origin of Mammals: New Approaches to the History of Palaeontology." *Zoological Journal of the Linnean Society, 82*: 7-16.

———. 1984b. "Robert E. Grant's Later Views on Organic Development: The Swiney Lectures on 'Palaeozoology,' 1853-1857," *Archives of Natural History, 11*: 395-413.

Dolby, R.G.A. 1976. "Debates over the Theory of Solution: A Study of Dissent in Physical Chemistry in the English-Speaking World in the Late Nineteenth and Early Twentieth Centuries." *Historical Studies in the Physical Sciences, 7*: 297-404.

Donovan, D. T. 1966. *Stratigraphy: An Introduction to Principles*. London: Thomas Murby.

Dott, Robert H., and R. L. Batten. 1976. *Evolution of the Earth*. 2d ed. New York: McGraw-Hill.

Douglas, Mary. 1973. *Natural Symbols: Explorations in Cosmology*. Harmondsworth: Penguin.

Dufrénoy, Ours-Pierre-Armand, and Léonce Élie de Beaumont. 1841. *Explication de la Carte Géologique de la France*, vol. 1. Paris: Imprimerie Royale.

Durkheim, Émile, and Marcel Mauss. 1963. *Primitive Classification*. Translated with an introduction by Rodney Needham. London: Cohen and West.

Egerton, Frank. 1970. "Refutation and Conjecture: Darwin's Response to Sedgwick's Attack on Chambers." *Studies in the History and Philosophy of Science, 1*: 176-183.

Élie de Beaumont, Léonce. 1829-1830. "Recherches sur quelques-unes des révolutions de la surface du globe." *Annales des Sciences Naturelles, 18*: 5-25, 284-416; *19*: 5-99, 177-240.

———. 1831. "Researches on Some of the Revolutions which have Taken

Place on the Surface of the Globe." *Philosophical Magazine*, 2d ser., *10*: 241-264.

———. 1852. *Notice sur les Systèmes de Montagnes.* 3 vols. Paris: Arthus Bertrand.

Ellenberger, François. 1978. "The First International Geological Congress: Paris, 1878." *Episodes: Geological Newsletter, International Union of Geological Sciences*, no. 2: 20-24.

Elles, Gertrude L. 1922. "The Bala Country: Its Structure and Rock-Succession." *QJGS, 78*: 132-172.

Elles, Gertrude L., and Ethel M. R. Wood. 1901-1918. *A Monograph of British Graptolites*. Edited by Charles Lapworth. London: Palaeontographical Society.

Eyles, Joan M. 1969. "William Smith: Some Aspects of his Life and Work." In C. J. Schneer (ed.), *Toward a History of Geology*, pp. 142-158. Cambridge, Mass.: M.I.T. Press.

———. 1978. "G. W. Featherstonhaugh (1780-1866), F.R.S., F.G.S., Geologist and Traveller." *JSBNH, 8*: 381-395.

Eyles, Victor A. 1937. "John MacCulloch, F.R.S., and his Geological Map: An Account of the First Geological Survey of Scotland." *Annals of Science*, 2: 114-129.

———. 1971. "Roderick Murchison, Geologist and Promoter of Science." *Nature, 234*: 387-389.

Farber, Paul L. 1976. "The Type Concept in Zoology During the First Half of the Nineteenth Century." *JHB, 9*: 93-119.

———. 1982a. "The Transformation of Natural History in the Nineteenth Century." *JHB, 15*: 145-152.

———. 1982b. *The Emergence of Ornithology as a Scientific Discipline: 1760-1850*. Dordrecht: D. Reidel.

Fitton, William H. 1828. [Presidential Address to the Geological Society of London, 15 Feb. 1828.] *PGS, 1*: 50-62.

———. 1829. [Presidential Address to the Geological Society of London, 20 Feb. 1829.] *PGS, 1*: 112-134.

———. 1836. "Observations on Some of the Strata Between the Chalk and the Oxford Oolite, in the South-east of England." *TGS*, 2d ser., *4*: 103-388*.

[Fitton, William H.] 1841. Review of R. I. Murchison, *The Silurian System. Edinburgh Review, 73*: 1-41.

Fleck, Ludwick. 1979. *Genesis and Development of a Scientific Fact*. Edited by T. J. Trenn and R. K. Merton. Chicago: University of Chicago Press.

Flett, John S. 1937. *The First Hundred Years of the Geological Survey of Great Britain*. London: H. M. Stationery Office.

[Forbes, Edward.] 1846. "Russian Geology." *Literary Gazette*, no. 1516: 122-124.

Forbes, Edward. 1848. "On Oldhamia, a New Genus of Silurian Fossils." *Journal of the Dublin Geological Society*, *4*: 20.

[Forbes, Edward.] 1852a. "The Poetry of Geology." *Literary Gazette*, no. 1835: 279; 1840: 370.

[Forbes, Edward.] 1852b. "The Future of Geology." *Westminster Review*, n. s., *2*: 67-94.

Forbes, Edward. 1854a. "Anniversary Address of the President." *QJGS, 10*: xxii-lxxxi.

[Forbes, Edward.] 1854b. Review of R. I. Murchison, *Siluria*. *Quarterly Review, 95*: 363-394.

[Forbes, Edward.] 1854c. Review of R. I. Murchison, *Siluria*. *Literary Gazette*, no. 1953: 581.

Foucault, Michel. 1970. *The Order of Things: An Archaeology of the Human Sciences*. London: Tavistock.

Gage, Andrew Thomas. 1938. *A History of the Linnean Society of London*. London: Taylor and Francis.

Garland, Martha McMackin. 1980. *Cambridge Before Darwin: The Ideal of a Liberal Education, 1800-1860*. Cambridge: Cambridge University Press.

Geertz, Clifford. 1973. *The Interpretation of Cultures*. New York: Basic Books.

Geikie, Archibald. 1871. "Sir Roderick Murchison." *Nature, 5*: 10-12.

———. 1875. *Life of Sir Roderick I. Murchison, Based on his Journals and Letters, with Notices of his Scientific Contemporaries and a Sketch of the Rise and Growth of Palaeozoic Geology in Britain*. 2 vols. London: John Murray.

———. 1882. *Text-book of Geology*. London: Macmillan.

———. 1895. *Memoir of Sir Andrew Crombie Ramsay*. London: Macmillan.

———. 1904. "Anniversary Address of the President." *QJGS, 60*: xlix-civ.

———. 1924. *A Long Life's Work: An Autobiography*. London: Macmillan.

Geison, Gerald L. 1978. *Michael Foster and the Cambridge School of Physiology: The Scientific Enterprise in Late Victorian Society*. Princeton: Princeton University Press.

Gerstner, Patsy A. 1979. "Henry Darwin Rogers and William Barton Rogers on the Nomenclature of the American Palaeozoic Rocks." In C. J. Schneer (ed.), *Two Hundred Years of Geology in America*, pp. 175-186. Hanover, N. H.: University Press of New England.

Gilbert, E. W., and A. Goudie. 1971. "Sir Roderick Impey Murchison, Bart., K.C.B., 1792-1871." *Geographical Journal, 137*: 505-511.

Gillespie, Neal C. 1979. *Charles Darwin and the Problem of Creation*. Chicago: University of Chicago Press.

Gillispie, Charles C. 1959. *Genesis and Geology: A Study in the Relations of Scientific Thought, Natural Theology, and Social Opinion in Great Britain, 1790-1850*. New York: Harper and Row.

———. 1981. *Science and Polity in France at the End of the Old Regime*. Princeton: Princeton University Press.

Gombrich, Ernst H. 1977. *Art and Illusion: A Study in the Psychology of Pictorial Representation*. London: Phaidon Press.

Goodchild, J. G. 1882. "Professor Robert Harkness, F.R.S., F.G.S." *Trans-*

actions of the Cumberland Association for the Advancement of Literature and Science, no. 8: 145-188.

Gooding, David. 1982. "Empiricism in Practice: Teleology, Economy, and Observation in Faraday's Physics." *Isis, 73*: 46-67.

Gordon, Elizabeth O. 1894. *The Life and Correspondence of William Buckland, D.D., F.R.S.* London: John Murray.

Greene, John C. 1959. *The Death of Adam: Evolution and its Impact on Western Thought.* Ames, Iowa: Iowa State University Press.

Greene, Mott T. 1982. *Geology in the Nineteenth Century: Changing Views of a Changing World.* Ithaca: Cornell University Press.

Greenly, Edward, and Howel Williams. 1930. *Methods in Geological Surveying.* London: Thomas Murby.

Greenough, George B. 1819. *A Critical Examination of the First Principles of Geology.* London: Longman.

———. 1820. *A Geological Map of England and Wales.* London: Longman.

———. 1840. *A Geological Map of England and Wales.* London: Geological Society.

———. 1841. "Address to the Royal Geographical Society of London; Delivered at the Anniversary Meeting on the 25th May, 1840." *Journal of the Royal Geographical Society of London, 10*: xliii-lxxxiii.

Guerlac, Henry. 1976. "Chemistry as a Branch of Physics: Laplace's Collaboration with Lavoisier." *Historical Studies in the Physical Sciences, 7*: 193-276.

Hall, A. Rupert. 1969. *The Cambridge Philosophical Society: A History, 1819-1969.* Cambridge: Cambridge Philosophical Society.

———. 1980. *Philosophers at War: The Quarrel between Newton and Leibniz.* Cambridge: Cambridge University Press.

———. 1983. "On Whiggism." *History of Science, 21*: 45-59.

Hall, James. 1847. *Palaeontology of New-York. Vol. 1, Containing Descriptions of the Organic Remains of the New-York System (Equivalent of the Lower Silurian Rocks of Europe).* Albany: printed by C. Van Benthuysen.

Hallam, Anthony. 1983. *Great Geological Controversies.* Oxford: Oxford University Press.

Hamilton, Beryl M. 1982. "The Influence of the Polarizing Microscope on Late Nineteenth Century Geology." *Janus, 69*: 51-68.

Hamilton, William J. 1856. "Anniversary Address of the President." *QJGS, 12*: xliii-cxix.

Hardy, Phillip Dixon (ed.). 1835. *Proceedings of the Fifth Meeting of the British Association for the Advancement of Science, Held in Dublin.* Dublin: printed by Phillip Dixon Hardy.

Harrison, J. F. C. 1979. *Early Victorian Britain, 1832-1851.* London: Fontana.

Harrison, J. M. 1963. "Nature and Significance of Geological Maps." In C. C. Albritton (ed.), *The Fabric of Geology*, pp. 225-232. Stanford, Calif.: Freeman, Cooper, and Co.

Hendrickson, Walter B. 1943. *David Dale Owen: Pioneer Geologist of the Middle West.* Indianapolis: Indiana Historical Bureau.

Henslow, John Stevens. 1822. "Geological Description of Anglesea." *Transactions of the Cambridge Philosophical Society, 1*: 359-452.

Herbert, Sandra. 1977. "The Place of Man in the Development of Darwin's Theory of Transmutation: Part II." *JHB, 10*: 155-227.

Herries Davies, Gordon L. 1983. *Sheets of Many Colours: The Mapping of Ireland's Rocks, 1750-1890*. Dublin: Royal Dublin Society.

Hesse, Mary. 1980. "The Strong Thesis of Sociology of Science." In *idem, Revolutions and Reconstructions in the Philosophy of Science*, pp. 29-60. Brighton: Harvester.

Heyck, T. W. 1980. "From Men of Letters to Intellectuals: The Transformation of Intellectual Life in Nineteenth-century England." *Journal of British Studies, 20*: 158-183.

Hicks, Henry. 1874. "On the Classification of the Cambrian and Silurian Rocks." *Proceedings of the Geologists' Association, 3*: 99-105.

Hicks, Henry, and John W. Salter. 1866. "Report on Further Researches in the Lingula-flags of South Wales. With Some Notes on the Sections and Fossils, by John W. Salter." *1865 British Association Report*, pt. i, pp. 281-286.

———. 1867. "Second Report on the 'Menevian Group' and the other Formations at St. David's, Pembrokeshire." *1866 British Association Report*, pt. i, pp. 182-186.

Hobsbawm, Eric. 1980. "The Revival of Narrative: Some Comments." *Past and Present*, no. 86: 3-8.

Hobsbawm, Eric, and Terence Ranger. 1983. *The Invention of Tradition*. Cambridge: Cambridge University Press.

Holland, C. H. 1974. "The Lower Palaeozoic Systems: An Introduction." In C. H. Holland (ed.), *Cambrian of the British Isles, Norden and Spitzbergen*, pp. 1-13. London: J. Wiley.

Holmes, Frederic L. 1974. *Claude Bernard and Animal Chemistry: The Emergence of a Scientist*. Cambridge: Harvard University Press.

———. 1981. "The Fine Structure of Scientific Creativity." *History of Science, 19*: 60-70.

Hopkins, William. 1853. "Anniversary Address of the President." *QJGS, 9*: xxii-xcii.

Horner, Leonard. 1846. "Anniversary Address of the President." *QJGS, 2*: 145-221.

———. 1847. "Anniversary Address of the President." *QJGS, 3*: xxiii-xc.

Horný, R. 1980. "Joachim Barrande (1799-1883), Life, Work, Collections." *JSBNH, 9*: 365-368.

Houghton, Walter E. 1957. *The Victorian Frame of Mind, 1830-1870*. New Haven: Yale University Press.

Houghton, Walter E. (ed). 1966–. *The Wellesley Index to Victorian Periodicals, 1824-1900*. Toronto: University of Toronto Press.

Hull, David L. 1973. *Darwin and his Critics: The Reception of Darwin's Theory of Evolution by the Scientific Community*. Cambridge, Mass.: Harvard University Press.

Hunt, Thomas Sterry. 1872a. "History of the Names Cambrian and Silurian in Geology." *Canadian Naturalist*, n.s., *6*: 281-312, 417-448.

———. 1872b. "History of the Names Cambrian and Silurian in Geology." *Nature, 6*: 15-17, 34-37, 53-55.

———. 1873. "History of the Names Cambrian and Silurian in Geology." *Geological Magazine, 10*: 385-395, 453-461, 504-510, 561-566.

———. 1878. "History of the Names Cambrian and Silurian in Geology." In *Chemical and Geological Essays*, 2d ed., pp. 349-425. Salem: S. E. Cassino.

Huxley, Leonard (ed.). 1918. *Life and Letters of Sir Joseph Dalton Hooker*. 2 vols. London: John Murray.

Inkster, Ian, and Jack Morrell (eds.). 1983. *Metropolis and Province: Science and British Culture, 1780-1850*. London: Hutchinson.

Itzkowitz, David C. 1977. *Peculiar Privilege: A Social History of English Foxhunting, 1753-1885*. Hassocks, Sussex: Harvester Press.

Johnson, Markes E. 1977. "Early Geological Explorations of the Silurian System in Iowa." *Proceedings of the Iowa Academy of Sciences, 84*: 150-156.

Jordanova, Ludmilla J., and Roy S. Porter (eds.). 1979. *Images of the Earth: Essays in the History of the Environmental Sciences*. Chalfont St. Giles: British Society for the History of Science.

Jukes, J. Beete. 1853. "On the Occurrence of Caradoc Sandstone at Great Barr, South Staffordshire." *QJGS, 9*: 179-181.

———. 1854. "Annual Address Delivered before the Geological Society of Dublin, February 8, 1854." *Journal of the Geological Society of Dublin, 6*: 61-108.

———. 1857. *The Student's Manual of Geology*. Edinburgh: Adam and Charles Black.

———. 1867. *Additional Notes on the Grouping of the Rocks of North Devon and West Somerset; with a Map and Section*. Dublin: Webb and Son.

Jukes, J. Beete, and Alfred R. Selwyn. 1848. "Sketch of the Structure of the Country Extending from Cader Idris to Moel Siabod, North Wales." *QJGS, 4*: 300-302.

Knorr, Karin D., Roger Krohn, and Richard Whitley (eds.). 1980. *The Social Process of Scientific Investigation*. Dordrecht: D. Reidel.

Kragh, Helge. 1980. "Anatomy of a Priority Conflict: The Case of Element 72." *Centaurus, 23*: 275-301.

Kubie, Lawrence S. 1953-1954. "Some Unsolved Problems of the Scientific Career." *American Scientist, 41*: 596-613; *42*: 104-112.

Kuhn, Thomas S. 1970. *The Structure of Scientific Revolutions*. 2d ed. Chicago: University of Chicago Press.

Lakatos, Imre. 1970. "Falsification and the Methodology of Scientific Research Programmes." In I. Lakatos and A. Musgrave (eds.), *Criticism and the Growth of Knowledge*, pp. 91-196. Cambridge: Cambridge University Press.

Lapworth, Charles. 1876. "On Scottish Monograptidae." *Geological Magazine*, 2d decade, *3*: 308-321, 350-360, 499-507, 544-552.

———. 1878. "The Moffat Series." *QGJS, 34*: 240-346.

———. 1879. "On the Tripartite Classification of the Lower Palaeozoic Rocks." *Geological Magazine*, n.s., 2d decade, *6*: 1-15.

Latour, Bruno, and Steve Woolgar. 1979. *Laboratory Life: The Social Construction of Scientific Facts*. Beverly Hills: Sage.

[Laudan], Rachel Bush. 1974. "The Development of Geological Mapping in Britain from 1795 to 1825." Ph.D. thesis, University of London.

Laudan, Rachel. 1976. "William Smith: Stratigraphy without Palaeontology." *Centaurus, 20*: 210-226.

———. 1977. "Ideas and Organizations in British Geology: A Case Study in Institutional History." *Isis, 68*: 527-538.

Law, John, and R. J. Williams. 1982. "Putting Facts Together: A Study of Scientific Persuasion." *Social Studies of Science, 12*: 535-558.

Lawrence, Phillip. 1978. "Charles Lyell Versus the Theory of Central Heat." *JHB, 11*: 101-128.

Logan, William E. 1861. "Remarks on the Fauna of the Quebec Group of Rocks, and the Primordial Zone of Canada." *The Canadian Journal of Industry, Science, and Art*, n.s., *6*: 40-46.

Lyell, Charles. 1830-1833. *Principles of Geology, Being an Attempt to Explain the Former Changes of the Earth's Surface, By Reference to Causes Now in Operation*. 3 vols. London: John Murray.

———. 1835. *Principles of Geology: Being an Inquiry How Far the Former Changes of the Earth's Surface are Referable to Causes Now in Operation*. 4th ed., 4 vols. London: John Murray.

———. 1838. *Elements of Geology*. London: John Murray.

———. 1841. *Elements of Geology*. 2d ed., 2 vols. London: John Murray.

———. 1845. *Travels in North America; With Geological Observations on the United States, Canada, and Nova Scotia*. 2 vols. London: John Murray.

———. 1849. *A Second Visit to the United States of North America*. 2 vols. New York: Harper and Brothers.

———. 1851. *A Manual of Elementary Geology: Or, the Ancient Changes of the Earth and its Inhabitants as Illustrated by Geological Monuments*. 3d ed. London: John Murray.

———. 1855. *A Manual of Elementary Geology: Or, the Ancient Changes of the Earth and its Inhabitants as Illustrated by Geological Monuments*. 5th ed. London: John Murray.

———. 1865. *Elements of Geology: Or the Ancient Changes of the Earth and its Inhabitants as Illustrated by Geological Monuments*. 6th ed. London: John Murray.

———. 1871. *The Student's Elements of Geology*. London: John Murray.

Lyell, Katherine M. (ed.). 1881. *Life, Letters and Journals of Sir Charles Lyell, Bart*. 2 vols. London: John Murray.

Lyell, Katherine M. (ed.). 1890. *Memoir of Leonard Horner, F.R.S., F.G.S., Consisting of Letters to his Family and from some of his Friends*. 2 vols. London: Women's Printing Society.

Lyons, Henry. 1944. *The Royal Society 1660-1940: A History of its Administration under its Charters*. Cambridge: Cambridge University Press.

McCartney, Paul J. 1977. *Henry De la Beche: Observations on an Observer*. Cardiff: Friends of the National Museum of Wales.

[Mackie, Samuel J.?] 1859. Review of R. I. Murchison, *Siluria*. *The Geologist*, *2*: 88-90.

MacLeod, Roy M. 1971. "Of Medals and Men: A Reward System in Victorian Science." *Notes and Records of the Royal Society of London*, *26*: 81-105.

MacLeod, Roy M., and Peter Collins (eds.). 1981. *The Parliament of Science: The British Association for the Advancement of Science, 1831-1981*. Northwood, Middlesex: Science Reviews.

Marcou, Jules. 1978. *Jules Marcou on the Taconic System in North America*. Edited by Hubert C. Skinner. New York: Arno.

Marr, John E. 1883. *The Classification of the Cambrian and Silurian Rocks: Being the Sedgwick Prize Essay for the Year 1882*. Cambridge: Deighton Bell.

Marshall, John G. 1840. "Description of a Section Across the Silurian Rocks in Westmoreland, from the Shap Granite to Casterton Fell." *1839 British Association Report*, pt. ii, p. 67.

Marston, V. Paul. 1984. "Science and Theology in the Work of Adam Sedgwick." Ph. D. thesis, Open University.

Mauskopf, Seymour H. 1976. "Crystals and Compounds: Molecular Structure and Composition in Nineteenth-Century French Science." *Transactions of the American Philosophical Society*. n.s., *66*: 1-82.

Mayr, Ernst. 1982. *The Growth of Biological Thought: Diversity, Evolution, and Inheritance*. Cambridge, Mass.: Harvard University Press.

Medawar, Peter. 1969. *The Art of the Soluble*. Harmondsworth: Penguin.

Merrill, George P. 1924. *The First One Hundred Years of American Geology*. New Haven: Yale University Press.

Merton, Robert K. 1973. *The Sociology of Science: Theoretical and Empirical Investigations*. Edited by Norman W. Storer. Chicago: University of Chicago Press.

Miller, David Philip. 1984. "The Social History of British Science: After the Harvest?" *Social Studies of Science*, *14*: 115-135.

Millhauser, Milton. 1954. "The Scriptural Geologists: An Episode in the History of Opinion." *Osiris*, *11*: 65-86.

———. 1959. *Just Before Darwin: Robert Chambers and Vestiges*. Middletown, Conn.: Wesleyan University Press.

Mills, Eric L. 1978. "Edward Forbes, John Gwyn Jeffreys, and British Dredging Before the Challenger Expedition." *JSBNH*, *8*: 507-536.

———. 1984. "A View of Edward Forbes, Naturalist." *Archives of Natural History*, *11*: 365-393.

Morrell, Jack B. 1971. "Individualism and the Structure of British Science in 1830." *Historical Studies in the Physical Sciences*, *3*: 183-204.

———. 1976. "London Institutions and Lyell's Career: 1820-1841." *BJHS*, *9*: 132-146.

———. 1984. Review of N. A. Rupke, *The Great Chain of History*. *Times Literary Supplement*, no. 4218: 106.

Morrell, Jack B., and Arnold Thackray. 1981. *Gentlemen of Science: Early Years of the British Association for the Advancement of Science*. Oxford: Clarendon Press.

———. 1984. *Gentlemen of Science: Early Correspondence of the British Association for the Advancement of Science*. London: Royal Historical Society.

Moxon, Charles. 1842a. "Sketch of the Progress of Geology &c., in 1842." *Geologist, 1*: v-xii.

———. 1842b. "Monthly Notice." *Geologist, 1*: 129-132.

Murchison, Roderick Impey. 1829a. "Geological Sketch of the North Western Extremity of Sussex, and the Adjoining Parts of Hants and Surrey." *TGS*, 2d ser., *2*: 97-107.

———. 1829b. "On the Coal-field of Brora in Sutherlandshire, and some other Stratified Deposits in the North of Scotland." *TGS*, 2d ser., *2*: 293-326.

———. 1833a. "Address to the Geological Society, Delivered on the Evening of the 15th of February 1833." *PGS, 1*: 438-464.

———. 1833b. "On the Sedimentary Deposits which Occupy the Western Parts of Shropshire and Herefordshire, and are Prolonged from N.E. to S.W. through Radnor, Brecknock, and Caermarthenshires, With Descriptions of the Accompanying Rocks of Intrusive or Igneous Characters." *PGS 1*: 470-477.

———. 1834a. "On the Structure and Classification of the Transition Rocks of Shropshire, Herefordshire, and Part of Wales, and on the Lines of Disturbance which have Affected that Series of Deposits, Including the Valley of Elevation at Woolhope." *PGS, 2*: 13-18.

———. 1834b. "On the Upper Greywacke Series of England and Wales." *Edinburgh New Philosophical Journal*, 1st ser., *17*: 365-368.

———. 1834c. *Outline of the Geology of the Neighbourhood of Cheltenham*. Cheltenham: Henry Davies, Montpellier Library.

———. 1835. "On the Silurian System of Rocks." *Philosophical Magazine*, 3d ser., *7*: 46-53.

———. 1836. "On the Geological Structure of Pembrokeshire, More Particularly on the Extension of the Silurian System of Rocks in the Coast Cliffs of the County." *PGS, 2*: 226-230.

———. 1839. *The Silurian System, Founded on a Series of Geological Researches in the Counties of Salop, Hereford, Radnor, Montgomery, Caermarthen, Brecon, Pembroke, Monmouth, Gloucester, Worcester, and Stafford; With Descriptions of the Coal-fields and Overlying Formations*. London: John Murray.

———. 1842. "Anniversary Address of the President." *PGS, 3*: 637-687.

———. 1843a. "Anniversary Address of the President." *PGS, 4*: 65-151.

———. 1843b. *Geological Map of England and Wales*. London: Society for the Diffusion of Useful Knowledge.

———. 1844. "Address to the Royal Geographical Society of London; De-

livered at the Anniversary Meeting on the 27th May, 1844." *Journal of the Royal Geographical Society of London, 14*: xlv-cxxviii.
———. 1845. "On the Palaeozoic Deposits of Scandinavia and the Baltic Provinces of Russia, and their Relations to Azoic or More Ancient Crystalline Rocks." *QJGS, 1*: 467-494.
———. 1847a. "On the Silurian and Associated Rocks in Dalecarlia, and on the Succession from Lower to Upper Silurian in Smoland, Öland, and Gothland, and in Scania." *QJGS, 3*: 1-46.
———. 1847b. "On the Meaning Originally Attached to the Term 'Cambrian System,' and on the Evidences since Obtained of its being Geologically Synonymous with the Previously Established Term 'Lower Silurian.' " *QJGS, 3*: 165-179.
[Murchison, Roderick Impey]. 1850. "Siberia and California." *Quarterly Review, 87*: 395-434.
Murchison, Roderick Impey. 1851. "On the Silurian Rocks of the South of Scotland." *QJGS, 7*: 139-178.
———. 1852a. "The Silurian System." *Literary Gazette*, no. 1835: 278-279.
———. 1852b. "The Silurian System." *Literary Gazette*, no. 1840: 369-370.
———. 1852c. "On the Meaning of the Term 'Silurian System' as Adopted by Geologists in Various Countries during the Last Ten Years." *QJGS, 8*: 173-184.
———. 1854. *Siluria. The History of the Oldest Known Rocks Containing Organic Remains, with a Brief Sketch of the Distribution of Gold over the Earth*. London: John Murray.
———. 1859. *Siluria. The History of the Oldest Fossiliferous Rocks and their Foundations; with a Brief Sketch of the Distribution of Gold over the Earth*. 3d ed. London: John Murray.
———. 1862. "Thirty Years Retrospect of the Progress in our Knowledge of the Geology of the Older Rocks." *American Journal of Science and Arts*, 2d ser., *33*: 1-21.
———. 1866. "Introduction. Observations on the Classification of the Silurian Rocks; Being an Introduction to the Brachiopoda of that System." In Thomas Davidson, *A Monograph of the British Fossil Brachiopoda. Part VII: The Silurian Brachiopoda*, pp. 19-31. London: Palaeontographical Society.
———. 1867. *Siluria. A History of the Oldest Rocks in the British Isles and other Countries; with Sketches of the Origin and Distribution of Native Gold, the General Succession of Geological Formations, and Changes of the Earth's Surface*. 4th ed. London, John Murray.
Murchison, Roderick Impey, Édouard de Verneuil, and Count Alexander von Keyserling. 1845. *The Geology of Russia in Europe and the Ural Mountains*. Vol. 1. London: John Murray; *Géologie de la Russie d'Europe et des Montagnes d'Oural*. Vol. 2. Paris: Bertrand.
Needham, Raymond, and Alexander Webster. 1905. *Somerset House, Past and Present*. London: T. Fisher Unwin.
Neve, Michael. 1980. "The Naturalization of Science." *Social Studies of Science, 10*: 375-391.

———. 1983. "Science in a Commercial City: Bristol 1820-60." In I. Inkster and J. Morrell (eds.), *Metropolis and Province,* pp. 179-204. London: Hutchinson.

Neve, Michael, and Roy Porter. 1977. "Alexander Catcott: Glory and Geology." *BJHS, 10*: 37-60.

Nicolson, Marjorie Hope. 1959. *Mountain Gloom and Mountain Glory: The Development of the Aesthetics of the Infinite.* Ithaca: Cornell University Press.

North, Frederick J. 1932. "From the Geological Map to the Geological Survey: Glamorgan and the Pioneers of Geology." *Transactions of the Cardiff Naturalists' Society, 65*: 45-115.

———. 1934. "Further Chapters in the History of Geology in South Wales." *Transactions of the Cardiff Naturalists' Society, 67*: 31-103.

O'Brien, Charles F. 1970. "Eozöon Canadense, the Dawn Animal of Canada." *Isis, 61*: 206-223.

O'Connor, J. G., and A. J. Meadows. 1976. "Specialization and Professionalization in British Geology." *Social Studies of Science, 6*: 77- 89.

Oldroyd, David R. 1972a. "Nineteenth Century Controversies Concerning the Mesozoic/Tertiary Boundary in New Zealand." *Annals of Science, 29*: 39-57.

———. 1972b. "Edward Daniel Clarke, 1769-1822, and his Rôle in the History of the Blow-pipe." *Annals of Science, 29*: 213-235.

———. 1979. "Historicism and the Rise of Historical Geology." *History of Science, 17*: 191-213, 227-257.

———. 1980. "Sir Archibald Geikie (1837-1924), Geologist, Romantic Aesthete, and Historian of Geology: The Problem of Whig Historiography of Science." *Annals of Science, 37*: 441-462.

D'Omalius D'Halloy, J. J. 1843. *Précis Elementaire de Géologie.* Paris: Arthus Bertrand.

D'Orbigny, Alcide. 1849-1852. *Cours Élementaire de Paléontologie et de Géologie Stratigraphiques.* 2 vols. Paris: Victor Masson.

Ospovat, Alexander M. 1969. "Reflections on A. G. Werner's 'Kurze Klassifikation.' " In C. J. Schneer (ed.), *Toward a History of Geology,* pp. 245-256. Cambridge, Mass.: M.I.T. Press.

———. 1976. "The Distortion of Werner in Lyell's *Principles of Geology.*" *BJHS, 9*: 190-198.

Ospovat, Dov. 1981. *The Development of Darwin's Theory: Natural History, Natural Theology, and Natural Selection, 1838-1859.* Cambridge: Cambridge University Press.

Outram, Dorinda. 1976. "Scientific Biography and the Case of Georges Cuvier: With a Critical Bibliography." *History of Science, 14*: 101-137.

———. 1978. "The Language of Natural Power: The '*Eloges*' of Georges Cuvier and the Public Language of Nineteenth Century Science." *History of Science, 16*: 153-178.

[Owen, David Dale.] 1844. "Review of the New York Geological Reports." *American Journal of Science and Arts, 46*: 144-157.

[Owen, David Dale.] 1852. *Report of a Geological Survey of Wisconsin, Iowa, and Minnesota; and Incidentally of a Portion of Nebraska Territory*. 2 vols. Philadelphia: Lippincott.

Owens, Robert M. 1971. "In Search of Welsh Trilobites." *Amgueddfa, Bulletin of the National Museum of Wales*, no. 9.

Page, Leroy E. 1969. "Diluvialism and its Critics in Great Britain in the Early Nineteenth Century." In C. J Schneer (ed.), *Toward a History of Geology*, pp. 257-271. Cambridge, Mass.: M.I.T. Press.

———. 1976. "The Rivalry Between Charles Lyell and Roderick Murchison." *BJHS*, *9*: 156-165.

Pantin, Carl F. A. 1968. *The Relations Between the Sciences*. Cambridge: Cambridge University Press.

Pattison, S. R. [n.d.] *Adam Sedgwick*. London: The Religious Tract Society.

Pengelly, Hester. 1897. *A Memoir of William Pengelly, of Torquay, F.R.S., Geologist, with a Selection from his Correspondence*. London: John Murray.

Perkin, Harold. 1969. *The Origins of Modern English Society, 1780-1880*. London: Routledge and Kegan Paul.

Phillips, John. 1836. *A Guide to Geology*. 3d ed. London: Longman.

[Phillips, John.] 1840a. "Organic Remains." *Penny Cyclopaedia*, *16*: 487-491.

[Phillips, John.] 1840b. "Palaeozoic Series." *Penny Cyclopaedia*, *17*: 153-154.

Phillips, John. 1841. *Figures and Descriptions of the Palaeozoic Fossils of Cornwall, Devon, and West Somerset; Observed in the Course of the Ordnance Survey of that District*. London: Longman.

———. 1848. "The Malvern Hills Compared with the Palaeozoic Districts of Abberley, &c." *MGS*, 2: 1-386.

———. 1855. *Manual of Geology, Practical and Theoretical*. 2d ed. London: Richard Griffin.

———. 1857. "Report on Cleavage and Foliation in Rocks, and on the Theoretical Explanations of these Phaenomena.—Part I." *1856 British Association Report*, pt. i, pp. 369-396.

———. 1873. "Sedgwick." *Nature*, *7*: 257-259.

Phillips, William, and Samuel Woods. 1822. "Sketch of the Geology of Snowdon, and the Surrounding Country." *Annals of Philosophy*, n.s., 4: 321-335, 401-424.

Pinch, Trevor J. 1980. "Theoreticians and the Production of Experimental Anomaly: The Case of Solar Neutrinos." In K. D. Knorr, R. Krohn, and R. Whitley (eds.), *The Social Process of Scientific Investigation*, pp. 77-106. Dordrecht: D. Reidel.

Porter, Roy. 1973. "The Industrial Revolution and the Rise of the Science of Geology." In M. Teich and R. Young (eds.), *Changing Perspectives in the History of Science*, pp. 320-343. London: Heinemann.

———. 1976. "Charles Lyell and the Principles of the History of Geology." *BJHS*, *9*: 91-103.

———. 1977. *The Making of Geology: Earth Science in Britain 1660-1815*. Cambridge: Cambridge University Press.

———. 1978. "Gentlemen and Geology: The Emergence of a Scientific Career, 1660-1920." *Historical Journal, 21*: 809-836.

———. 1979. "Creation and Credence: The Career of Theories of the Earth in Britain, 1660-1820." In B. Barnes and S. Shapin (eds.), *Natural Order*, pp. 97-123. Beverly Hills: Sage.

———. 1982a. "The Natural Science Tripos and the 'Cambridge School of Geology,' 1850-1914." *History of Universities, 2*: 193-216.

———. 1982b. "Charles Lyell: The Public and Private Faces of Science." *Janus, 69*: 29-50.

———. 1983. *The Earth Sciences: An Annotated Bibliography*. New York: Garland.

Portlock, Joseph E. 1843. *Report on the Geology of the County of Londonderry, and of Parts of Tyrone and Fermanagh*. Dublin: Andrew Milliken.

———. 1857. "The Anniversary Address of the President." *QJGS, 13*: xxvi-cxlv.

Ramsay, Andrew C. 1846. "On the Denudation of South Wales and the Adjacent Counties of England." *MGS, 1*: 297-335.

———. 1853. "On the Physical Structure and Succession of Some of the Lower Palaeozoic Rocks of North Wales and Part of Shropshire." *QJGS, 9*: 161-176.

———. 1863. "The Anniversary Address of the President." *QJGS, 19*: xxxvi-lii.

———. 1866. "The Geology of North Wales." *MGS, 3*: 1-238.

———. 1878. *The Physical Geology and Geography of Great Britain: A Manual of British Geology*. 5th ed. London: Edward Stanford.

Ramsay, Andrew C., and William Talbot Aveline. 1848. "Sketch of the Structure of Parts of North and South Wales." *QJGS, 4*: 294-299.

Rappaport, Rhoda. 1982. "Borrowed Words: Problems of Vocabulary in Eighteenth-century Geology." *BJHS, 15*: 27-44.

Ravetz, Jerome R. 1971. *Scientific Knowledge and its Social Problems*. New York: Oxford University Press.

———. 1973. "Tragedy in the History of Science." In M. Teich and R. Young (eds.), *Changing Perspectives in the History of Science*, pp. 204-222. London: Heinemann.

Rehbock, Philip F. 1979. "The Early Dredgers: 'Naturalizing' in British Seas, 1830-1850." *JHB, 12*: 293-368.

———. 1983. *The Philosophical Naturalists: Themes in Early Nineteenth Century British Biology*. Madison, Wis.: University of Wisconsin Press.

Reingold, Nathan. 1976. "Definitions and Speculations: The Professionalization of Science in America in the Nineteenth Century." In A. Oleson and S. C. Brown (eds.), *The Pursuit of Knowledge in the Early American Republic*, pp. 33-69. Baltimore: Johns Hopkins University Press.

Rogers, Henry Darwin. 1858. *The Geology of Pennsylvania*. 2 vols. Edinburgh: W. Blackwood and Sons.

Rogers, Mrs. William B. (ed). 1869. *Life and Letters of William Barton Rogers.* 2 vols. Boston: Houghton Mifflin.

Roscoe, Thomas. 1836. *Wanderings and Excursions in North Wales.* London: C. Tilt.

Rothblatt, Sheldon. 1968. *The Revolution of the Dons: Cambridge and Society in Victorian England.* New York: Basic Books.

Rudler, F. W. 1887. "Fifty Years' Progress in British Geology." *Proceedings of the Geologists' Association, 10*: 234-272.

Rudwick, Martin J. S. 1963. "The Foundation of the Geological Society of London: Its Scheme for Co-operative Research and its Struggle for Independence." *BJHS, 1*: 325-355.

———. 1971. "Uniformity and Progression: Reflections on the Structure of Geological Theory in the Age of Lyell." In D. H. D. Roller (ed.), *Perspectives on the History of Science and Technology*, pp. 209-227. Norman: University of Oklahoma Press.

———. 1974. "Darwin and Glen Roy: A 'Great Failure' in Scientific Method?" *Studies in History and Philosophy of Science, 5*: 97-185.

———. 1975. "Caricature as a Source for the History of Science: De la Beche's Anti-Lyellian Sketches of 1831." *Isis, 65*: 534-560.

———. 1976a. *The Meaning of Fossils: Episodes in the History of Palaeontology.* 2d ed. New York: Science History Publications.

———. 1976b. "The Emergence of a Visual Language for Geological Science, 1760-1840." *History of Science, 14*: 149-195.

———. 1976c. "Levels of Disagreement in the Sedgwick-Murchison Controversy." *Journal of the Geological Society of London, 132*: 373-375.

———. 1978. "Charles Lyell's Dream of a Statistical Palaeontology." *Palaeontology, 21*: 225-244.

———. 1979a. "The Devonian: A System Born from Conflict." In *The Devonian System: Special Papers in Palaeontology*, no. 23, pp. 9-21.

———. 1979b. "Transposed Concepts from the Human Sciences in the Early Work of Charles Lyell." In L. Jordanova and R. Porter (eds.), *Images of the Earth*, pp. 67-83. Chalfont St. Giles: British Society for the History of Science.

———. 1980. "Social Order and the Natural World." *History of Science, 18*: 269-285.

———. 1982a. "Cognitive Styles in Geology." In Mary Douglas (ed.), *Essays in the Sociology of Perception*, pp. 219-241. London: Routledge and Kegan Paul.

———. 1982b. "Charles Darwin in London: The Integration of Public and Private Science." *Isis, 73*: 186-206.

———. 1985. *The Great Devonian Controversy: The Shaping of Scientific Knowledge Among Gentlemanly Specialists.* Chicago: University of Chicago Press.

Rupke, Nicholaas A. 1983. *The Great Chain of History: William Buckland and the English School of Geology (1814-1849).* Oxford: Clarendon Press.

Ruse, Michael. 1976. "The Scientific Methodology of William Whewell." *Centaurus, 20*: 227-257.

———. 1979. *The Darwinian Revolution.* Chicago: University of Chicago Press.

Sabine, Edward. 1863. [Presidential Address.] *Proceedings of the Royal Society of London, 13*: 22-39.

Salter, John W. 1856. "On Fossil Remains in the Cambrian Rocks of the Longmynd and North Wales." *QJGS, 12*: 246-251.

———. 1857. "On Annelide-Burrows and Surface-Markings from the Cambrian Rocks of the Longmynd." *QJGS, 13*: 199-206.

———. 1865. "On Some New Forms of Olenoid Trilobites from the Lowest Fossiliferous Rocks of Wales." *1864 British Association Report*, pt. ii, p. 67.

———. 1866. "Appendix. On the Fossils of North Wales." *MGS, 3*: 239-381.

———. 1867. "On the May Hill Sandstone." *Geological Magazine, 4*: 201-205.

———. 1873. *Catalogue of the Collection of Cambrian and Silurian Fossils Contained in the Geological Museum of the University of Cambridge.* Cambridge: Cambridge University Press.

Salter, John W., and William Talbot Aveline. 1854. "On the 'Caradoc Sandstone' of Shropshire." *QJGS, 10*: 62-75.

Sarjeant, William A. S. 1980. *Geologists and the History of Geology: An International Bibliography from the Origins to 1978.* 5 vols. London: Macmillan.

Sarjeant, William A. S., and Anthony P. Harvey. 1979. "Uriconian and Longmyndian: A History of the Study of the Precambrian Rocks of the Welsh Borderland." In W. O. Kupsch and W.A.S. Sarjeant (eds.), *History of Concepts in Precambrian Geology. Geological Association of Canada Special Paper, 19*, pp. 181-224.

Schaffer, Simon. 1980. "Herschel in Bedlam: Natural History and Stellar Astronomy." *BJHS, 13*: 211-239.

———. 1983. "Natural Philosophy and Public Spectacle in the Eighteenth Century." *History of Science, 21*: 1-43.

Schneer, Cecil J. (ed.). 1969a. *Toward a History of Geology.* Cambridge, Mass.: M.I.T. Press.

Schneer, Cecil J. 1969b. "Ebenezer Emmons and the Foundations of American Geology." *Isis, 60*: 439-450.

———. 1978. "The Great Taconic Controversy." *Isis, 69*: 173-191.

Schneer, Cecil J. (ed.). 1979. *Two Hundred Years of Geology in America.* Hanover, N.H.: University Press of New England.

Schwartz, Joel S. 1980. "Three Unpublished Letters to Charles Darwin: The Solution to a Geometrico-geological Problem." *Annals of Science, 37*: 631-637.

Secord, James A. 1982. "King of Siluria: Roderick Murchison and the Imperial Theme in Nineteenth-Century British Geology." *Victorian Studies, 25*: 413-442.

———. 1983. "Historical Geology in the Ascendant." *Times Higher Education Supplement*, no. 582: 10.

———. 1985a. "Natural History in Depth." *Social Studies of Science, 15*: 181-200.

———. 1985b. "John W. Salter: The Rise and Fall of a Victorian Palaeontological Career." In A. Wheeler and J. Price (eds.), *From Linnaeus to Darwin*, pp. 61-75. London: Society for the History of Natural History.

———. 1985c. "Darwin and the Breeders: A Social History." In D. Kohn (ed.), *The Darwinian Heritage*, pp. 519-542. Princeton: Princeton University Press.

———. 1986. "The Geological Survey of Great Britain as a Research School, 1839-1855." *History of Science, 24.*

———. (forthcoming). "A Romance of the Field: Roderick Murchison's Geological Discovery of 1831."

Sedgwick, Adam. 1820. "On the Physical Structure of those Formations which are Immediately Associated with the Primitive Ridge of Devonshire and Cornwall." *Transactions of the Cambridge Philosophical Society, 1*: 89-146.

———. 1821. *A Syllabus of a Course of Lectures on Geology.* Cambridge: J. Hudson.

———. 1830. [Anniversary address of the President, 19 Feb. 1830.] *PGS, 1*: 187-212.

———. 1831. "Address to the Geological Society, Delivered on the Evening of the 18th of February 1831." *PGS, 1*: 281-316.

———. 1832. *A Syllabus of a Course of Lectures on Geology.* 2d ed. Cambridge: J. Hudson.

———. 1834. [Presidential Address.] 1833 British Association Report, pp. xxvii-xxxii.

———. 1835a. "Remarks on the Structure of Large Mineral Masses, and Especially on the Chemical Changes Produced in the Aggregation of Stratified Rocks During Different Periods after their Deposition." *TGS*, 2d ser., *3*: 47-68.

———. 1835b. [Address delivered 14 Aug. 1835.] In P. D. Hardy (ed.), *Proceedings of the Fifth Meeting of the British Association*, p. 118. Dublin: P. D. Hardy.

———. 1836. "Extrait d'une lettre de M. le professeur A. Sedgwick à M. Élie de Beaumont." *Bulletin de la Société Géologique de France, 7*: 152-155.

———. 1837. *A Syllabus of a Course of Lectures on Geology.* 3d ed. Cambridge: John W. Parker.

———. 1838. "A Memoir Entitled, a Synopsis of the English Series of Stratified Rocks Inferior to the Old Red Sandstone—With an Attempt to Determine the Successive Natural Groups and Formations." *PGS, 2*: 675-685.

———. 1841. "Supplement to a 'Synopsis of the English Series of Stratified Rocks Inferior to the Old Red Sandstone,' with Additional Remarks on the Relations of the Carboniferous Series and Old Red Sandstone of the British Isles." *PGS, 3*: 541-564.

———. 1843a. "Letter III." In J. Hudson (ed.), *A Complete Guide to the Lakes*, pp. 215-245. 2d ed. Kendal: J. Hudson.

———. 1843b. "Outline of Geological Structure of North Wales." *PGS, 4*: 212-224.

———. 1844. "On the Older Palaeozoic (Protozoic) Rocks of North Wales." *PGS, 4*: 251-266.

———. 1845a. "On the Older Palaeozoic (Protozoic) Rocks of North Wales." *QJGS, 1*: 5-22.

———. 1845b. "On the Geology of North Wales." *QJGS, 1*: 214.

[Sedgwick, Adam.] 1845c. Review of *Vestiges of the Natural History of Creation. Edinburgh Review, 82*: 1-85.

Sedgwick, Adam. 1845d. "On the Comparative Classification of the Fossiliferous Strata of North Wales, with Corresponding Deposits of Cumberland, Westmoreland, and Lancashire." *PGS, 4*: 576-584.

———. 1846. "On the Classification of the Fossiliferous Slates of Cumberland, Westmoreland and Lancashire (Being a Supplement to a Paper Read to the Society, March 12, 1845)." *QJGS, 2*: 106-131.

———. 1847. "On the Classification of the Fossiliferous Slates of North Wales, Cumberland, Westmoreland and Lancashire (Being a Supplement to a Paper Read to the Society, March 12, 1845)." *QJGS, 3*: 133-164.

———. 1848. "On the Organic Remains Found in the Skiddaw Slate, with Some Remarks on the Classification of the Older Rocks of Cumberland and Westmoreland, &c." *QJGS, 4*: 216-225.

———. 1850. *A Discourse on the Studies of the University of Cambridge*. The Fifth Edition, with Additions, and a Preliminary Dissertation. London: John W. Parker.

———. 1851. "On the Geological Structure and Relations of the Frontier Chain of Scotland." *1850 British Association Report*, pt. ii, pp. 103-107.

———. 1852a. "On the Slate Rocks of Devon and Cornwall." *QJGS, 8*: 1-19.

———. 1852b. "On the Classification and Nomenclature of the Older Palaeozoic Rocks of Great Britain." *Literary Gazette*, no. 1833: 234.

———. 1852c. "The Silurian System." *Literary Gazette*, no. 1838: 338-340.

———. 1852d. "On the Classification and Nomenclature of the Lower Palaeozoic Rocks of England and Wales." *QJGS, 8*: 136-168.

———. 1852e. "The Silurian System." *Literary Gazette*, no. 1843: 417-419.

———. 1853a. "Letter II—(5th of the Series)." In J. Hudson (ed.), *A Complete Guide to the Lakes*, pp. 236-258. 4th ed. Kendal: J. Hudson.

———. 1853b. "On a Proposed Separation of the So-called Caradoc Sandstone into Two Distinct Groups; viz. (1) May Hill Sandstone; (2) Caradoc Sandstone." *QJGS, 9*: 215-230.

———. 1854a. "On the Classification and Nomenclature of the Older Palaeozoic Rocks of Britain." *1853 British Association Report*, pt. ii, pp. 54-61.

———. 1854b. "On the May Hill Sandstone, and the Palaeozoic System of England." *Philosophical Magazine*, 4th ser., *8*: 301-317, 359-370.

———. 1854c. "On the May Hill Sandstone, and the Palaeozoic System of England." *Philosophical Magazine*, 4th Ser., *8*: 472-506.

———. 1855. "Introduction." In A. Sedgwick and F. McCoy, *A Synopsis of the Classification of the British Palaeozoic Rocks . . .*, pp. v-xcviii. London: John W. Parker.

———. 1857. "Remarks on a Passage in the President's Address Delivered at the Anniversary Meeting of the Geological Society of London, on the 15th of February, 1856." *Philosophical Magazine*, 4th ser., *13*: 176-182.

———. 1873. "Preface." In John W. Salter, *A Catalogue of the Collection of the Cambrian and Silurian Fossils Contained in the Geological Museum of the University of Cambridge*, pp. ix-xxxiii. Cambridge: Cambridge University Press.

Sedgwick, Adam, and Frederick McCoy. 1855. *A Synopsis of the Classification of the British Palaeozoic Rocks, with a Systematic Description of the British Palaeozoic Fossils in the Geological Museum of the University of Cambridge*. London: John W. Parker.

Sedgwick, Adam, and Roderick Impey Murchison. 1830. "A Sketch of the Structure of the Austrian Alps, &c. &c." *Philosophical Magazine*, 2d ser., *8*: 81-134.

———. 1835a. "On the Geological Relations of the Secondary Strata in the Isle of Arran." *TGS*, 2d ser., *3*: 21-36.

———. 1835b. "On the Structure and Relations of the Deposits Contained between the Primary Rocks and the Oolitic Series in the North of Scotland." *TGS*, 2d ser., *3*: 125-160.

———. 1835c. "A Sketch of the Structure of the Eastern Alps; with Sections through the Newer Formations on the Northern Flanks of that Chain, and through the Tertiary Deposits of Styria, &c., &c." *TGS*, 2d ser., *3*: 301-420.

———. 1836. "On the Silurian and Cambrian Systems, Exhibiting the Order in which the Older Sedimentary Strata Succeed each Other in England and Wales." *1835 British Association Report*, pt. ii, pp. 59-61.

———. 1839. "Classification of the Older Stratified Rocks of Devonshire and Cornwall." *Philosophical Magazine*, 3d ser., *14*: 241-260.

———. 1840. "On the Distribution and Classification of the Older or Palaeozoic Deposits of the North of Germany and Belgium, and their Comparison with Formations of the same Age in the British Isles." *TGS*, 2d ser., *6*: 221-301.

Seymour, W. A. (ed.). 1980. *A History of the Ordnance Survey*. Folkestone, Kent: Dawson.

Shapin, Steven. 1974. "The Audience for Science in Eighteenth Century Edinburgh." *History of Science*, *12*: 95-121.

———. 1982. "History of Science and its Sociological Reconstructions." *History of Science*, *20*: 157-211.

Shapin, Steven, and Barry Barnes. 1979. "Darwin and Social Darwinism: Purity and History." In B. Barnes and S. Shapin (eds.), *Natural Order*, pp. 125-142. Beverly Hills: Sage.

Sharpe, Daniel. 1842. "Sketch of the Geology of the South of Westmoreland." *PGS*, *3*: 602-608.

———. 1843. "On the Bala Limestone." *PGS*, *4*: 10-14.

———. 1846. "Contributions to the Geology of North Wales." *QJGS*, *2*: 283-316.

Sheets-Pyenson, Susan. 1982. "Geological Communication in the Nineteenth Century: The Ellen S. Woodward Autograph Collection at McGill University." *Bulletin of the British Museum (Natural History), Historical Series, 10*: 179-226.

Silliman, Benjamin. 1841. "Miscellanies: British Association—Mr. Murchison—His Journey to the Ural Mountains—Opinions, &c." *American Journal of Science and Arts, 41*: 207-208.

Skinner, Quentin. 1969. "Meaning and Understanding in the History of Ideas." *History and Theory, 8*: 3-53.

Smiles, Samuel. 1891. *A Publisher and his Friends: Memoir and Correspondence of the Late John Murray*. 2 vols. London: John Murray.

Smith, Crosbie W. 1985. "Geologists and Mathematicians: The Rise of Physical Geology." In P. M. Harman (ed.), *Wranglers and Physicists*, pp. 49-83. Manchester: Manchester University Press.

Smyth, Warington W. 1846. "Note on the Gogofau, or Ogofau, Mine, near Pumpsant, Caermarthenshire." *MGS, 1*: 480-484.

Speakman, Colin. 1982. *Adam Sedgwick, Geologist and Dalesman, 1785-1873: A Biography in Twelve Themes*. Heathfield, East Sussex: Broad Oak Press.

Stafford, Robert A. 1984. "Geological Surveys, Mineral Discoveries, and British Expansion, 1835-71." *Journal of Imperial and Commonwealth History, 12*: 5-32.

Stone, Lawrence. 1979. "The Revival of Narrative: Reflections on a New Old History." *Past and Present*, no. 85: 3-24.

Sulloway, Frank J. 1979. *Freud: Biologist of the Mind: Beyond the Psychoanalytic Legend*. New York: Basic Books.

Tayler, John James. 1846. "A Sketch of the Life and Character of John Eddowes Bowman." *Memoirs of the Literary and Philosophical Society of Manchester*, 2d ser., *7*: 45-85.

Teich, Mikuláš, and Robert Young (eds.). 1973. *Changing Perspectives in the History of Science: Essays in Honour of Joseph Needham*. London: Heinemann.

Tennyson, Alfred. 1899. *The Works of Alfred, Lord Tennyson, Poet Laureate*. 8 vols. London: Macmillan.

Thackray, John C. 1972. "Essential Source-material of Roderick Murchison." *JSBNH, 6*: 162-170.

———. 1976. "The Murchison-Sedgwick Controversy." *Journal of the Geological Society of London, 132*: 367-372.

———. 1977. "T. T. Lewis and Murchison's Silurian System." *Transactions of the Woolhope Naturalists' Field Club, 42*: 186-193.

———. 1978. "R. I. Murchison's *Silurian System* (1839)." *JSBNH, 9*: 61-73.

———. 1979. "R. I. Murchison's *Geology of Russia* (1845)." *JSBNH, 8*: 421-433.

———. 1981. "R. I. Murchison's *Siluria* (1854 and later)." *Archives of Natural History, 10*: 37-43.

Torrens, Hugh S. 1982. "The Reynolds-Anstice Shropshire Geological Collection, 1776-1981." *Archives of Natural History, 10*: 429-441.

———. 1983. "Arthur Aikin's Mineralogical Survey of Shropshire, 1796-

1816, and the Contemporary Audience for Geological Publications." *BJHS, 16*: 111-153.

Toulmin, Stephen, and June Goodfield. 1965. *The Discovery of Time.* New York: Harper and Row.

Tozer, E. T. 1984. *The Trias and its Ammonoids. The Evolution of a Time Scale. Geological Survey of Canada Miscellaneous Report, 35.*

Trimmer, Joshua. 1844. "Agriculture and Geology.—The Palaeozoic System." *Journal of Agriculture*, n.s., *1*: 421-440.

Turner, Frank M. 1978. "The Victorian Conflict between Science and Religion: A Professional Dimension." *Isis, 69*: 356-376.

———. 1980. "Public Science in Britain, 1880-1919." *Isis, 71*: 589-608.

Verneuil, Édouard de. 1847. "Note sur le parallélisme des roches des dépôts palaeozoïques de l'Amérique septentrionale avec ceux de l'Europe. . . ." *Bulletin de la Société Géologique de France*, 2d ser., *4*: 646-710.

Vucinich, Alexander. 1965. *Science in Russian Culture: A History to 1860.* London: Peter Owen.

Waterston, Charles D. 1982. "John Farey's Mineral Survey of South-East Sutherland and the Age of the Brora Coalfield." *Annals of Science, 39*: 173-185.

Watts, William W. 1939. "The Author of the Ordovician System; Charles Lapworth." *Proceedings of the Geologists' Association, 50*: 235-286.

Weaver, Thomas. 1824. "Geological Observations on Part of Gloucester and Somersetshire." *TGS*, 2d ser., *1*: 317-368.

Weindling, Paul J. 1979. "Geological Controversy and its Historiography: The Prehistory of the Geological Society of London." In L. Jordanova and R. Porter (eds.), *Images of the Earth*, pp. 248-271. Chalfont St. Giles: British Society for the History of Science.

Weld, Charles R. 1848. *History of the Royal Society, with Memoirs of the Presidents.* 2 vols. London: John W. Parker.

Wheeler, Alywne, and J. Price (eds.). 1985. *From Linnaeus to Darwin: Commentaries on the History of Biology and Geology.* London: Society for the History of Natural History.

Whewell, William. 1836. *Thoughts on the Study of Mathematics as a Part of a Liberal Education.* 2d ed. Cambridge: Deighton.

———. 1837. *History of the Inductive Sciences from the Earliest to the Present Time.* 3 vols. London: John W. Parker.

———. 1838. "Address to the Geological Society." *PGS, 2*: 624-649.

———. 1839. "Address to the Geological Society, Delivered at the Anniversary on the 15th of February, 1839." *PGS, 3*: 61-98.

———. 1857. *History of the Inductive Sciences, from the Earliest to the Present Time.* 3d ed., 3 vols. London: John W. Parker.

Whitaker, William. 1900. "The Anniversary Address of the President." *QJGS, 56*: li-lxxxviii.

Whitley, Richard D. 1983. "From the Sociology of Scientific Communities to the Study of Scientists' Negotiations and Beyond." *Social Science Information, 22*: 681-720.

Wilson, David B. (ed.). 1976. *Catalogue of the Manuscript Collections of Sir George Gabriel Stokes and Sir William Thomson, Baron Kelvin of Largs, in Cambridge University Library*. Cambridge: Cambridge University Library.

Wilson, George, and Archibald Geikie. 1861. *Memoir of Edward Forbes, F.R.S.* Cambridge: Macmillan.

Wilson, Leonard G. 1967. "The Emergence of Geology as a Science in the United States." *Journal of World History, 10*: 416-437.

Wilson, Leonard G. (ed.). 1970. *Sir Charles Lyell's Scientific Journals on the Species Question*. New Haven: Yale University Press.

Wilson, Leonard G. 1972. *Charles Lyell, The Years to 1841: The Revolution in Geology*. New Haven: Yale University Press.

———. 1980. "Geology on the Eve of Charles Lyell's First Visit to America." *Proceedings of the American Philosophical Society, 124*: 168-202.

Winsor, Mary P. 1969. "Barnacle Larvae in the Nineteenth Century: A Case Study in Taxonomic Theory." *Journal of the History of Medicine and Allied Sciences, 24*: 294-309.

———. 1976. *Starfish, Jellyfish, and the Order of Life: Issues in Nineteenth Century Science*. New Haven: Yale University Press.

Woodward, Horace B. 1908. *The History of the Geological Society of London*. London: Longmans.

———. 1911. *History of Geology*. London: Watts.

Yeo, Richard. 1979. "William Whewell, Natural Theology and the Philosophy of Science in Mid Nineteenth Century Britain." *Annals of Science, 36*: 493-516.

———. 1984. "Science and Intellectual Authority in Mid-nineteenth Century Britain: Robert Chambers and *Vestiges of the Natural History of Creation*." *Victorian Studies, 28*: 5-31.

Young, G. M. 1964. *Victorian England: Portrait of an Age*. 2d ed. New York: Oxford University Press.

Young, Robert. 1973. "The Historiographic and Ideological Contexts of the Nineteenth-Century Debate on Man's Place in Nature." In M. Teich and R. Young (eds.), *Changing Perspectives in the History of Science*, pp. 344-438. London: Heinemann.

Zaslow, Morris. 1975. *Reading the Rocks: The Story of the Geological Survey of Canada, 1841-1972*. Toronto: Macmillan.

Zittel, Karl Alfred von. 1901. *History of Geology and Palaeontology to the End of the Nineteenth Century*. Translated by Maria M. Ogilvie-Gordon. London: Walter Scott.

Index

Secord, James A.
Controversy in Victorian geology.

Bibliography: p.
Includes index.
1. Geology, Stratigraphic—Cambrian—History.
2. Geology, Stratigraphic—Silurian—History.
3. Geology—Wales—History. I. Title.
QE656.S4 1986 551.7′23 85-43310
ISBN 0-691-08417-3
4